Classical and Molecular Genetics

(The second edition)

Md. Mohan Mia

AMERICAN ACADEMIC PRESS

AMERICAN ACADEMIC PRESS

By AMERICAN ACADEMIC PRESS

201 Main Street

Salt Lake City

UT 84111 USA

Email manu@AcademicPress.us

Visit us at http://www.AcademicPress.us

ISBN: 979-8-3370-8949-2

Distributed to the trade by National Book Network Suite 200, 4501 Forbes Boulevard, Lanham, MD 20706

10　9　8　7　6　5　4　3　2　1

Preface to the Second Edition

In the Name of Allah, the Most Gracious, the Most Merciful.

Since the publication of the first edition of *Classical and Molecular Genetics*, the field of genetics has continued to advance rapidly, bringing forth new concepts, refined methodologies, and broader applications in agriculture, medicine, and biotechnology. The encouraging feedback from students, teachers, and fellow researchers has been both humbling and motivating, reinforcing the need for an updated, accessible, and concise textbook that bridges classical principles with modern molecular insights.

This second edition retains the original aim of presenting genetics in a clear, logical, and student-friendly manner, avoiding unnecessary repetition while ensuring that core concepts are thoroughly explained. At the same time, several chapters have been revised to include recent discoveries, improved illustrations, and updated examples that reflect current research trends.

I have remained mindful of the challenges faced by learners in grasping both the foundational aspects of Mendelian genetics and the intricate details of molecular mechanisms. My approach has been to present information progressively, linking theory with application, and showing how classical concepts remain deeply relevant in the era of genomics and bioinformatics.

I am grateful to colleagues, reviewers, and students whose valuable suggestions have shaped this edition. It is my hope that this book will continue to serve as a reliable resource for undergraduate and postgraduate students of biological science, as well as anyone seeking to understand the fascinating science of heredity and variation.

May this effort benefit all who seek knowledge, and may Allah bless it with acceptance.

Professor Dr. Md. Mohan Mia
Sylhet Agricultural University
Sylhet, Bangladesh

Dedication

For those who by their wisdom, love and encouragement have been
made from me what I am today
To the candle of given and sacrifice my father Md. Mosan Mia
To the angel of love and tenderness my mother Afia Begum
To my lovely soul mate Krishibid Sharifa Akter Begum
To my children that have a great part in my success
Dr. Mashkura Afroz Tuli and Shadman Shahriar
To my son-in-law Dr. Abu Sahal Md. Fahim
To my daughter-in-law Nupur Begum
To all of them I dedicate my book

Dr. Md. Mohan Mia

Contents

CHAPTER THREE   MODIFICATIONS OF MENDELIAN RATIOS

CHAPTER FOUR   CHROMOSOMAL BASIS OF INHERITANCE

V

VIII

XIII

CHAPTER ONE   INTRODUCTION

In 1905, the English biologist William Bateson used the word 'genetics' (from the Greek *gennō*; "*to give birth*"), in a letter to his colleague Adam Sedgwick. In 1906, he proposed that the 'new science of heredity based on Mendel's laws' be named as genetics at the "Third International Conference on Plant Hybridization" in London and this proposal was enthusiastically approved. Genetics is a young science. Genetics is the study of the biological (biochemical basis) of inheritance. Inheritance is how traits, or characteristics, are passed on from generation to generation. This passing can occur through sexual reproduction or asexual reproduction. The characteristics are passed onto offspring as genetic information. It is the most fundamental life science. Because, the things we study in biology, including most of the physical characteristics, behaviors, and capabilities of living things, are inherited. Genetics is often described as a biological science, which deals with the principles of heredity and variation. It is the scientific discipline that deals with the similarities and differences among related individuals. In the very early ages though improvement of the races of plants and animals were conducted by the Babylonians and Assyrians, it was not known what exactly caused the characters to be passed from one generation to the next. The science of Genetics helps us to differentiate between heredity and variations and seeks to account for the resemblances and differences due to heredity, their source and development.

Heredity

The term heredity comes from the Latin word *hereditatem*, which means "condition of being an heir", was first used in the 1530s. Heredity or inheritance or biological inheritance refers to the transmission of characters from parents to their offspring. The children or offspring closely resemble their parents and to some extent their grandparents and great grandparents. An elephant always gives birth only to a baby elephant and not some other animal. A mango seed forms only a mango plant and not any other plant. Thus, the offspring resemble their parents. Heredity explains how offspring in a family resemble their parents. The process of heredity occurs among all living things including animals, plants, bacteria, protists and fungi. Through this process a cell or organism acquires the characteristics of its parent cell or organism. Heredity is the conservative factor in nature, results from transmission of genes from parents to offspring.

Variation

Though the offspring in a species may resemble very closely to their parents, but they never resemble exactly with them. The offspring of a particular set of parents differ from each other and from their parents in many respects and to different degrees. The differences among the individuals of a species for a character constitute the variation. Variation refers to the differences shown by individuals of the same species and also by offspring (siblings) of the same parents. It explains why offspring even though born to the same parents differ from each other. Variation may be shown in physical appearance, metabolism, fertility, mode of reproduction, behaviour, learning and mental ability, and other characters. Variation can be discontinuous or continuous. Some of the features of the different organisms in a species show continuous variation, and some features show discontinuous variation.

Discontinuous variation

In discontinuous variation individuals fall into a number of distinct classes or categories, and the characteristics cannot be measured across a complete range. Human blood group is an example of discontinuous variation. In the ABO blood group system, only four blood groups are possible (A, B, AB or O). Humans can only belong to one of these categories. There are no other possibilities and there are no values in between (Image 1.1). So, this is discontinuous variation. Other examples of discontinuous variation are: gender, tongue rolling, eye colour, *etc.* A bar graph is used to represent discontinuous variation. Such data is called discrete (or categorical) data. Chi-squared statistical calculations work well in this case. Discontinuous variation is controlled by alleles of a single gene or small number of genes. The environment has little effect on this type of variation.

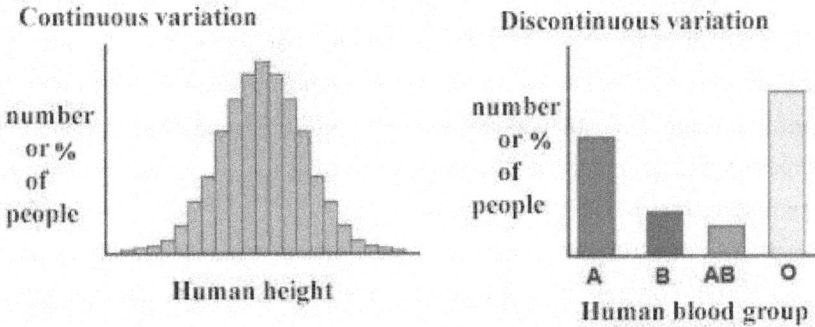

Image 1.1 Discontinuous and continuous variation.

Continuous variation

In continuous variation there is a complete range of measurements from one extreme to the other, no distinct classes or categories. Characteristics can be measured and fall within a range between two extremes. Human height is an example of continuous variation. Height ranges from that of the shortest person in the world to that of the tallest person, any height is possible between these values (Image 1.1). Other examples of continuous variation include weight, hand span, heart rate, finger length, foot length, milk yield in cows, *etc*. A line graph is used to represent continuous variation. Continuous variation is the combined effect of many genes and is often significantly affected by environmental influences. Milk yield in cows, for example, is determined not only by their genetic make-up but is also significantly affected by environmental factors such as pasture quality and diet, weather, and the comfort of their surroundings.

Causes of Variation

Variation in the phenotype is caused by heredity (genetic), by the environment, or by a combination of the two. Hereditary variation refers to differences in inherited traits. Examples of inherited variation in humans include eye colour, hair colour, skin colour, lobed or lobeless ears, gender, dimples, freckles, *etc.* Characteristics of animal and plant species can be affected by factors such as climate, diet, accidents, culture and lifestyle. These are environmental variation. Examples of environmental variation are: language and religion, flower colour in hydrangeas (a genus of more than 70 species of flowering plants native to Asia and the Americas) – these plants produce blue flowers in acidic soil and pink flowers in alkaline soil. Environmental variation is temporary and not inherited to next generation.

Origins/Sources of Genetic Variation

Genetic variation refers to differences in the genetic makeup of individuals in a population. Three primary sources of genetic variation or genetic diversity in a population are mutation, sexual reproduction, and gene flow.

Mutation. Mutation produces new alleles of gene. Mutation is the ultimate source of all genetic variation. They may impact an individual gene or an entire chromosome.

Sexual reproduction. Sexual reproduction cannot produce new alleles but it promotes genetic variation by producing different gene combinations through meiosis and random fertilization.

❖ Meiosis produces two types of recombination
  1. Interchromosomal recombination. Independent assortment of non-homologous chromosomes during metaphase I of meiosis.
  2. Intrachromosomal recombination, *i.e.*, crossing-over in prophase I of meiosis.
❖ Random fertilization of gametes. A single egg is fertilized by a single sperm in a random manner.

Because of independent assortment of chromosomes and random fertilization of gametes, a human offspring represents any of over (one of 8.38 million possible chromosome combination for each sperm × one of 8.38 million possible chromosome combination for each ovum) 70 trillion zygote possibilities. Genetic variation produced in sexual life cycles is the raw material for evolution by natural selection.

Gene flow. The movement of genes into a population from other populations with different gene frequencies.

Modern Concept of Genetics

The word genetics comes from genes. Genetics is the study of genes including the structure of genetic materials, what information is stored in the genes, how the genes are expressed and how the genetic information is transferred. Genetics is of central importance to biology because genes are the principal determinants of all life processes, from cell structure and function to reproduction of the organism. Genes play central roles in determining all aspects of the lives of all living organisms. Genes not only control what we look like physically, but also influence our behavior, our personalities, our susceptibility to diseases, our aging processes, and, indeed, our longevity.

Genetics is primarily concerned with the understanding of biological properties that are transmitted from parent to offspring. The subject matter of genetics includes the phenomenology of heredity, the molecular nature of the genetic material, how genes control metabolism and development, and the distribution and behavior of genes in populations.

Branches of Genetics

Genetics is a broad field with several branches, each focusing on different aspects of heredity and variation. Based on the various ways genetics is analyzed, it can be subdivided into four major sub disciplines for the convenience of discussion on its scope and significance: transmission genetics, molecular genetics, population genetics, and quantitative genetics. Each of these areas focuses on a different aspect of heredity.

Transmission genetics

Transmission genetics (also called Mendelian or classical, genetics), which remains the foundation for all other areas in genetics, is primarily concerned with the study of the transmission of traits (genes) from generation to generation. The unit of study for transmission genetics is the individual. Classical genetics began with Mendel's study of inheritance in garden peas and continues with studies of inheritance in many different plants and animals, and provides a starting point for the study of genetics.

Molecular genetics

Molecular genetics is the branch of genetics that studies the chemical and physical nature of genes and the mechanisms by which genes control development, growth and physiology. In molecular genetics we focus on the cell. During the last few decades, molecular genetics has developed at such a fast rate, that no aspect of genetics is complete without a discussion and explanation at the molecular level. Many recent achievements, including isolation and characterization of genes, have facilitated greatly the recent advances in the field of genetics.

Population genetics

Population genetics studies the heredity in groups of individuals for traits that are determined by one or only a few genes. Therefore, in population genetics the unit of study is the population. Population genetics is the study of the distributions and changes in the frequencies of alleles and genotypes in populations over space and time, and the forces which maintain or change the frequencies of particular alleles and genotypes in populations.

Quantitative genetics

Quantitative genetics concerned with the study of the inheritance of metric characters in groups of individuals.

Population genetics and quantitative genetics both focus on the genetic basis of phenotypic variation among the individuals in a population. Population genetics traditionally deals with frequencies of alleles and genotypes, whereas quantitative genetics concentrates on the ways that individual variation in genotype and environment contribute to variance in phenotype.

The above classification is arbitrary, and the four areas are inter-related, there is no sharp boundary between them.

There are many different branches of genetics depending upon the organism you wish to work upon, such as microbial genetics, plant genetics, animal genetics, human genetics, *etc*. Animal genetics is a branch of genetics that studies the hereditary material of animals for theoretical and practical applications, such as increased production, conservation, and disease research. Depending upon its applications, genetics is divided into various branches as immunogenetics, behavioral genetics, clinical genetics, epigenetics, conservation genetics, developmental genetics, ecological genetics, evolutionary genetics, genetic engineering, genetic epidemiology, archaeogenetics, genetics of intelligence, genomics, medical genetics, psychiatric genetics, statistical genetics, *etc.*

The Development of the Field of Genetics

Genetics has shown us that many of the differences between organisms are the result of differences in the genes they carry. The history of genetics dates from the classical era with contributions by Hippocrates, Aristotle and Epicurus. While breeding experiments with domesticated animals and cultivated plants were done for centuries, the principles of heredity were first written about by Gregor Johann Mendel in the 1860s. The importance of his findings was not recognized in his lifetime and it was only around 1900, when Mendel's principles of heredity were rediscovered, that the development of the subject of genetics truly began. Table 1.1 presents a time line of important milestones in the history of genetics.

Table 1.1 Timeline of Important Milestones in the History of Genetics

1866	Gregor Johann Mendel	Published the results of his experiments with pea plants and demonstrated that heredity is transmitted in discrete units.
1866	Ernst Heinrich Philipp August Haeckel	Proposed that the nucleus contained the factors responsible for the transmission of hereditary traits.
1869	Friedrich Miescher	Isolated DNA from pus cells for the first time and called it "nuclein".
1871	Ernst Heinrich Philipp August Haeckel	Proved that the genetic material is indeed located in the nucleus.
1875	Eduard Adolf Strasburger	First observed chromosome during cell division.
1879	Walter Flemming	Described the splitting of dark stained rod-like structures during cell division which he called as chromatin.

1882	Walter Flemming	Described the whole process of mitosis.
1887	Van Beneden, Theodor Boveri	First discovered that the number of chromosomes in a given species remains constant.
1887	August Weismann	Proposed a universal theory of chromosome behavior that predicts meiosis in sex cells.
1888	Henrich Wilhelm Von Waldeyer	Coined and described chromosome as a condensed form of material inside the nucleus.
1889	Francis Galton	Published the book "Natural Inheritance". The field of biometry is formally founded.
1889	August Weismann	Proposes that inheritance occurs through germ cells, separating hereditary material from somatic cells.
1900	Hugo de Vries, Carl Correns, Erich von Tschermak	Independently rediscovered Mendel's work while doing their own work on the laws of inheritance.
1901	Thomas Harrison Montgomery	Studied spermatogenesis in various species of Hemiptera and finds that maternal chromosomes only pair with paternal chromosomes during meiosis.
1901	Hermann Emil Fischer, Ernest Fourneau	Reported the synthesis of the first dipeptide, glycylglycine.
1901	Hugo de Vries	Adopted the term 'mutation' to describe sudden, spontaneous, drastic alterations in the hereditary material of *Oenothera*.
1901	William Bateson	Working with chicken, showed that Mendelian principles were applicable to animals as well as plants.
1901	Karl Landsteiner	Pioneered ABO blood groups in humans.
1902	William Bateson	Coined the terms genetics, F_1, F_2, allelomorph (later shortened to allele), homozygote, heterozygote, and epistasis.

1902	Clarence Erwin McClung	Presented the concept that specific chromosomes are responsible for determining sex in a number of animals.
1902	Sir Archibald Edward Garrod	The first person to associate Mendel's theories with a human disease alkaptonuria.
1902	Hermann Emil Fischer, Franz Hofmeister	Demonstrated that proteins are polypeptides.
1902	Carl Erich Correns	Discussed the time and place of segregation.
1902	Walter Sutton, Theodor Boveri	Proposed the chromosome theory of heredity independently.
1903	Wilhelm Ludvig Johannsen	Introduced and defined the concepts of phenotype, genotype, and selection.
1903	William Castle	Recognized the relationship between allele and genotype frequencies for the first time.
1905	William Bateson, Edith Rebecca Saunders, Reginald Crundall Punnett	Investigated the exceptions to Mendel's rules, leading to the discovery of two new genetic principles: linkage and gene interaction.
1905	William Bateson	Coined the term genetics.
1905	Edmund Beecher Wilson, Nellie Maria Stevens	Independently discovered that separate X and Y chromosomes determine sex.
1905	Nettie Stevens	Observed the sex chromosomes X and Y.
1905	Lucien Claude Jules Cuénot	Extended Mendel's discoveries to animals and discovered the first lethal allele: the yellow coat color allele in mice.
1906	Craig W. Woodworth, William Ernest Castle	Introduced *Drosophila melanogaster* as a new experimental material for genetic studies.
1906	Robert Heath Lock	Suggested the relation between linkage and exchange of parts between homologous chromosomes.
1907	Friedrick Laibach	First suggested Arabidopsis as a model organism.

1908	Godfrey Harold Hardy, Wilhelm Weinberg	Formulated the Hardy-Weinberg principle independently.
1908	Archibald Edward Garrod	Discussed the biochemical genetics of man (or any other species).
1908	Erwin Baur	First clearly demonstrated a lethal gene in *Antirrhinum*.
1909	William Bateson	Published his book entitled *Mendel's Principles of Heredity: A Defense*. This book was the first English textbook on genetics and became the main popularizer of Mendel's ideas after their rediscovery.
1909	Wilhelm Ludvig Johannsen	Introduced the term 'gene' to describe the Mendelian unit of heredity. He coined the terms phenotype and genotype.
1909	Frans Alfons Janssens	Proposed the chiasma-type hypothesis.
1909	Herman Nilsson-Ehle	Explained the quantitative variation of seed coat color in wheat and oat by the interaction of multiple genetic factors. This is the birth of quantitative genetics.
1910	William Bateson, Reginald Punnett	Discovered genetic linkage. They also coined the term "epistasis" to describe the interaction between two different genes.
1910	Thomas Hunt Morgan	Demonstrated that specific gene resides on specific chromosome. He established the sex liked inheritance of white eyes in *Drosophila melanogaster* and awarded the Nobel Prize in Physiology or Medicine in 1933.
1910	Ludwig Karl Martin Leonhard Albrecht Kossel	Awarded the Nobel Prize in 1910 in Physiology or Medicine for his discovery of the five nucleotide bases; adenine, cytosine, guanine, thymine and uracil.

1911	Thomas Hunt Morgan	Proposed that the genes for white eyes, yellow body, and miniature wings in *Drosophila* are linked together on the X chromosome.
1911	Alexis Carrel, Montrose Thomas Burrows	Developed the technique of *in vitro* tissue culture.
1911	Thomas Hunt Morgan, Alfred Henry Sturtevant	Discovered the position of linked genes on a chromosome.
1912	Thomas Hunt Morgan	Reported the first sex-linked recessive lethal gene in *Drosophila*.
1912	Sir William Henry Bragg, Sir William Lawrence Bragg	Developed the X-ray crystallography technique and both are awarded the Nobel Prize for Physics in 1915.
1912	Thomas Hunt Morgan, Clara Julia Lynch	First published the case of autosomal linkage in *Drosophila*.
1913	Alfred Henry Sturtevant	Constructed the first genetic map of a chromosome.
1914	Calvin Bridges	Observed non-disjunction in sex chromosomes.
1915	Thomas Hunt Morgan, Alfred Henry Sturtevant, Hermann Joseph Muller, Calvin Blackman Bridges	Published the mechanism of Mendelian heredity.
1915	Frederick William Twort	First discovered a virus capable of infecting and destroying bacteria in Great Britain.
1916	Thomas Hunt Morgan	Formally named Mendel's two laws.
1917	Félix Hubert d'Hérelle	Discovered the virus infecting and destroying bacteria in France. He coined the term *bacteriophage*, meaning "bacteria eater," to describe the agent's bactericidal ability.
1917	Calvin Blackman Bridges	Discovered the first chromosome deficiency in *Drosophila*.

1918	Ronald Aylmer Fisher	Began the study of quantitative genetics by partitioning phenotypic variance into a genetic and an environmental component.
1919	Calvin Blackman Bridges	Discovered the chromosomal duplications in *Drosophila*.
1919	Thomas Hunt Morgan	Showed that the number of chromosomes equals the number of linkage groups.
1920	Hans Winkler	Coined the term 'genome' as a conjunction between gene and chromosome.
1921	Calvin Blackman Bridges	Published analysis resulted from the study of triploidy, genic balance, and sex determination in *Drosophila*.
1921	Alfred Sturtevant	First observed chromosomal inversion in *Drosophila melanogaster*.
1923	Calvin Blackman Bridges	Discovered chromosomal translocations in *Drosophila*.
1923	Arthur Edwin Boycott, Cyril Diver	Demonstrated maternal inheritance for shell coiling direction in water snail, *Limnea peregra*.
1925	Felix Bernstein	Proposed multiple allele interpretation of human ABO blood groups.
1925	Alfred Henry Sturtevant	First demonstrated position effect in Drosophila Bar-eye.
1927	Herman Joseph Muller	Induced mutation by X-rays.
1927	Leo Loeb, Sewall Wright	Demonstrated genetics of transplant specificity in mammals.
1927	Karl Landsteiner, Philip Levine	Discovered the MN and the P blood groups in human.
1927	Bernard Ogilvie Dodge	First used Neurospora as a genetic organism.
1928	Frederick Griffith	Discovered that hereditary material from dead bacteria can be incorporated into live bacteria.
1929	Phoebus Aaron Levene	Discovered that DNA is made up of nucleotides.

1930	Ronald Aylmer Fisher	Published *Genetical Theory of Natural Selection*, a formal analysis of the mathematics of selection.
1930	Arne Tiselius	Developed electrophoresis as a technique to separate proteins.
1931	Phoebus Aaron Levene, Lawrence Wade Bass	Introduced the chemical component and the basic structure of DNA.
1931	Harriet Baldwin Creighton, Barbara McClintock	Obtained physical evidence for recombination.
1932	Sewall Wright	Stressed the importance of "genetic drift" due to chance in small populations during evolution.
1932	Ernst Ruska, Max Knoll	Produced first electron microscope.
1933	Emil Heitz, Hans Bauer, and independently Theophilus Shickel Painter	Discovered the nature of salivary gland chromosomes.
1933	Theophilus Shickel Painter	First used salivary gland chromosomes of Drosophila in cytogenetic studies.
1934	Dorothy Crowfoot Hodgkin	Used X-rays crystallography to illuminate the structure of protein.
1935	Calvin Bridges	Published the first cytogenetic map of the Drosophila salivary gland chromosome.
1935-36	Otto Heinrich Warburg, Hans von Euler-Chelpin	Isolated pyrimidine nucleotides and determined their structure and action.
1937	Albert Francis Blakeslee, Amos G. Avery	Used colchicine to produce artificial polyploidy in plant cells.
1937	Arne Wilhelm Kaurin Tiselius	Developed electrophoresis technique.
1938	William Thomas Astbury, Florence Ogilvy Bell	First used X-ray crystallography to analyze the structure of DNA.

1940	Karl Landsteiner, Alexander Solomon Wiener	Discovered the Rh blood factor.
1941	George Wells Beadle, Edward Lawrie Tatum	Showed that genes act by regulating distinct chemical events. They proposed that each gene directs the formation of one enzyme.
1941	Kenneth Mather	Coined the term 'polygene' and described polygenic traits in various organisms.
1942	Salvador Edward Luria, Thomas Foxen Anderson	Obtained the first high-quality electron micrograph of a bacteriophage.
1943	William Astbury	Obtained the first X-ray diffraction pattern of DNA, which reveals that DNA must have a regular periodic structure. He suggested that nucleotide bases are stacked on top of each other.
1943	Salvador Edward Luria, Max Ludwig Henning Delbrück	Demonstrated that bacteria have normal genetic systems and thus could be used to study genetic processes.
1944	Oswald Theodore Avery, Colin McLeod, Maclyn McCarty	Identified DNA (not proteins) as the genetic material.
1945	John von Neumann	Developed the concept of a computer.
1946	Max Delbrück, William T. Bailey, Alfred Day Hershey	Demonstrated Bacteriophage genetic recombination.
1946	Joshua Lederberg, Edward Tatum	Discovered genetic recombination (conjugation) in bacteria and awarded the Nobel Prize in 1958.
1948	Hermann Joseph Muller	Described the phenomenon of dosage compensation.
1948	Barbara McClintock	Discovered transposons in maize and awarded the Nobel Prize in 1983 in Physiology or Medicine for her experiments detailing the evidence for transposons.

1949	James Van Gundia Neel, E.A. Beet	Independently provided genetic evidence that the sickle-cell disease is inherited as a simple Mendelian autosomal recessive.
Early 1950s	Salvadore Luria, Jean Weigle, Giuseppe Bertani	Discovered the first restriction enzymes, but the enzymes they found were all type I enzymes that cleaved DNA randomly from a recognition site.
1950	Erwin Chargaff	Discovered base equivalencies in DNA. Thus, the amount of adenine (A) is always equal to the amount of thymine (T), and the amount of guanine (G) is always equal to the amount of cytosine (C).
1951	Rosalind Elsie Franklin, Maurice Wilkins, Raymond Gosling	Conducted X-ray diffraction studies that provided images of the helical structure of DNA.
1952	Alfred Hershey, Martha Chase	Proved that the genetic material of phages and all other organisms to be DNA.
1953	James Dewey Watson, Francis Harry Compton Crick	Discovered the double helical structure of DNA using Chargaff's data and the X-ray images recorded by Rosalind Franklin and Maurice Wilkins.
1954	George Gamow	Suggested that DNA contains a code that is responsible for the production of proteins.
1955	Arthur Kornberg and colleagues	Isolated the first DNA polymerase.
1956	Vernon Ingram	Discovered that a specific chemical alteration in a hemoglobin protein is the cause of sickle cell disease.
1956	Joe Hin Tjio	Established the correct chromosome number in humans to be 46.
1956	Alfred Gierer, Gerhard Schramm	Concluded that RNA is the genetic material of tobacco mosaic virus (TMV).
1957	Heinz L. Fraenkel-Conrat, Beatrice Singer	Confirmed that TMV uses RNA as its genetic material.

1958	Mathew Meselson, Franklin Stahl	Proved the semiconservative model for DNA replication.
1958	Arthur Kornberg	Synthesized DNA *in vitro* for the first time and won the Nobel Prize in 1959.
1959	Jerome Lejeune and his colleagues	Discovered that Down Syndrome is caused by trisomy 21.
1959	Severo Ochoa	Discovered the first RNA polymerase.
1961	Francois Jacob, Jacques Monad	Hypothesized the existence of an intermediary between DNA and its protein products, which they called messenger RNA.
1961	Marshall Nirenberg, Har Gobind Khorana and colleagues	Worked out the complete genetic code, showing that three DNA bases code for one amino acid. In 1968 Nirenberg won the Nobel Prize in Physiology or Medicine for his seminal work on the genetic code. He shared the award with Har Gobind Khorana (University of Wisconsin), who mastered the synthesis of nucleic acids, and Robert Holley (Cornell University), who discovered the chemical structure of transfer-RNA.
1961	Sidney Brenner, Francois Jacob, Matthew Meselson	Discovered mRNA that carries genetic information from DNA in the nucleus to the protein-making machinery in the cytoplasm.
1961	Robert Guthrie	Developed a method to test newborns for the metabolic defect, phenylketonuria (PKU).
1962	James Dewey Watson, Francis Harry Compton Crick, Maurice Wilkins	Won the Nobel Prize for their discovery of the structure of DNA.
1964	Howard Temin	Using RNA viruses showed that the direction of DNA to RNA transcription can be reversed.
1967	W.M. Fitch, E. Margoliash	Developed the first phylogenetic tree.
1969	Jonathan Beckwith	Isolated the first bacterial gene.

1970	Howard Temin, David Baltimore	Independently discovered reverse transcriptase.
1970	Werner Arber, Hamilton Othanel Smith, Daniel Nathans	Discovered type II restriction enzymes and awarded the 1978 Nobel Prize for Physiology or Medicine
1972	Paul Berg	Constructed the first recombinant DNA molecule *in vitro*.
1973	Herbert Boyer, Stanley Cohen	For the first time cut a kanamycin-resistant gene from one bacterium and inserted it into a plasmid of another bacterium transferring the kanamycin-resistant trait. This was the first genetically modified organism developed by mankind.
1973	Walter Fiers and his team	Determined the sequence of a gene: the gene for bacteriophage MS2 coat protein.
1974	Rudolf Jaenisch	First created a transgenic mouse by injecting the SV40 virus in early mouse embryos.
1974	Roger David Kornberg	Proposed the concept of chromosome ultra-structure in the form of nucleosome.
1975	César Milstein, Georges Köhler, Niels Jerne	Published a principle for producing monoclonal antibodies and awarded the Nobel Prize for Medicine in 1984.
1975	Edwin Southern	Developed gel transfer hybridization to detect specific DNA sequences.
1975	Frederick Sanger and colleagues; and Alan Maxam and Walter Gilbert	Both groups developed rapid DNA sequencing methods.
1976	Genentech (The First Genetic Engineering Company)	Produced the first human protein insulin (the first recombinant DNA drug) in a bacterium.
1976	Walter Fiers and his team	Determined the complete nucleotide-sequence of bacteriophage MS2 RNA.

1977	Walter Gilbert, Frederick Sanger	Devised methods for DNA sequencing.
1977	Frederick Sanger	Determined the base sequence of an entire viral genome (ØX174).
1977	Richard J. Roberts, Phillip A. Sharp	Independently discovered introns in eukaryotic genes.
1977	Richard J. Roberts and Phillip A. Sharp	Discovered splicing of adenovirus RNAs and awarded the Nobel Prize in 1993
1979	M. Goodman, J. Czelusniak, G.W. Moore, AE Romero-Herrera, G. Matsuda	First described the parsimony procedure for developing phylogenetic trees.
1980	Frederick Sanger, Wally Gilbert and Paul Berg	Won the Nobel Prize for Chemistry for pioneering DNA sequencing methods.
1981	Felsenstein	Published the maximum likelihood procedure to develop phylogenetic trees.
1981	Thomas Wagner and his team	Performed the first transgenic animal experiment.
1982	GenBank	Established GenBank to be a database of all DNA sequences.
1982	Genentech	Introduced world's first genetically engineered drug 'insulin' on the market for humans.
1983	Kary Banks Mullis	Invented the polymerase chain reaction (PCR) at the Cetus Corporation in California, USA for which he was awarded the Nobel Prize in chemistry in 1993.
1983	James Gusella and his team	Identified the location of the gene responsible for Huntington's disease at Massachusetts General Hospital, USA.
1983	Sidney Altman, Thomas Cech	Established the existence of catalytic RNAs and awarded the Nobel Prize in 1989.
1985	Alec Jeffreys	Developed a method for DNA profiling.

1986	Leroy Hood, Lloyd Smith, Michael Hunkapiller, Tim Hunkapiller	Released the first automated DNA sequencer.
1987	E.S. Lander, P. Green, J. Abrahamson, A. Barlow, M.J. Daly, S.E. Lincoln, L.A. Newberg	Released MAPMAKER, a computer program for developing genetic linkage maps from molecular marker data.
1989	Francis Collins, Lap-Chee Tsui	Sequenced the human gene that encodes the CFTR protein.
1990	James Dewey Watson and many others (funded by US Government)	Launched the 'Human Genome Project (HGP)' to map and sequence the entire human genome.
1991	Hood and Hunkapillar	Introduced a new automated DNA sequencing technique.
1992	A French team	Built a low-resolution, microsatellite genetic map (Second-Generation Genetic Map) of the entire human genome.
1993	Huntington's Disease Collaborative Research Group	Identified the Huntington's disease gene.
1993	Victor Ambros and colleagues	Discovered that microRNA can regulate gene expression by base-pairing to an mRNA.
1994	The Food and Drug Administration (FDA), USA	Approved the sale of the first genetically modified food (*Flavr Savr* Tomato).
1995	Craig Venter, Hamilton Smith	Determined the base sequences of the genomes of two bacteria *Hemophilus influenzae* and *Mycoplasma genitalium*.
1996	An international consortium of scientists	Determined the base sequence of the genome of brewer's yeast *Saccharomyces cerevisiae*, the first eukaryotic genome to be sequenced.

1996	David J. Lockhart and colleagues as well as Joseph DeRisi and colleagues	Presented DNA microarrays, which allowed the simultaneous examination of thousands of genes.
1996	Ian Wilmut and colleagues	Cloned the first cloned mammal – the sheep Dolly from an adult sheep udder cell.
1996	William F. Dietrich, Joyce Miller, Robert Steen, Mark A. Merchant, Deborah Damron-Boles, Zeeshan Husain, Robert Dredge, Mark J. Daly, Kimberly A. Ingalls, Tara J. O'Connor, Cheryl A. Evans, Margaret M. DeAngelis, David M. Levinson, Leonid Kruglyak, Nathan Goodman, Neal G. Copeland, Nancy A. Jenkins, Trevor L. Hawkins, Lincoln Stein, David C. Page, Eric S. Lander	Completed mouse genetic map.

1997	F.R. Blattner, G. Plunkett 3rd, C.A. Bloch, N.T. Perna, V. Burland, M. Riley, J. Collado-Vides, J.D. Glasner, C.K. Rode, G.F. Mayhew, J. Gregor, N.W. Davis, H.A. Kirkpatrick, M.A. Goeden, D.J. Rose, B. Mau, Y. Shao	Published the complete sequence of *Escherichia coli* (4.7 Mbp) and shown to contain 4,500 genes.
1998	S.T. Cole, R. Brosch, J. Parkhill, T. Garnier, C. Churcher, D. Harris, S.V. Gordon, K. Eiglmeier, S. Gas, C.E. Barry 3rd, F. Tekaia, K. Badcock, D. Basham, D. Brown, T. Chillingworth, R. Connor, R. Davies, K. Devlin, T. Feltwell, S. Gentles, N. Hamlin, S. Holroyd, T. Hornsby, K. Jagels, A. Krogh, J. McLean, S. Moule, L. Murphy, K. Oliver, J. Osborne, M.A. Quail, M.A. Rajandream, J. Rogers, S. Rutter, K. Seeger, J. Skelton, R. Squares, S. Squares, J.E. Sulston, K. Taylor, S. Whitehead, B.G. Barrell	Published the complete genome sequence of *Mycobacterium tuberculosis*.

1998	Andrew Fire, Craig Mello	Discovered the method of RNA interference: genes are silenced with small RNA bits.
1998	James Thomson	Succeeded in isolating stem cells from human embryos and cultivating them as embryonic stem cells (ES cells) in the laboratory.
1998	John Sulston, Bob Waterston	Published the genome of the nematode *Caenorhabditis elegans*. This is the first genome of a multicellular organism to be sequenced. The genome contains 19,100 genes.
1999	Human Genome Project	Fully sequenced chromosome 22 as the first human chromosome.
2000	Drosophila Genome Project Group	Completed the full genome sequence of the model organism *Drosophila melanogaster*.
2000	Many investigators	Sequenced the genome of *Arabidopsis thaliana* (a plant model organism).
2001	Human Genome Project and Celera Genomics	Published Draft Human Genome.
2001	The Food and Drug Administration (FDA), USA	Approved the first targeted gene therapy.
2002	Malaria Sequencing Consortium	Published the sequence of *Plasmodium falciparum* genome (23 Mbp). The genome consists of 5300 genes.
2002	Mosquito Sequencing Consortium	Published the malaria-parasite-carrying mosquito genome sequence (278 Mbp). It is shown to contain 13,600 genes, similar to the number found in Drosophila

2002	The International Mouse Genome Sequencing Consortium	Published the mouse genome sequence (2500 Mbp). The number of genes is estimated to be 30,000. The mouse is the first mammal to have its full genome sequence completed. The mouse genome is 14 percent smaller than the human genome, but over 95 per cent of the mouse genome is similar to human.
2002	International HapMap Project	The International HapMap Project, which was designed to identify genetic variations contributing to human disease through the development of a haplotype map of the human genome, began.
2002	Eckhard Wimmer	Synthesized the first complete genome of a poliovirus in a test tube. The work is considered to be a milestone for the emerging field of synthetic biology.
2002	Malcolm J. Gardner and many others	The genome sequence of the parasite *Plasmodium falciparum* that causes the majority of human malaria is completed.
2003	British Columbia Cancer Agency	Released the SARS-associated coronavirus genome sequence (30 Kbp).
2003	Francis Collins and many others	Published full sequence of the human genome and confirms that humans have approximately 20,000-25,000 genes.
2003	Consortium of more than 30 research groups	The ENCODE (Encyclopedia of DNA Elements) project was launched by the National Human Genome Research Institute and aims to identify and characterise all the genes in the human genome.
2004	Solexa (now Illumina)	Developed the sequencing-by-synthesis technology. This is the first of the genome sequencing technologies developed.
2005	International HapMap Consortium	HapMap (Haplotype map) report was published in *Nature*.

2005	The Chimpanzee Genome Project	The chimpanzee genome is completed.
2006	Shinya Yamanaka	First introduced the induced pluripotent stem cell (iPSC) technology.
2006	Andrew Fire and Craig Mello	Won the Nobel prize for discovery of RNA interference (RNAi).
2008	1000 Genomes Project Consortium	The 1000 Genomes Project (1KGP) was launched.
2009	The Wellcome Trust, Sanger Institute	The first comprehensive analysis of cancer genomes is published, including lung cancer and malignant melanoma.
2010	Hamilton Smith and colleagues	Succeeded in incorporating a synthetic genome into a bacterium. The bacterium with the transplanted genome develops and passes it on to its daughter cells.
2010	Pacific Biosciences	Released single molecule DNA sequencer.
2010	Craig Venter and his team	Created the world's first synthetic life form (a 1.08-mega-base pair *Mycoplasma mycoides* JCVI-syn1.0 genome).
2012	Jennifer Doudna, Emmanuelle Charpentier, and their teams	Discovered CRISPR Genome Editing Tool.
2012	The ENCODE (Encyclopedia of DNA Elements) Project Consortium	ENCODE study published 30 research papers describing the active regions of the human genome including confirmation that the human genome contains 20,687 protein-coding genes.
2013	DNA Worldwide, Eurofins Forensic	Discovered identical twins have differences in their genetic makeup.
2013	Wellcome Trust, Sanger Institute	The Zebrafish genome sequence was completed
2015	AquaBounty	The first GMO Salmon was sold in Canadian Markets.
2015	Junjiu Huang	First used CRISPR to edit human embryos.

2017	The Food and Drug Administration (FDA), USA	Approved the first CAR T-cell Therapy for Cancer.
2018	He Jiankui affair	Announced the birth of the first gene-edited babies using CRISPR-Cas9.
2019	Andrew Anzalone and his colleagues	Developed the gene editing technique 'prime editing'.
2020	The Telomere-to-Telomere (T2T) Consortium	Produced Telomere-to-Telomere assembly of human X chromosome
2023	The Human Pangenome Reference Consortium	Released its first *pangenome draft* based on 47 diploid assemblies, capturing far more global genetic diversity than a single reference genome.
2025	The Human Pangenome Reference Consortium	Released Data Release 2, which includes high-quality phased genomes from over 200 individuals, nearly five times more than in Release 1.

Scope of Genetics

Genetics is considered to be a central subject in biology. In the last few decades, the science of genetics has pervaded all aspects of biology. While on the one hand, genetics is used for the study of the mechanism of heredity and variation, on the other hand it has provided tools for the study of the fundamental biological processes examined and taught in areas, like physiology, biochemistry, ecology, pathology, microbiology, *etc.* Genetic insight helps up understanding humanity and human society. Genetics has many practical applications which are of great value to human beings. The modern science of genetics influences all aspects of our lives. The science of genetics also had a tremendous impact in applied areas including medicine, agriculture, forestry, fisheries, law and religion. The recent upsurge of biotechnology has added further to the significance of the science of genetics, so that the products of genetics have also become a subject of discussion for Trade Related Aspects of Intellectual Properties (TRIPs) under the aegis of General Agreement on Tariffs and Trade (GATT). Significance of genetics also stems from the fact that the genetic material containing information for hereditary traits consists of nucleic acids only, across the entire spectrum of life on the earth. Genetics has innovative work in the scientific field, with the technology strategies genetics has the key technology that can offer solutions for the challenges of the present day.

No field of science has changed the world more, in the past 50 years than genetics. The scientific and technological advances in genetics have transformed agriculture, medicine and forensic science *etc*. In broad sense we discuss the scope of genetics in some major portion of life.

Genetics and agriculture

From Mendel's principles to the application of the most modern molecular technologies, genetics has impacted all of agriculture. The contribution of genetics to increased food production is one of the premier success stories of science in the twentieth century. The Green Revolution, which dramatically increased crop productivity on a global scale, is a genetic success story about the breeding of highly productive strains of certain crop species. Studies of the genetic composition of economically important plants have enabled plant breeders to institute rational programs for developing new varieties. Some food crops (oranges, potatoes, wheat, and rice) have been genetically altered to withstand insect pests, resulting in a higher crop yield. Tomatoes and apples have been modified so that they resist discoloration or bruising. Our knowledge of genetics has also been applied to animal improvement with impressive result, such as large increases in meat, milk, and egg production in domestic animals. Genetic makeup of cows has been modified to increase their milk production, and cattle raised for beef have been altered so that they grow faster. Through the application of genetics, scientists have been able to produce domestic animals with superior qualities.

Genetics in health care

Genetic technologies have a rapidly expanding role in health care. Human beings are at risk for several thousand different inherited diseases. Inherited genetic diseases are caused by abnormal forms of a single or a group of genes that are passed on from one generation to the next. Some single genes responsible for inherited genetic diseases, such as the ones responsible for familial Alzheimer's disease, familial breast cancer, Duchenne muscular dystrophy and cystic fibrosis are being isolated and characterized at molecular level. More complex human hereditary disorders, those which are caused by the interaction of several genes, also interacting with the environment, such as heart diseases, hypertension, diabetes, various forms of cancer and infections, are also being investigated using a molecular approach. Genetic techniques are used in medicine to diagnose and treat inherited human disorders. Cells from embryonic tissues reveal certain genetic abnormalities, including enzyme deficiencies, which may be present in newborn babies, thus permitting early treatment. Scientific advancements in genetic medicine are enabling better means to diagnose disease, to predict risk, and to direct and personalize treatment. Gene therapy is used in treating some devastating conditions, including some forms of cancer and cystic fibrosis. Gene therapy is based on modification of defective genotypes by adding functional genes made through recombinant DNA technology. Genetically engineered vaccines are being tested for possible use against HIV. For many diseases, an accurate diagnosis can be made more quickly and accurately through a study of one's family history than through elaborate and expensive laboratory tests. Genetics is also important in preventive medicine. In many cases, it is possible to anticipate the development of a disease or other body abnormalities based on family history. Thus, appropriate steps can be taken to prevent its occurrence. Genetic testing to predict individual risk of developing certain diseases, such as breast or colon cancer, can greatly reduce the risk of developing disease by timely intervention strategies.

Genetics and law

The collection and analysis of DNA is an important tool in law enforcement. Genetic technology – specifically the development of 'DNA fingerprinting', has led to the widespread use of genetic testing for identification. DNA recovered from semen, blood, skin cells, or hair found at a crime scene can be analyzed in a laboratory and compared with the DNA of a suspect. Genetic testing has also been used to exonerate those wrongly accused of a crime by demonstrating that their DNA does not match the evidence found at the crime scene. Beyond the criminal context, DNA fingerprinting has been used to identify victims of disasters and to confirm or disprove paternity and maternity.

Genetics and biotechnology

The recent upsurge of biotechnology has added further to the significance of the science of genetics. It is estimated that the biotechnology industry, which has genetics as its basis, will become one of the top moneymaking industries in the coming decades. Gene cloning has made available large quantities of substances previously in short supply *e.g.*, insulin, interferon, growth hormone, growth factors, blood clotting factors VIII and IX, plasma protein α1 antitrypsin, cyclosporine, interleukin 2, *etc*. Genetic engineering allows scientists to replicate, improve or even completely change the natural functions of the organisms we rely on for food. A genetically engineered version of an enzyme or a protein can often be produced more efficiently than its natural counterpart. Genetically engineered traits like disease and insect resistance are now common in crops like corn and soybeans. Foods could be genetically engineered to have fewer calories, increased vitamins and minerals, or have allergens removed. The very first whole food genetically engineered was the *Flavr Savr* Tomato. Agricultural breeders also use biotechnology to move genes across taxonomic barriers, combining genetic material from species that would not cross-breed naturally. For example, Bt corn has been modified by inserting a gene from the bacterium *Bacillus thuringiensis* that kills harmful insects so that farmers do not need to use insecticide.

CHAPTER TWO   FUNDAMENTALS
OF MENDELIAN GENETICS

—————— ⇒ ⇒ ⇒ ⋅ ✳ ⋅ ⋐ ⋖ ⋗ ——————

Modern genetics is based on the concept of gene. Geneticists are concerned with the transmission of genes from generation to generation, with the physical structure of genes, with the variation in genes, and with the ways in which genes dictate the features of a species. Mendel was the first to recognize the existence of genes. The concept of gene (but not the word) was first proposed by Gregor Mendel in 1865. There have been several explanations on the possible mechanism of inheritance of traits from the parent to the offspring before the rediscovery of Mendel's laws of inheritance in 1900. Early philosophers and scholars had forwarded their own opinion and theory to explain the phenomenon of inheritance. All these theories are collectively known as pre-Mendelian view or theory.

Pre-Mendelian Concept about Heredity

The pre-Mendelian theories are briefly presented below.

Moist Vapour Theory

To explain why children looks like their father, The Greek philosopher and mathematician Pythagoras (580-500 BC) proposed that each organ in the body of male released some kind of moist vapour during coitus, and that the new individual is formed by the combination of vapours from different organs. Children look like their mother because they develop in mother's body.

Fluid Theory

Greek pre-Socratic philosopher and physiologist Empedocles (494-434 BC) proposed that each body part produces a fluid. The fluid of different body parts of the two parents mixes up and is used in the formation of an embryo. Any defect in the descent process and fluid mixing results one or both parents' characteristics to be missing. Child phenotype changes according to the ratio of the fluids.

Reproductive Blood Theory

The famous Greek philosopher and scientist Aristotle (384-322 BC) proposed that the males produce extremely pure reproductive blood containing the nutrients from all body parts. Females also produce reproductive blood but this is impure. The two reproductive bloods coagulate in the body of the female and form the embryo. Due to the purity of reproductive blood, male contribute more characteristics than the female. This belief lasted very long. He also held that the female semen just provides the inert substance while the vital ingredient required for the formation of an embryo was provided by the male semen.

Preformation Theory, popular in the late 17th to late 18th century

Two Dutch biologists, Swammerdam and Bonnet (1679) proposed that the organism is already present or preformed in the ovum (according to the ovists) or spermatozoon (according to the animalculists) as a tiny structure called homunculus that unfold during development. Homunculus can be male or female. Malpighi (1673) promoted the preformation theory. Many people of that time, including Hartosoeker (1694) and Dalepatius (1694), held this belief. It was also supported by Roux as recently as 1888, but rejected by many scientists because this could not be proved scientifically.

William Harvey (1578-1657) cut of the uterus of recently mated dears and looked for tiny dears to prove preformation hypothesis. But he couldn't find it. He saw embryo development after few days of mating. He hypothesized that semen solely has a vitalizing function and that all animals develop from eggs.

Theory of Epigenesis

German physician and naturalist Caspar Friedrich Wolff (1734-1794), one of the founders of Embryology, proposed that structures arise during development that are not already preformed (Wolff 1759, 1764). Only after fertilization the zygote begin to differentiate into different organs and body parts, leading to the growth of adult tissues and organs. This is one of the most accepted and true theory of that time.

Theory of Pangenesis

French mathematician and astronomer Pierre Louis Moreau de Maupertuis (1698-1759) proposed that heredity is controlled by minute particles which come from all parts of the body of both mother and father to the reproductive organs. The particles combine in the embryo and multiply there. Certain particles of one parent can dominate over those of other so that the offspring resembles one parent in certain characteristics and the second parent in others. This theory prevailed for many centuries and was accepted by Charles Darwin. In 1868 English naturalist Charles Darwin (1809-1882), the father of evolution, modified the above theory and proposed that each parts of the body generate its own type of minute particles called the gemmules or pangene. These particles are carried by the blood to the reproductive organ and are deposited in the gametes, which again carry them to the next generation. After fertilization the gemmules reproduce the same kind of organ, cell, tissue, *etc.* in the next generation. The pangenes from both the parents mix-up and as a result mixed characteristics appear in the offspring. By this mechanism, acquired characters would also be inherited because as the body parts changed, so did the pangenes or gemmules they produced. A defective gemmule will lead to the development of defective organ in an individual. This theory was rejected too because it lacked scientific basis.

Theory of Acquired Characters

Famous French biologist Lamarck (1774-1829) proposed that a new character once acquired by an individual shall pass on to its progeny. It means if a man develops a strong muscle by exercise, all his children would have strong muscles. On the other hand, if a person becomes weak all his children would be weak. Lamarck takes the example of the long neck of a giraffe. This theory was very popular with the name Lamarckism but failed to give a physical basis of the theory. This concept was totally rejected by Weismann on the basis of experiment on mice performed through 22 generations. He cut the tail of mice for successive generations and always got the baby mice with tail.

Germplasm Theory, concept of the physical basis of heredity

In 1883 German evolutionary biologist August Friedrich Leopold Weismann (1834-1914) proposed that organism's body contains two types of cells, namely somatic cells and germ cells. The somatic cells form the body and its various organ systems, while the germ cells form the gametes. The somatic cells contain the 'somatoplasm' and germ cells contain the 'germplasm'. The germplasm can form somatoplasm, but somatoplasm cannot form germplasm. Germplasm is the essential element of germ cells (eggs and sperm) and is the hereditary material that is passed from generation to generation. Any change in the germplasm will lead to change in the next generation. This theory was accepted in a broad sense. This view contradicted Lamarck's theory, which was a prevalent theory of heredity of the time. Although the details of the germplasm theory have been modified, its premise of the continuity of hereditary material is the basis of the modern understanding of inheritance.

All these early theories presume that the characteristics of the two parents blended or got mixed during coitus to form the characteristics of the offspring. Hence these ideas came to be known as **'blending theories of inheritance'**. Once blended like two liquids in solution, the hereditary material is inseparable and the offspring's traits are some intermediate between two parents. For example, suppose a red-flowered plant were crossed with a yellow-flowered plant of the same species. According to the blending theory, the red and yellow hereditary material would blend to produce orange-flowered offspring – like red and yellow paint blend to make orange paint. Based on this hypothesis, all offspring of orange-flowered plants would also have orange flowers.

There are some obvious problems associated with the theory of blending inheritance, one of which is that offspring are not always an intermediate blend of their parent's characteristics. A second problem with this idea is that if the characteristics of the offspring always existed qualitatively between those of the parents, eventually all members of a species would become homogenous, displaying the same characteristics. The blending hypothesis was eventually discarded because it could not explain how traits that disappear in one generation can reappear in later ones (atavism).

In the nineteenth century, most biologists worked by observing and describing nature. Gregor Mendel was one of the first to apply an experimental approach that revealed the basic rules underlying patterns of inheritance. He was able to observe the mathematical patterns of inheritance from one generation to the next generation. For seven years, Mendel bred pea plants and recorded inheritance patterns in the offspring. Based on his results, Mendel proposed the '**particulate theory of inheritance**' in 1865. He described the unit of heredity as a particle that does not change. He proposed that characteristics are determined by paired discrete units called particles/factors (that we now call genes) that are passed on intact from parent to offspring through gametes. Each trait, such as color, is controlled by two factors, one from the male parent and one from the female parent. The two factors/alleles do not mix together (blend) or contaminate each other but they just remain together. These heritable factors retain their identity generation after generation, even if they do not appear in each generation. In other words, genes are more like marbles of different colors than paints. Just as marbles retain their individual colors, genes retain their own identities. Mendel implies that mother transmits a red colored marble and father contributes a yellow colored one. The offspring transmits one of the two marbles – totally intact – to the next generation. Factors/particles/genes are thus the physical basis of inheritance.

Mendel did not know what genes are like, or where they are located. Today we might say discrete alleles are passed from parents to offspring and maintain their integrity across generations. Mendelian principles apply to humans as well as to peas and all other eukaryotes.

Mendel's great contribution was that he replaced the **blending theory** with the **particulate theory**. Mendel first presented his findings in 1865, but they were not accepted then and remained unknown for many years. Their rediscovery in 1900 by de Vries of Holland, Carl Correns of Germany and Tschermak of Austria independently, led to the beginning of modern genetics.

Gregor Mendel, Austrian Monk, the Father of Modern Genetics (July 20, 1822-January 6, 1884)

The first person to gain some understanding of the principles of heredity was Gregor Johann Mendel, whose work on heredity is considered to be the foundation of modern genetics. Mendel was born into an ethnic German poor farming family in July 20, 1822 on his family's farm in the village of

Heinzendorf in the district of Moravia, Austria, now known as Hyncice in the Czech Republic. Mendel was an Austrian monk, lifelong learner, teacher and, scientist with interests in astronomy and plant breeding. Since he was a child, he had contact with agricultural activities. Mendel was a keen gardener and Beekeeper during his childhood at the farm. At the end of high school, he was admitted as a novice at the Augustinian monastery of St. Thomas in 1843 in the city of Brünn, Austria, now Brno of the Czech Republic. The monastery had a botanical garden and library. It was a centre for science, religion and culture. He was fortunate in having an abbot who encouraged intellectual and research pursuits, and spent his early years mastering mathematics, physics and chemistry. Later, he became particularly interested in the study of hybrids, a "hot topic" in biology at the time.

Mendel worked as a substitute high school teacher. In 1850 he failed the oral part, the last of three parts, of his exams to become a certified high school teacher in natural science. His monastery was dedicated to teaching science and to scientific research. To enable him to further his education, the abbot arranged for Mendel to attend the University of Vienna to get a teaching diploma. In 1851 he was sent to the University of Vienna. He studied at the University of Vienna from 1851 to 1853 where he was influenced by physicist Christian Doppler who encouraged experimentation and the application of mathematics to science and a botanist Franz Unger who aroused Mendel's interest in the causes of variation in plants. In his studies at the University of Vienna, Mendel paid much attention to physics, mathematics, chemistry, entomology, and plant anatomy and physiology, and acquired the theoretical background and the skill required to perform experiments and undertake independent research. Mendel did not perform well. Mendel's examiner failed him with the comments, "he lacks insight and the requisite clarity of knowledge". He returned to the monastery as a failure in 1853 where he was an 'unaccredited' teacher for many years. He devoted all his free time to his long-term research program on plant hybridization

In the summer of 1854, Mendel grew thirty-four strains of common garden peas (*Pisum sativum*); he tested them for constancy in 1855. In 1856 he began his breeding experiments with different varieties of garden peas and systematically analyzed the results of his tests in an attempt to learn something about the mechanism of heredity. Between 1856 and 1863 Mendel carefully analysed seven pairs of contrasting characters. Mendel confirmed his findings by testing his crosses through at least four generations. As a result of his creativity, Mendel discovered some fundamental principles of genetics. The real importance of his research was to demonstrate the inadequacy of blending inheritance and propose a new model involving the transmission of discrete units. In 1865, Gregor Mendel gave two lectures (delivered in German) on February 8 and March 8 and presented the results of his experiments on pea plants, to the Brünn Society for the Study of Natural Science. Mendel turned these lectures into 48-page paper, "Experiments in Plant Hybridization" published in the 1866 issue of the proceedings of that society. That paper reported research done by Mendel from 1854-1863 involving almost 28,000 plants, of which he "carefully examined" 12,835 plants. The Brünn Society had a considerable exchange list, and its proceedings were sent to more than 120 libraries. Mendel also had about 40 reprints of his work which he sent to biologists throughout Europe, but Mendel's contemporaries did not appreciate his findings. Mendel sent a copy of his major paper to Karl Nageli, who was professor of botany at Munich and a major figure of his time in biology. He was also interested in heredity and was actively working on it. He completely failed to appreciate Mendel's work and made some rather pointless criticisms of it in his reply to Mendel's letter. Mendel was elected abbot of the monastery in 1868, and his scientific work effectively came to an end. Until his death in January 6, 1884, he spent much of his time engaged in administrative conflicts with the local government. After his death, Mendel's personal papers were burned by the new abbot. Luckily, some of the letters and documents generated by Mendel were kept in the monastery archives. His observations led to laws regarding the transmission of hereditary characteristics from generation to generation. Mendel's observations led him to coin two terms which are still used in present-day genetics: dominance and recessiveness.

Rediscovery of Mendel's Principles

The importance of Mendel's experiments with respect to the field of genetics was unfortunately paid no attention and ignored for 34 years. In 1900, 16 years after his death, Gregor Mendel's pea plant research finally made its way into the wider scientific community. The Dutch botanist and geneticist Hugo Marie de Vries, German botanist and geneticist Karl Franz Joseph Correns and Austrian agronomist Erich von Tschermak (three turn-of-the-century scientists), independently rediscovered Mendel's work. When they reviewed the literature before publishing their own results, they were startled to find Mendel's old papers spelling out those laws in detail. Each man announced Mendel's discoveries and his own work as confirmation of them. They realized that the rules originally deduced by Mendel were also fulfilled in several animal and plant species. Therefore, they were accorded the status of "laws." Hugo de Vries republished the Mendel's work in 'Flora'. Its rediscovery prompted the foundation of the discipline of genetics.

Other Experiments

Mendel began his studies on heredity using mice. But his bishop did not like one of his friars studying animal sex, so Mendel switched to plants. Mendel also worked with honeybees to determine genetic traits in animals. He also studied astronomy and meteorology. The majority of his published works were related to meteorology.

The Second Type of Mendelian Inheritance

Mendel conducted research with more than 20 plant species to verify the laws of inheritance he discovered using pea plant. All of them followed the rules except "hawkweed". Hawkweeds, or Hieracium species, are native to Eurasia. In his experiments on hawkweed, Mendel found that neither the first rule nor the second rule he had observed in the pea plant held true. The F_1 hybrids from crossings between "true breeding" strains were not morphologically uniform in appearance. Mendel observed that many of the plants were identical to the maternal parent, others resembled the pollen donor, and there were even a few in between. The F_2 progeny derived from examined F_1 plants were uniform and resembled their parent in morphological appearance. Uniform progeny continued to be stably maintained in subsequent generations and did not segregate for any characters. Mendel published his findings on hawkweed in 1869. He argued that in nature there must be two distinct types of inheritance in plants: the *Pisum* type where trait segregation occurred, and the *Hieracium* type where trait segregation was largely absent (Nogler, 2006). But he went no further.

Rediscovery of *Hieracium*-Type Heredity

From 1904 to 1910, two researchers, Ostenfeld and Rosenberg, worked on hawkweed, and even repeated Mendel's earlier experiments. They observed that *Hieracium* is a genus of plants of the daisy family, *Asteraceae* in which some species reproduce by the normal sexual process. Seeds are formed by fusion of male and female gametes. But there are other species of the *Hieracium* genus that reproduce wholly or partially asexually, by a phenomenon whereby seeds develop from either non-reproductive somatic cells close to female gametes, or from female gametes that do not need to be fertilized by male gametes. These seeds and plants are effectively clones of the female parent. The phenomenon is termed "apomixis" (from the Greek *apo*, "lack," and *mixis*, "mingling"). The hawkweed plants that Mendel chose for his experiments happened to be totally or partly apomictic.

The absence of male gametes in the formation of seeds explains why, for example, unlike the case with peas, Mendel's first generation of hybrids in hawkweed was not uniform. Many of the hybrids inherited the traits of the female parents selected by Mendel, which were those involved in seed formation by apomixis. And hybrids displaying traits of male parents or a mixture of both parents were formed by "normal" sexual reproduction.

What we might call *Hieracium*-type heredity is due to the fact that two kinds of seeds and plants are formed. One kind originates from sexual reproduction, involving male and female gametes. While a second kind of seeds and plants originate by asexual reproduction, from female gametes alone or from somatic cells. The former kind follows Mendel's laws while the later does not.

In the end, therefore, Mendel was right to conclude that *Hieracium*, or hawkweed, is an interesting exception to the rules of heredity he had observed in the pea and other plants. It was later discovered that the exception was due to hawkweed's special mode of reproduction: a mixture of "normal" sexual reproduction and a form of asexual reproduction. It is now known that in nature there are a wide range of animals and plants that display this special *Hieracium*-type heredity. It occurs in about 50 families and 150 genera of plants. For animals, *apomixis* is termed *parthenogenesis* (from the Greek *parthenos*, "virgin," and *genesis*, "creation").

Genetic Terminology

Asexual reproduction: A single individual is the sole parent and passes copies of all its genes to its offspring. In asexual reproduction one parent produces genetically identical copies of the parent by mitosis. Each offspring in asexual reproduction is called a clone.

Clone: A clone is a large population of identical molecules, bacteria, or cells that arise from a common ancestor. Researchers are interested in cloning individual genes so that these genes can be studied in detail.

Sexual reproduction: The process by which organisms produce offspring through the fertilization of gametes from a male and a female parent. This is the default form of reproduction in nature. Sexual reproduction is at the heart of genetics. During sexual reproduction the female gamete (egg) is fertilized by the male gamete (sperm), which means that the egg and sperm fuse together. In sexual reproduction two parents give rise to offspring that have unique combinations of genes inherited from the two parents. Sexual reproduction is the most successful form of reproduction within eukaryotes. The variation that sexual reproduction creates among offspring is very important to the survival and reproduction of the population.

Perfect flower: A **perfect flower** is one that contains both male and female reproductive parts within a single flower (Image 2.1). In contrast, **imperfect flowers** possess only one type of reproductive organ – they are either **male** or **female** flowers.

Stamen: The **male reproductive part** of a flower is called the **stamen**. It consists of an **anther**, where pollen is produced, supported by a slender stalk called the **filament** (2.1).

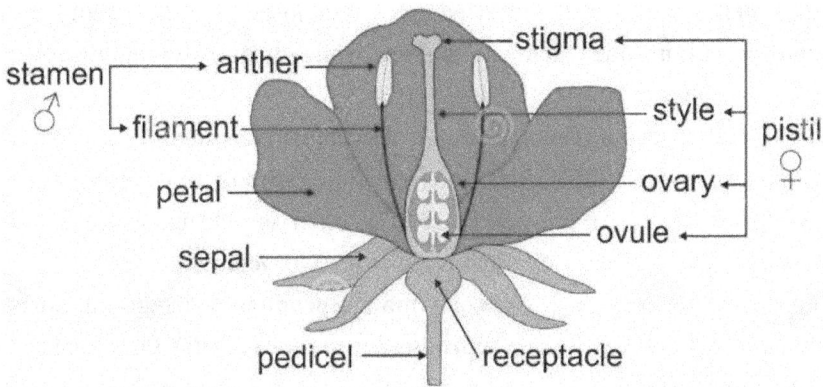

Image 2.1 Parts of a perfect flower (source:
https://www.dreamstime.com/stock-illustration-parts-flower-educational-
illustration-angiosperm-morphology-image58036575).

Anther: The anther is the part of the stamen that produces and bears pollen
(Image 2.1). It typically contains **four pollen sacs**, each responsible for
generating pollen grains. Each pollen grain is a **single cell containing two
male gametes**. During cross-pollination, pollen is collected from the male
parent. Both male gametes participate in **double fertilization**, a process
unique to flowering plants, resulting in the formation of a **zygote** and
endosperm. When mature, the anther splits open to release the pollen.

Pistil: The **female reproductive structure** of a flower is called the **carpel** or
pistil (Image 2.1). It consists of three main parts: the **stigma**, the **style**, and
the **ovary**.

Stigma: The **stigma** is the part of the pistil that receives pollen in a flower
(Image 2.1). During cross-pollination, the pollen from the male parent is
placed on the stigma of the female parent.

Style: The **style** is the part of the pistil that connects the stigma to the ovary
and **transports the sperm cells** from the pollen to the ovule for fertilization
and subsequent development of the egg (Image 2.1).

Ovule: The **ovule** is a structure within the ovary of a flower that develops into
a seed after fertilization (Image 2.1). It is the site where **meiosis, female
gamete development**, and **fertilization** occur.

Angiosperms or Angiospermae: Angiosperms are a group of plants whose
main characteristic is the presence of **true flowers**. They are divided into

monocotyledons and dicotyledons. Examples of monocotyledon angiosperms include corn and palms, while dicotyledon angiosperms include sunflower, apple, and rose.

Dioecious/Unisexual/Gonochoric: A dioecious individual is a plant or animal that possesses only male or only female reproductive organs, and its sex remains the same throughout life (*i.e.*, non sex changing species). In dioecious species, each individual is either male or female.

In plants, some individuals have only male reproductive organs (stamens), while others have only female reproductive organs (pistils). Only about 5% of all plants are dioecious. Examples of dioecious plants include papaya, date palm, spinach, and asparagus. Different cases of sex determination are observed in dioecious plants.

A wide range of organisms in the animal kingdom is dioecious. Examples of dioecious species include humans, dogs, cats, *etc*.

Monoecious / Bisexual / Hermaphroditic species: Individuals that possess both male and female reproductive organs, allowing them to perform both male (sperm/pollen production) and female (egg/ovule production) functions, either simultaneously or sequentially (*i.e.*, in sex-changing species). In animals, hermaphroditism is most common in invertebrates, such as flukes, worms, snails, slugs, and barnacles. It also occurs in some fish species but is rare among other vertebrates.

In plants, monoecious species are of two types:

1. **Monoecious plants with unisexual flowers:** These plants have separate male and female flowers on the same individual. They comprise only about 5% of all plants. Examples include maize, coconut, cucurbits (squash, cucumber, pumpkins, watermelon), oak, and pine.

2. **Monoecious plants with bisexual (hermaphroditic) flowers:** These plants have both male and female reproductive structures in the same flower. This is a very large group, including most native trees, shrubs, and vegetables. About 90% of all plants fall into this category. Examples include apple trees, cherry trees, roses, olive trees, tomatoes, beans, cabbages, strawberries, and hibiscus.

In monoecious plants, there are **no special sex chromosomes**; instead, the formation of **ova and pollen** occurs through **histological differentiation**.

Intersex: Usually reserved for individuals of intermediate or indeterminate sexual differentiation. This state is not normal and the affected individuals are often sterile.

Gamete: A mature reproductive cell that is specialized for sexual fusion. Each gamete is haploid and fuses with a cell of similar origin but of opposite sex to produce a diploid zygote.

Fertilization: Fertilization is the process of joining the male gamete with the female gamete. A single egg is fertilized by a single sperm in a random manner (Image 2.2).

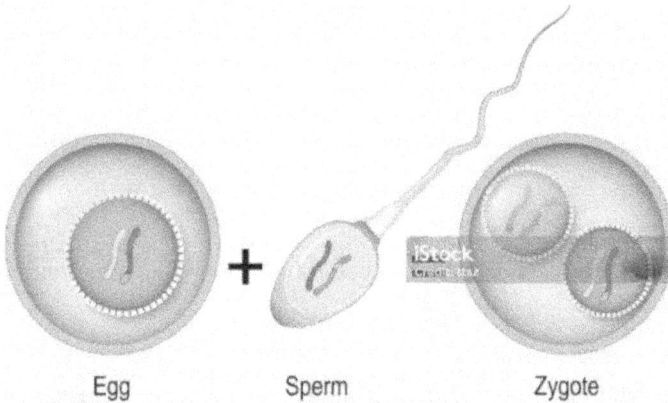

Egg   Sperm   Zygote

Image 2.2 Fertilization (source:
https://www.istockphoto.com/vector/fertilization-zygote-is-egg-plus-sperm-gm687728436-126650857?searchscope=image%2Cfilm).

Pollination: Pollination is the transfer of pollen from the **stamen** (male part) to the **stigma** (female part) of a flower. **Self-pollination** occurs when pollen is transferred to the stigma of the **same flower** or to a flower on the **same plant**. **Cross-pollination** occurs when pollen is deposited on the stigma of a **different plant** of the same species.

Self-fertilization (also selfing): The fusion of male and female gametes produced by the same (hermaphrodite or bisexual) individual.

Zygote: The cell produced by the fusion of male and female gametes (Image 2.2). All animals originate from the union of a single haploid cell from the female parent (ovum or egg) and a single haploid cell from the male parent (spermatozoa or sperm). The result of this union is a zygote (diploid cell), which develops into a new animal with a full set of chromosomes.

Mutation Breeding: A technique in which seeds or plant cuttings are exposed to **mutagens** (such as radiation or chemicals) to induce **random mutations**, some of which may result in **new and beneficial traits**.

Chromosome Doubling: A process that **inhibits normal cell division** to produce cells with **twice the usual amount of DNA**. It can occur naturally or be induced using **radiation** or **chemical treatments**.

Monozygotic/identical twins: Twins derived from a single fertilized ovum. Differences between monozygotic twins later in life are virtually always the result of environmental influences rather than genetic inheritance.

Dizygotic/fraternal twins: Twins produced from two separate ova that are separately fertilized. Fraternal twins may look similar but are not genetically identical.

Cross: A mating between two individuals, leading to the fusion of gametes.

Back Cross: The back cross is a cross of an F_1 hybrid with any one of the parental genotypes.

Test Cross: A test cross is a cross between an individual with an unknown genotype with a homozygous recessive genotype.

Reciprocal cross: A cross with the phenotype of each sex reversed as compared with the original cross. If the results of reciprocal crosses are the same, the interpretation is that the trait is not dependent on the sex of the organism.

Backcross population: A backcross population consists of individuals produced by crossing a **hybrid organism** with **one of its parents**. In plant breeding, backcrossing is commonly used to **introduce or retain a specific trait** from one parent while maintaining most of the **genetic background** of the other parent.

Gene (Mendelian factor): Genes are the basic physical and functional units of inheritance. The inherited traits are determined by genes that are passed from parents to offspring. Genes code for proteins that determine virtually all characteristics of living organisms. Humans have estimated 20,687 protein-coding genes.

Candidate Gene: A gene that is suspected of being responsible for a particular trait of interest.

Locus: A gene is indeed located at a specific position on a chromosome This specific location on a chromosome is called a locus (plural, loci). Because chromosomes are paired there are two such loci and thus every animal/plant has two alleles of a gene (Image 2.3). Each chromosome contains many genes, and each gene occupies a unique locus.

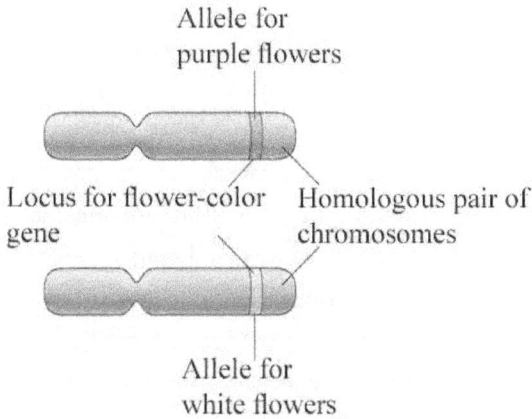

Image 2.3 Gene and locus.

Alleles or Allelomorphs: A gene can exist in several alternative forms owing to differences in DNA sequence. Alleles are the alternative forms of the same gene present at the same locus on the homologous chromosomes that code for alternative forms of the protein (Image 2.3). An allele can be either dominant or recessive. Each individual carries two alleles of each gene, one from each parent. For example, sheep and cattle can be polled or horned. One gene codes for this trait and the two alleles of the gene are polled or horned. The polled allele is dominant and is, therefore, denoted by *P*, while the horned allele is recessive and is denoted by *p*.

The terms genes and alleles are potentially confusing. When alleles like *P* and *p* are examined at the DNA level using modern technology, it is generally found that they are identical for most of their sequence, and differ only at one or a few nucleotides out of the thousands that make up the gene.

Genetic polymorphism: Genetic polymorphism is the occurrence of more than one allele at a given locus in the same population. If the frequency of the most common allele is less than 0.99 the gene is considered polymorphic;

otherwise, it is monomorphic. Many genes in natural population are polymorphic.

Character: Character means a specific property of an organism; geneticists use this term as a synonym for characteristic or trait.

Trait: A trait is a variation of a particular character. For example, one plant might have the trait of red flowers, while another might have the trait of yellow flowers.

Inherited trait: The characteristic that is transmitted through genes from parents to offspring is called inherited (hereditary) trait (*e.g.*, eye color and dimples in humans).

Acquired trait: Acquired traits are characteristics that we are not born with, but that are obtained during the lifetime of an individual as a result of their experiences (*e.g.*, pierced ears, dyed hair, and scars in humans).

Phenotype: Visible or otherwise measurable properties of an individual plant/animal are called its phenotype. More generally, a phenotype is any characteristic of an organism that can be determined; its appearance, physiology, behavior, *etc.* Round seeds, wrinkled seeds, yellow seeds, and green seeds are all phenotypes. Phenotype is determined by interaction of genes and environment. Genes provide potential, but environment determines whether that potential is realized. Any organism is a by-product of both its genetic makeup and the environment.

Genotype: Genotype is the particular combination of alleles (genetic makeup) responsible for creating the phenotype of an organism. With respect to seed shape in peas, *WW*, *Ww*, and *ww* are examples of the possible genotypes for the *W* and *w* alleles. Because gametes contain only one allele of each gene, *W* and *w* are example of genotypes of gametes.

Genotyping: Genotyping is the process by which differences in the genotype of individuals are analyzed using biological assays. Genotyping techniques includes PCR, DNA Microarray, Allele-specific Oligonucleotide (ASO) Probes, DNA Hybridization, *etc.*

Homozygous (pure): An organism with two identical (same) alleles for a character is said to be homozygous for the gene controlling that character. A homozygous organism may be homozygous dominant (*AA*), or homozygous recessive (*aa*) (Image 2.4). The terms homozygous and heterozygous cannot apply to gametes, which contain only one allele of each gene. Homozygote always produces single type of gametes.

Heterozygous/hybrid: An organism that has two different alleles for a gene is said to be heterozygous for the gene controlling that character (Image 2.4). A monohybrid is heterozygous for one pair of alleles (*e.g.*, Aa). Individual heterozygous for two alleles of a gene produce two kinds of gametes.

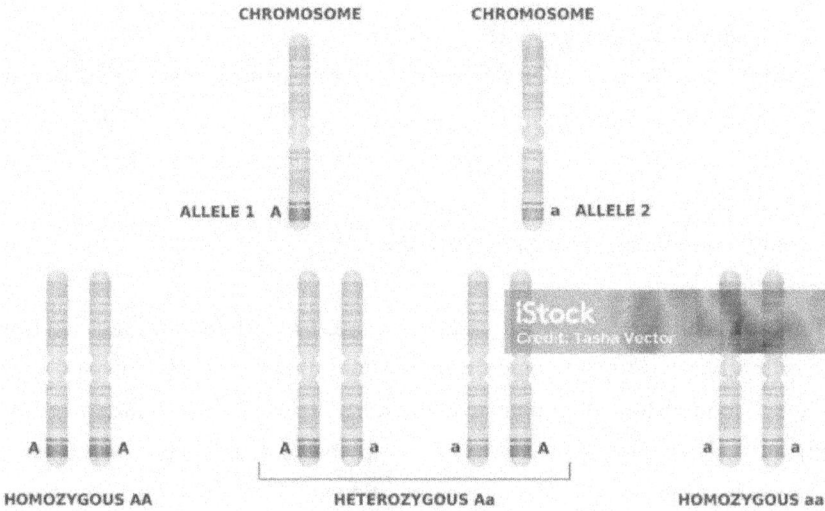

Image 2.4 Homozygous and heterozygous genotype (source: https://www.istockphoto.com/vector/homozygous-and-heterozygous-homozygous-has-same-allele-for-a-particular-trait-gm2065041082-564190010).

Carriers: Carriers are heterozygous individuals for a recessive genetic disorder but are phenotypically normal.

Double heterozygote (dihybrid): An individual that is heterozygous at two loci (YyRr) under investigation.

Haplotype: The word "haplotype" is derived from the word "haploid", and from the word "genotype". The term haplotype is used in genetics to loosely mean "half of a genotype". A haplotype is a set of alleles (or DNA variants) on a single chromosome inherited together from one parent. This group of genes was inherited together because of genetic linkage. For two loci A and B with two alleles each (A_1 and A_2, and B_1 and B_2, respectively) there are four possible gametes (or haplotypes) present in the population: A_1B_1, A_1B_2, A_2B_1, A_2B_2. A diploid individual's genotype will be something like A_1B_1/A_1B_2: it has two haplotypes, one inherited from each parent. In general, if there are **n loci**, each with **2 alleles**, then the maximum possible haplotypes = 2^n. The frequency of a haplotype in a population can be counted as the number of gametes bearing a particular combination of genes. The term "haplotype" can also refer to the inheritance of a set of single nucleotide polymorphisms (SNPs) found on the same chromosome.

Generation notation: When discussing different generations in genetics, the original members of a mating are referred to as the parental (P) generation. The first generation of progeny from the parental generation is referred to as the first filial generation or F_1 generation. When individuals from the F_1 generation are mated with each other, their offspring are referred to as the F_2 generation.

Dominant allele: The allele that is expressed where an individual is heterozygous for a given gene. In most situations, the normal, or wild type, allele is completely dominant over mutant alleles. In some instances, the mutant allele is dominant over the wild type; that is, the wild type is recessive. **Recessive allele:** The allele that is 'hidden' (not expressed) when an individual is heterozygous for a given gene.

Dominant and recessive phenotype: The phenotype expressed in hybrids is said to be dominant and the other recessive. For example, in crosses of purebred short-winged fruit flies with purebred long-winged fruit flies the progeny will all have long wings. Here long wings are dominant to short wings.

Wild-type (standard/common type): An organism that displays the usual phenotype for that species is called the wild type. Wild type allele appears most frequently in a population. Wild type allele generally confers the greatest fitness, and produces a fully functional gene product. In Drosophila, red eyes and full size (long) wings are wild type.

Mutant type: An organism whose usual phenotype has changed as the result of a mutation is called the mutant type. Mutant alleles are usually recessive (white and miniature alleles in Drosophila) but not always. Polycystic Kidney disease, and Huntington's disease alleles are mutant but dominant.

Monohybrid crosses: A monohybrid cross is a mating between two individuals that differ in a single characteristic or that are heterozygous for a single trait or gene (*e.g., Aa × Aa*). The simplest form of a cross is a monohybrid cross, which analyses a single trait and its associated variations.

Dihybrid crosses: A dihybrid cross is a cross between two organisms that differ in two characteristics or that are heterozygous for two different characters or genes (*e.g., Aa Bb × Aa Bb*).

Backcross: When F_1 individuals are crossed with one of the two parents from which they were derived.

Testcross: The mating of the individual to be tested (F_1) with a recessive homozygote. A testcross is designed to reveal whether an organism that displays the dominant phenotype is homozygous or heterozygous.

Pure-breeding (true-breeding) line: This refers to organisms in which all offspring produced by selfing or crossing within population show the same form of the character.

The Punnett-square and the forked-line method: There are two good methods for determining the theoretical outcome of any cross. One of these is the **Punnett-square**. The Punnett square is a diagram/chart showing the expected ratios of possible genotypes and phenotypes in the offspring of two parents. It shows parental gametes and the genotypes of next generation. The male gamete is normally indicated at the top and the female gamete is indicated in the vertical margin. Reginald Punnett developed the method to analyze data from genetic crosses that is particularly useful where small numbers of genes are involved. When higher order crosses (trihybrid and tetrahybrid crosses) are done, the dimensions of the Punnett square increase and this method of analyzing genetic data becomes unwieldy. The other method for determining the outcome of a cross is the **forked-line method**. This method takes advantage of the 'product law', and exploits the fact that a cross can be broken down into a set of monohybrid crosses. The forked line method is especially useful when three or more characteristics are crossed simultaneously.

Consanguineous mating: A mating between close relatives. Consanguineous mating increases the chance of mating between two carriers of the same rare allele. Consanguineous mating often produces offspring that have rare recessive, and often deleterious, traits. Most societies and cultures have laws or taboos against marriages between close relatives.

Genetic distance: A measure of the **genetic relatedness** or difference between populations, indicating how closely or distantly they are related at the DNA level.

Gene therapy: Gene therapy is the insertion, alteration, or removal of genes within an individual's cells and biological tissues to treat disease. It is a technique for correcting defective genes that are responsible for disease development.

Bioinformatics: Bioinformatics, also known as **Computational Biology**, is an interdisciplinary field that combines **molecular biology** and **computer science**. It involves the application of **computational and statistical techniques** to manage and analyze biological data. Key problems addressed in bioinformatics include **protein and nucleotide sequence alignment, protein structure prediction, evolutionary modeling, gene expression analysis,** and **prediction of protein-protein interactions**. The field has greatly enhanced **genomic and proteomic research**. Bioinformatics exercises often consist of **interrelated modules** designed to introduce students to modern techniques in this area.

Evolution: Genetic change in a population of organisms that occurs over time. Biological evolution refers to populations and not to individuals and that the changes must be passed on to the next generations.

Epigenetics: The discovery of **DNA** has greatly enhanced our understanding of life, but DNA alone does not explain how genes are regulated. The term **epigenetics** was coined in the 1940s by **Conrad Hal Waddington**, a developmental biologist, paleontologist, geneticist, embryologist, and philosopher. The prefix "**epi**" means "on" or "above" in Greek, so epigenetics refers to changes **"above" or "on top of" genetics**.

While **genetics** studies heritable changes in gene expression caused by direct alterations in the DNA sequence – such as **mutations, deletions, insertions,** and **translocations** – **epigenetics** studies heritable changes in gene expression **without altering the DNA sequence**. Epigenetics involves mechanisms that **turn genes on or off**, allowing different cell types to express specific sets of genes despite having the same genome.

The Epigenome: The **epigenome** encompasses all chemical modifications that regulate gene expression within the genome. These modifications ensure that each cell produces only the proteins necessary for its function. For example, proteins promoting **bone growth** are not produced in **muscle cells**. Epigenetic patterns vary among individuals, between tissues, and even among cells within the same tissue. Environmental factors, such as **diet** and **pollutant exposure**, can influence the epigenome. Some epigenetic modifications are **heritable**, being passed from cell to cell or, in some cases, across generations.

Major Epigenetic Mechanisms: Epigenetic mechanisms primarily include **DNA methylation** and **post-translational modifications (PTMs) of histone proteins.**

1. DNA Methylation

DNA methylation involves the addition of a **methyl group** to the C5 position of cytosine bases in **CpG islands** within gene promoters, forming **5-methylcytosine (5mC)**. This is catalyzed by **DNA methyltransferases (DNMTs)**. DNA methylation can:

- ❖ **Repress gene expression** by blocking transcription factor binding or recruiting repressive proteins.
- ❖ **Silence genes** such that no protein is produced.
- ❖ Be **heritable** through cell divisions.
- ❖ Be reversed by **demethylation**, which can activate genes.

2. Histone Modifications

Histone proteins can undergo chemical modifications that affect **chromatin structure** and gene expression. These modifications influence how tightly DNA is wrapped around histones:

- ❖ **Tightly packed nucleosomes (heterochromatin):** Genes are **silenced**.
- ❖ **Loosely packed nucleosomes (euchromatin):** Genes are **active**.

Common histone modifications include:

1. **Acetylation:** Addition of acetyl groups by **histone acetyltransferases (HATs)** reduces positive charge on histones, loosening DNA-histone interactions and **activating gene expression**.

2. **Methylation:** Addition of methyl groups by **histone methyltransferases (HMTs)** can **activate or repress genes** depending on the site and type of methylation.

    Example: H3K4 methylation = active gene; H3K9 methylation = silenced gene.

3. **Phosphorylation**: Addition of phosphate groups can influence interactions with other histone modifications, initiating downstream regulatory effects.

4. **Ubiquitylation:** Addition of ubiquitin molecules affects chromatin structure and gene expression, playing a role in **DNA damage response**.

These modifications act as **molecular switches**, regulating gene accessibility and transcriptional activity.

Paramutation: A **paramutation** is an epigenetic interaction between two alleles at a single locus, where one allele (**paramutagenic**) induces a heritable change in the expression of the other allele (**paramutant**). Neutral alleles remain unaffected. The altered expression of a paramutant can be **maintained in offspring**, even if the paramutagenic allele is no longer present.

Non-coding RNA (ncRNA): While DNA provides instructions for making both **coding and non-coding RNA**, only **coding RNA** is translated into proteins. **Non-coding RNA** regulates gene expression by interacting with coding RNA or by **recruiting proteins** that modify histones, thereby turning genes **on or off**.

Transcriptome: The total amount of genetic information which has been transcribed by the cell. This information will be stored as RNA. The transcriptome is unique to a cell type and is a measure of the gene expression. Different cells within an organism will have different transcriptomes. Cell types can be identified by their transcriptome.

Proteome: Proteome is the cell's complete protein output. This reflects all the mRNA sequences translated by the cell. Cell types have different proteomes and these can be used to identify a particular cell.

Proteomics: Proteomics is the study of all of the proteins in an organism (proteome) and their interactions with their environment. The nature of proteins makes proteomics far more complex than genomics. Study in proteomics includes investigations in quantitation, structure, function, interactions, and modifications of proteins. In industry, protein separation and purification of proteins are also important aspects of proteomics.

Cytogenetics: The branch of genetics that studies **chromosomes** and **chromosomal abnormalities**, including their structure, function, and inheritance.

Symbolization

Symbolization/Symbolism is not the important part of genetics and no universal notation has been established. But symbols are useful for providing a language to describe important aspects of genetics. To symbolize the genotype of an organism, the post-Mendelian geneticists have evolved the following three methods of symbolism:

Classical method: The first letter of the dominant trait represents the alleles. A capital letter signified a dominant and a lowercase letter a recessive member of a pair of alleles. Thus, according to this system, the genotype of homozygous tall pea plant will be TT and that of homozygous dwarf plant as tt. The genotype of heterozygote will be Tt.

Modified classical method: Capital and small letters are commonly used to designate dominant and recessive alleles, but the first letter of the recessive trait represents the alleles. According to this system, the genotype of homozygous tall, heterozygous tall, and dwarf pea plant will be DD, Dd, and dd, respectively.

Wild type symbolism: In this method the lowercase letter corresponding to the initial letter of the mutant phenotype designates the mutant allele. In the same time the wild type allele is designated with the same small letter with a + as a superscript. According to this system, the genotype of homozygous tall, heterozygous tall, and dwarf pea plant will be d^+d^+, d^+d, and dd, respectively.

Mendel's Selection of the Experimental Plant

Mendel made careful selection of garden pea (*Pisum sativum*) for his experiments for the following reasons:

1. Pea plants have a number of easily observable traits with contrasting forms.
2. Pure varieties of pea plants were available in which mating of plants can be controlled. Pea flowers contain both male and female parts, called stamen and pistil, and usually self-pollinate. Peas can be easily cross-pollinated by removing stamens before pistils of flower mature by the process called emasculation and cross with stigma of desired pea plant.
3. Cultivation of pea plant is easy, economic and required small space.
4. Generation interval of pea plant is relatively short, thus many generation can be obtained in a single growing season, and produce many offspring in one cross.
5. The hybrids of pea plant produced by cross pollination are perfectly fertile.

Among 20-30 different characters, Mendel used following seven pairs of contrasting characters in his hybridization crosses:

1. Seed shape – Round : wrinkled
2. Seed color – Yellow : green
3. Pod shape – Inflated : constricted
4. Pod color – Green : yellow
5. Flower color – Purple : white
6. Flower and pod position – Axial : terminal
7. Stem length – Tall : dwarf

Mendel's first step was to identify pea plants that were true-breeding. When self-fertilize, a true-breeding plant produces offspring identical in appearance to itself generation after generation. For instance, Mendel identified a purple-flowered pea plant that, when self-fertilized, always produced offspring plants that had purple flowers. Garden pea (*Pisum sativum*) is known to be the first model organism in genetics.

Mendel's Laws

Mendel himself did not postulate any genetic principle or laws. It was Karl

Correns, the discoverer of Mendel's work who thought that Mendel's discovery could be represented by the two laws of heredity. These laws of heredity are 'law of segregation' and the 'law of independent assortment'. These two principles form the basis of our knowledge of transmission genetics. Mendel's first law, the principle of segregation, states that alleles of a gene exist in pairs but in gamete formation the two alleles of the same gene segregate (separate) from each other so that each gamete receives (carries) only one allele of each gene. Nondisjunction is an exception of law of segregation.

Mendel's second law, the principle of independent assortment, states that segregating alleles of different gene assort independently of one another during gamete formation, *i.e.*, every possible combination of alleles for every gene is equally likely to occur into gametes. Therefore, traits are transmitted to offspring independently of one another.

Independent assortment can be observed when studying more than one character at a time. In fact, Mendel's law of independent assortment is not always correct – this law is accurate for genes that are on separate chromosomes, but not necessarily for genes that are on the same chromosome.

The Monohybrid Cross

Gregor Mendel chose common garden pea to study inheritance of several phenotypes, but we will concentrate on only one of these, stem height. He made a cross between two pure-breeding lines of plants, one of which had tall stems and the other dwarf stems. Tall plants were six to seven feet high, while dwarf in the same environment measured nine to eighteen inches. The hybrids (F_1) produced in this cross were all tall. Thus, tall was dominant to dwarf. He then allowed these tall hybrid plants to self-fertilize to produce the F_2 generation. Some plants were tall and some were dwarf. The ratio of tall to dwarf plants was close to 3 : 1. This 3 : 1 F_2 ratio was observed to occur in the monohybrid cross for each of the seven pairs of contrasting characters originally selected (Table 2.1). When the recessive progeny of F_2 (which were forming 25% of total F_2 progeny) were self-pollinated, they always produced F_3 offspring with only recessive characters. From F_2 dominant progeny, ⅓ yielded F$_3$ offspring with only dominant characters, while remaining ⅔ yielded the F$_3$ offspring displaying the dominant and recessive characters in 3 : 1 ratio (Figure 2.1).

Table 2.1 Results of Mendel's crosses in which parents differed for one character

Parental Phenotypes	F$_1$	F$_2$	F$_2$ Ratio
Round × wrinkled seeds	All round	5474 round; 1850 wrinkled	2.96 : 1
Yellow × green seeds	All yellow	6022 yellow; 2001 green	3.01 : 1
Purple × white petals	All purple	705 purple; 224 white	3.15 : 1
Inflated × pinched pods	All inflated	882 inflated; 299 pinched	2.95 : 1
Green × yellow pods	All green	428 green; 152 yellow	2.82 : 1
Axial × terminal flowers	All axial	651 axial; 207 terminal	3.14 : 1
Long × short stems	All long	787 long; 277 short	2.84 : 1

These experimental data led Mendel to suggest that inheritance of each character is controlled by discrete units called particles/factors, which we now call genes. These factors existed in several alternative forms (now call alleles). Each individual had two factors/alleles, receiving one allele from the gamete of the male parent and the other allele from the gamete of the female parent. When two different alleles for a single characteristic are present, one allele is dominant to the other. The recessive allele remains hidden or masked by the dominant allele. Hence a plant with dominant trait might carry, in unchanged form, the allele for the recessive trait. During gamete formation, two alleles segregate randomly from each other so that each gamete receives one or the other with equal likelihood. Although this elegant interpretation seems simple to us today, no one in Mendel's time appreciated it, because it was in complete opposition to the conventional view that heredity was due to blending of fluids from both parents. Clearly no blending had occurred in Mendel's experiments, neither in the F$_1$ where only one phenotype was expressed nor in the F$_2$ where both were expressed separately. He was the first scientist to use this type of quantitative analysis in a biological experiment. Mendel's data is concerned with the proportions of offspring.

P	Tall	×	Dwarf
F₁	All Tall		
F₂	¾ tall, ¼ dwarf		
F₃	Tall	⅓ pure-breeding tall	
		⅔ impure tall	
	Dwarf	pure-breeding dwarf	

Figure 2.1 Results of Mendel's monohybrid cross.

Parents	DD (Tall)	×	dd (Dwarf)
Gametes	D		d
F₁	Dd (Tall)		
Gametes	½ D, ½ d		
F₂		Male Gametes	
		½ D	½ d
	Female Gametes	½ D: ¼ DD tall	¼ Dd tall
		½ d: ¼ Dd tall	¼ dd dwarf

Figure 2.2 Genetic basis of the principle of segregation in F₂ progeny.

From our modern knowledge of genes and gene structure we may now reconstruct Mendel's experiment and explanation by making use of symbols. The pure-breeding tall plants carried two dominant alleles for tallness, D. The dwarf plants carried two recessive alleles for dwarfness, d (Figure 2.2). When these parents produce gametes, the DD individuals produce gametes with the D allele, and the dd individuals produce gametes with the d allele. Fertilization results in all progeny (the F₁) being heterozygotes, or Dd, and having the dominant tall phenotype. When F₁ tall plants (Dd) were selfed, separation of allele pairs of both the male and female occurred. Half the gametes carried the D allele and half the d allele. These gametes can combine in four different, equally frequent ways. Overall, then, $^1/_4$ of the progeny are DD, $^1/_2$ are Dd, and $^1/_4$ are dd, but because of the dominance of allele D, $^3/_4$ are the tall phenotype (*DD* or *Dd*) and ¹/₄ are the dwarf phenotype (*dd*). In other words, there are three of the dominant types to every one of the recessive type, giving a 3 : 1 ratio.

The validity of this hypothesis was strengthened when individual plants from each of the F_2 classes were self-fertilized. All dwarf plants were found to be pure-breeding. One-third of the tall plants were pure-breeding and two-thirds give tall and dwarf plants in the ratio 3 : 1.

Summary chart

Phenotypes	Genotypes	Genotypic ratio	Phenotypic ratio
Tall	*DD*	1	3
	Dd	2	
Dwarf	*dd*	1	1

The 3 : 1 ratio is referred to as the monohybrid ratio, and is the basis for all patterns of inheritance in higher organisms.

There have subsequently been many examples of the 3 : 1 ratio for the inheritance of characters controlled by a single gene in many different species. A few of these are listed in Table 2.2.

Table 2.2 Examples of characters controlled by single gene

Species/breed	Dominant Phenotype	Recessive Phenotype
Cattle	Black coat color (Angus/Holstein)	Red coat color
	Solid color	Irregular white spotting
	Hornless (Polled)	Horned
	White face	Solid color
	Cloven hooves	Mule feet
Horse	Black coat color	Red (chestnut) coat color
	Chestnut mane and tail	Flaxen mane and tail
	Smooth hair	Curly hair
Goat	Polled	Horned
Sheep	White fleece	Black fleece
	Brown eyes	Blue/Green eyes

	Hairy fleece	Wooly fleece
Swine	Black hair (Hampshires)	Red hair
	White belt	No belt
	Erect ears	Drooping ears
	Mule foot	Cloven hoof
Dogs	Wire hair	Smooth hair
	Black hair	Liver color
	Red hair	Yellow hair
	Solid color	White spotting
Cats	Short hair	Long hair
	Black hair	Brown hair
	Agouti (wild color)	Nonagouti
	Color	White
Chickens	White skin color (wild-type)	Yellow skin color
	Blue eggshell	White eggshell (wild-type)
	Feathered shanks	Clean shanks
	White shank color	Yellow shank color
	Black feathers	White/Red feathers
	Rose comb	Single comb
	Pea comb	Single comb
	Normal feather	Silky feather
	Close feathering	Loose feathering
	White plumage in White Leghorn	Color plumage
	Color plumage in White Wyandottes	White plumage
	Black skin colour gene found in Silkie breed	Other skin color
	Black feather color	White feather color
	Normal feathers (wild-type)	Silkie feathers, Woolly feathers
	Yellow down	Snow-white down

	Normal hock	Vulture hocks
	Broodinance	Nonbroodiness
	Multiple spur	Single spur (wild-type)
	Cleft palate	Normal palate
	Single spur	Spurlessness
Guinea pigs	Short hair	Long hair
	Black eyes	Red eyes
Mice	Black coat color	White coat color
	Pale ears	Normal ears
Fruit flies	Normal wings	Vestigial wings
	Red eye	White eye
Humans	Brown eye (iris) color	Green or blue eye color
	Green eye (iris) color	Blue eye color
	Large eyes	Small eyes
	Horizontal slanting eyes	Upward slanting eyes
	Broad/bushy eyebrows	Slender/fine eyebrows
	Long eyelash	Short eyelash
	Almond-shaped eyes	Round eyes
	Rounded face shape	Square face shape
	Very prominent chin	Less prominent chin
	Round chin	Square chin
	Cleft chin	Normal chin
	Abundant body hair	Little/no body hair
	Black hair	Brown hair
	Hair on back of hand	No hair
	Broad/thick lips	Thin lips
	Broad nose	Narrow nose
	Rounded nose	Pointed nose
	Rounded nostrils	Pointed nostrils
	Hairy ears	Ears without hair

Free (detached) earlobes	Attached earlobes
Dwarfism	Normal growth
Presence of freckles on the forehead	Absence of freckles on the forehead
Presence of freckles on the cheeks	Absence of freckles on the cheeks
Face freckles	No face freckles
Straight thumb	Curved thumb
Normal skin pigmentation	Albinism
Widow's peak	Straight hairline
Presence of cheek dimples	No cheek dimples

Testcross: Detection of Heterozygotes

A more common method to determine whether an individual with the dominant phenotype is homozygous or heterozygous is to perform a testcross by crossing the individual with one that is homozygous recessive. As shown in Figure 2.3 the heterozygote F_1 can produce two classes of gametes, carrying either the dominant or the recessive allele. The parent with the recessive phenotype can only produce gametes with recessive alleles, and so the progeny of the cross have the dominant and recessive phenotypes in equal numbers, a 1 : 1 phenotypic ratio.

Parents	Dd (Tall)	×	dd (Dwarf)
Gametes	$\frac{1}{2}D, \frac{1}{2}d$		d
Progeny	$\frac{1}{2}Dd$ (Tall; Dominant) , $\frac{1}{2}dd$ (Dwarf; Recessive)		

Figure 2.3 Results of a testcross of Dd heterozygotes.

The phenotype of the progeny of the testcross indicates the genotype of the individual tested. If the progeny all show the dominant phenotype; the individual was homozygous dominant. If there is approximately 1 : 1 ratio of progeny with dominant and recessive phenotypes, the unknown individual was heterozygous. The recessive individual is called a tester. Because the tester contributes only recessive alleles, the genotypes of the unknown individual can be deducted from progeny phenotypes.

Dihybrid Cross and Inheritance of Two Genes

Parents		*WW GG* round yellow	×	*ww gg* wrinkled green
Gametes		*WG*		*wg*
F₁ progeny		*Ww Gg* round yellow (double heterozygote)		
Gametes		¼*WG*, ¼ *Wg*, ¼*wG*, ¼ *wg*		

F₂			Male gametes			
			WG	*Wg*	*wG*	*wg*
	Female gametes	*WG*	*WW GG* round yellow	*WW Gg* round yellow	*Ww GG* round yellow	*Ww Gg* round yellow
		Wg	*WW Gg* round yellow	*WW gg* round green	*Ww Gg* round yellow	*Ww gg* round green
		wG	*Ww GG* round yellow	*Ww Gg* round yellow	*ww GG* wrinkled yellow	*ww Gg* wrinkled yellow
		wg	*Ww Gg* round yellow	*Ww gg* round green	*ww Gg* wrinkled yellow	*ww gg* wrinkled green

Figure 2.4 Punnett square showing results of Mendel's dihybrid cross.

Instead of studying a single pair of contrasting traits, Mendel conducted dihybrid crosses considering two distinct and contrasting characters at a time. In one of his experiments, he crossed true-breeding plants that differed in two seed characteristics – seed shape (round or wrinkled) and seed/cotyledon color (yellow or green). A cross between a pure-breeding line of round yellow peas with a pure-breeding line of wrinkled green peas yielded an F_1 generation which consisted only of round yellow seeds. This indicated that round phenotype is completely dominant over wrinkled and yellow color is completely dominant over green. When these F_1 individuals were allowed to self-fertilize, four types of progeny were produced in the F_2, with the ratio of 9 : 3 : 3 : 1 (nine plants with round yellow seeds, three plants with round green seeds, three plants with wrinkled yellow seeds and one plant with wrinkled green seeds). From this experiment, Mendel observed that the pairs of traits in the parental generation assorted independently of one another, which resulted in a phenotypic ratio of 9 : 3 : 3 : 1. Seed shape is determined by a single gene that has alleles for round, W, or wrinkled seeds, w. Again, seed color is determined by two alleles of a single gene (alleles for yellow, G, or green seeds, g). The experiment is set out in Figure 2.4.

For the F_2 generation, the law of segregation requires that each gamete receive either a W allele or a w allele along with either a G allele or a g allele. The law of independent assortment states that during gamete formation, segregating alleles of different gene assort independently of each other i.e., a gamete into which a w allele sorted would be equally likely to contain either a G allele or a g allele. Thus, there are four equally likely genetically different gametes that can be formed as follows: WG, Wg, wG, and wg. These four types of gametes of F_1 dihybrids can unite randomly in 16 different genotypic combinations in the process of fertilization in F_2. From these genotypes, we infer a phenotypic ratio of 9 round yellow : 3 round green : 3 wrinkled yellow : 1 wrinkled green. Mendel obtained ratios close to this for several different combinations of pairs of genes. The 9 : 3 : 3 : 1 ratio is simply two 3 : 1 ratios combined, and shows that the alleles of the two genes assort independently of each other. This is demonstrated more easily by looking at one gene at a time. Take the seed shape gene first; in the F_2 generation a ratio of 3 round to 1 wrinkled would be expected. Now look at the seed color gene in those seeds that have the wrinkled phenotype; these should have a ratio of 3 yellow to 1 green. The same is true for round seeds, these should also have a ratio of 3 yellow to 1 green. If the ratio for the two phenotypes is multiplied across the 9 : 3 : 3 : 1 ratio is obtained.

Phenotypes	Genotypes	Genotypic ratio	Phenotypic ratio
Round yellow	*WW GG*	1	9
	Ww GG	2	
	WW Gg	2	
	Ww Gg	4	
Round green	*WW gg*	1	3
	Ww gg	2	
Wrinkled yellow	*ww GG*	1	3
	ww Gg	2	
Wrinkled green	*ww gg*	1	1

Summary of Mendel's Ratios

Monohybrid ratio:

Phenotypic ratio   3 : 1

Genotypic ratio   1 : 2 : 1

Testcross ratio 1 : 1

Dihybrid ratio:

Phenotypic ratio $9 : 3 : 3 : 1$; *i.e.*, $(3 : 1)^2$

Genotypic ratio $1 : 2 : 1 : 2 : 4 : 2 : 1 : 2 : 1$; *i.e.*, $(1 : 2 : 1)^2$

Testcross ratio $1 : 1 : 1 : 1$; *i.e.*, $(1 : 1)^2$

If n is the number of heterozygous gene pair, there are 2^n phenotypes/gametes and 3^n genotypes.

Phenotypic ratio $(3 : 1)^n$

Genotypic ratio $(1 : 2 : 1)^n$

Testcross ratio $(1 : 1)^n$

Later studies have shown that there are some important exceptions to Mendel's principle of independent assortment, but otherwise, these principles are recognized as the basis of inheritance. The $9 : 3 : 3 : 1$ F_2 ratio depends on the following conditions:

❖   The two genes must not act on the same character; *i.e.*, there is no epistasis.

❖ The two genes are not situated close together on the same chromosome; *i.e.*, there is no linkage. If the two genes lie close together on the same chromosome, the four classes of gametes are not produced at equal frequencies.

Where there is complete dominance at both gene loci and no epistasis and no linkage, there are four phenotypes, with a 9 : 3 : 3 : 1 F_2 ratio.

Mendel's Contributions

Gregor Johann Mendel set the framework for genetics long before chromosomes or genes had been identified; at a time when meiosis was not well understood. Gregor Mendel's contribution to science was that he discovered the basics of hereditary laws. He:

❖ Developed pure lines.
❖ Predicted the existence of genes as particles of inheritance.
❖ Discovered patterns of inheritance.
❖ Established the laws of heredity.
❖ Discovered genes come from both parents.
❖ Discovered one allele dominant to another.
❖ Discovered recessive allele expressed in absence of dominant allele.
❖ Recognized the mathematical patterns of inheritance from one generation to the next

Reasons for Mendel's Success

Before Mendel, several researchers experimented with different plant varieties and produced fertile hybrids through artificial pollination. However, they failed to explain the mechanism of heredity. The main reasons for their failure were:

1. **Studying too many characters simultaneously:**
   They considered many traits at the same time in which the parents

differed. This created confusion, making it impossible to trace individual characters through successive generations or maintain complete numerical records.

2. **Belief in blending inheritance:**
   They assumed that the hereditary traits of two parents became thoroughly mixed in the offspring, which hindered a correct understanding of heredity.

Mendel recognized these problems and carefully designed experiments to overcome them. The following factors contributed to his success:

- ❖ **Careful selection of pea plants:** Mendel chose plants suitable for studying inheritance.
- ❖ **Studying one character at a time:** He analyzed each of the seven pairs of contrasting traits individually in parent plants and their hybrid offspring across generations. He classified traits as **dominant** or **recessive**.
- ❖ **Limiting the number of generations:** He focused on the **F2** and **F3** generations to observe patterns of inheritance.
- ❖ **Controlled crosses:** Mendel ensured precise control over which plants were crossed.
- ❖ **Pedigree records:** He meticulously recorded the traits of each generation.
- ❖ **Luck in trait selection:** All seven pairs of traits he selected showed **complete dominance**, without complications such as **linkage** or **epistasis**.
- ❖ **Application of mathematics:** Mendel used statistical and probability methods to analyze his breeding results quantitatively.
- ❖ **Scientific method and detailed records:** He followed strict experimental protocols and recorded every individual with a particular trait.
- ❖ **Large sample sizes:** His experiments involved sufficient numbers of plants to produce reliable data.
- ❖ **Pure breeding parents:** He used pure lines and confirmed their purity through self-fertilization over multiple generations.

In summary: Mendel's success was due to careful planning, systematic study, statistical analysis, and attention to experimental detail, combined with favorable biological traits in his chosen plants.

Single Gene Genetic Disorders in Humans

Mendelian inheritance refers to the inheritance of traits controlled by a single gene with two alleles, one of which may be completely dominant to the other. Single gene disorders, also known as monogenic disorders, are conditions caused by mutations in the DNA sequences of a single gene. Around 10,000 known human genetic disorders are caused by inheriting an altered gene. Individually, single gene disorders are each very rare, but as a whole, they affect about one per cent of the population. The four basic modes of inheritance exist for single-gene disorders: autosomal dominant, autosomal recessive, X-linked dominant, and X-linked recessive. In 1903, an English physician Archibald Garrod first connects human disease alkaptonuria with Mendel's law of inheritance.

Autosomal Recessive

Most human genetic disorders result from homozygosity for recessive mutant alleles. Both parents of an affected person are usually carriers; not typically seen in every generation. Examples of this type of disorder are albinism, cystic fibrosis, spinal muscular atrophy, medium-chain acyl-CoA dehydrogenase deficiency, phenylketonuria, sickle cell anaemia, Tay-Sachs disease, Niemann-Pick disease, spinal muscular atrophy, adenosine deaminase deficiency, Smith-Lemli-Opitz syndrome, beta thalassemia, maple syrup urine disease, *etc*.

Albinism. The word "albino" comes from the Latin word "albus," which means white. Albinism is a group of rare autosomal recessive genetic disorder present at birth. It is characterized by reduced or absent melanin pigmentation and caused by mutations in genes involved in the biosynthesis of melanin pigment, with an overall estimated frequency of about 1 per 20000 people worldwide. Melanin is the pigment responsible for the color of the skin, hair and eyes. It's also involved in optic nerve development. Melanin absorbs UV light and is important in protecting the skin against harmful UV radiation from the sun. Symptoms vary based on how much melanin someone makes. Signs and symptoms may include: very pale skin, very light-blond or white hair, light-blue eyes that can appear red in certain lighting, vision problems, *etc.* Mutations in at least 12 different genes have been identified in different types of albinism in humans. Albinism is known to affect mammals, birds, fish, reptiles, and amphibians.

Cystic Fibrosis (CF). Cystic fibrosis (CF) is one of the most common genetic disorders caused by a mutation in the CFTR (cystic fibrosis transmembrane conductance regulator) gene located on the long arm of chromosome 7. There are more than 180 known mutations in the CFTR gene that causes cystic fibrosis and disrupt the normal work of the chloride channels. The dominant allele codes for a membrane protein that is responsible for the movements of negatively charged particles known as chloride ions into and out of cells. These movements are lacking or are defective in recessive homozygotes; as a result, abnormally, thick and sticky mucus is produced on the outside of the cell. The production and accumulation of the mucus damages the lungs, pancreas, digestive tract and sex organs. The cells most seriously affected by this are the lung cells. This mucus clogs the airways in the lungs, and increases the risk of infection by bacteria. The thick mucus also blocks ducts in the pancreas, so digestive enzymes can't get into the intestines. Without these enzymes, the intestines cannot properly digest food. People who have the disorder often do not get the nutrition they need to grow normally. Finally, cystic fibrosis affects the sweat glands. Too much salt is lost through sweat, which can disrupt the delicate balance of minerals in the body. Over time, it can cause chronic coughing, wheezing and inflammation, and develop into permanent lung damage, the formation of scar tissue (fibrosis), and cysts in the lungs. Therefore, cystic fibrosis patients who don't receive treatment have shortened lifespan. It is a common genetic disease within the white population in the United States. The disease occurs in 1 in 2,500 to 3,500 white newborns. Cystic fibrosis is less common in other ethnic groups, affecting about 1 in 17,000 African Americans and 1 in 31,000 Asian Americans.

Phenylketonuria. Phenylketonuria (PKU) is a rare metabolic disorder caused by a mutation in the phenylalanine hydroxylase (PAH) gene at 12q24.1 that results in a deficiency of the enzyme phenylalanine hydroxylase. The absence of that enzyme activity prevents the conversion of the amino acid phenylalanine to the amino acid tyrosine. Phenylalanine is one of the most essential amino acids. It is required to make our own proteins, but excess amounts are harmful. The absence of phenylalanine hydroxylase results in the accumulation of the phenylalanine in the body. The accumulated phenylalanine in phenylketonurics is converted by a secondary pathway to phenylpyruvic acid. This substance drastically affects the cells of the central nervous systems and produces severe brain damage. This damage can cause epilepsy, behavioral problems, and stunt the growth of the baby. Other symptoms include skin rashes, such as eczema; a musty body odor in the breath, skin or urine; a small head (microcephaly); hyperactivity; intellectual disability; delayed development; mental health disorders; fair skin; or even death. Norwegian doctor Asbjørn Følling discovered PKU in 1934. If the condition is detected early and treatment is begun, individuals with PKU can lead healthy lives. The occurrence of PKU varies among ethnic groups and across geographic regions worldwide. In the United States, PKU occurs in 1 in 25,000 newborns.

Sickle-Cell Disease (SCD). Normal red blood cells have a biconcave, disc-like shape and contain enormous amounts of hemoglobin. Sickle-cell disease is an autosomal recessive disorder first described by J. B. Herrick in 1910 (Herrick, 1910) that affects the red blood cells, which use hemoglobin to transport oxygen from the lungs to the rest of the body. It is the most common serious genetic diseases in childhood. The hemoglobin molecule has two parts: an alpha and a beta. The β chains consist of 146 amino acids, numbered from the N-terminal end. The disease is caused by a mutation in the sixth codon of the β-globin gene located on chromosome 11. The altered codon specifies a different amino acid (valine rather than glutamic acid), and this causes a structural abnormality of the β-globin molecule that produce a new form of haemoglobin called sickle hemoglobin (hemoglobin S, Hb^S), and behaves very differently to normal haemoglobin (Hb^A). Hb^S causes the red blood cells to develop abnormally into a crescent, or sickle, shape. This change in shape makes the sickle red blood cells less efficient at transporting oxygen through the bloodstream. The sickle shape causes red blood cells to pile up, causing blockages and damaging vital organs and tissue. The sickle cells also block the flow of blood through vessels, resulting in lung tissue damage. It also causes damage to the spleen, kidneys and liver. These sickle cells can block veins, arteries, and capillaries and prevents oxygen from reaching the spleen, liver, kidneys, lungs, heart, or other organs, causing pain and a lot of damage to the organs. Unlike normal red blood cells, which can live for 120 days, sickle-shaped cells live only 10 to 20 days. Individuals who are homozygous for the sickle-cell allele often have anemia, which is why this disease is commonly referred to as sickle-cell anemia that require hospitalization. Central nervous system injury is the most debilitating frequent complication of SCD and includes stroke, silent cerebral infarct (SCI), and cognitive impairment. Other symptoms include delayed growth, and jaundice. Organ damage and other complications often shorten patient's lives by about 30 years. Sickle cell disease is the most common inherited blood disorder in the United States, affecting an estimated 100,000 Americans. The disease is estimated to occur in 1 in 500 African Americans and 1 in 1,000 to 1,400 Hispanic Americans. Sickle cell disease primarily affects humans.

Other point mutations in and around the β-globin gene result in decreased or, in some instances, no production of β-globin; β-thalassemia is the result of these mutations.

Tay-Sachs Disease. Tay-Sachs disease (TSD) (also known as GM2 gangliosidosis or hexosaminidase A deficiency or the child-killer disease) is a rare genetic disorder that causes progressive damage and, ultimately, the death of nerve cells (neurons) in the brain and spinal cord (central nervous system). In the most common and severe form of Tay-Sachs disease (known as infantile Tay-Sachs disease), symptoms like developmental delays, hearing and vision loss usually start to show up at about 3 to 6 months of age in previously healthy babies. As the disease progresses, development slows and muscles begin to weaken. Over time, this leads to seizures, blindness and degeneration of motor and mental performance. Children with this form of Tay-Sachs disease typically live only a few years. Tay-Sachs disease is caused by a mutation of the HEXA gene on chromosome 15. The HEXA gene gives instructions to make enzyme beta-hexosaminidase A that break down the fatty substance GM2 ganglioside. A mutation in the HEXA gene leads to a deficiency of enzyme beta-hexosaminidase A that causes the accumulation of harmful quantities of GM2 ganglioside in the nerve cells, eventually damages nerve cells in the brain and spinal cord. Severity and age of onset of the disease relates to how much enzyme is still produced. The prevalence of the disease is 1 case per 320 000 live births. There is no known cure or treatment. Tay-Sachs disease primarily affects humans.

Spinal Muscular Atrophy (SMA). Spinal muscular atrophy (SMA) is a progressive lethal motor neuron disease. The most common form of SMA (types 1-4) is caused by a mutation in the survival motor neuron 1 (SMN1) gene on chromosome 5. A mutation in the SMN1 gene leads to a deficiency of SMN protein that supports normal muscle movement and control of the limbs, abdomen, head and neck, chest and breathing muscles. SMA leads to progressive muscle weakness and atrophy; especially in the muscles of the torso, upper legs and upper arms. Symptoms can begin prior to six months of age, in childhood and, more rarely, in adulthood. In the most common form of the disease, lifespan is often less than two years of age. The incidence of spinal muscular atrophy is about 1 in 10,000 live births with a carrier frequency of approximately 1 in 50. SMA primarily affects humans.

Galactosemia. Galactosemia means "galactose in the blood". Galactosemia is a rare genetic disorder where galactose fails to convert to glucose. Classic galactosemia occurs when liver enzyme galactose-1-phosphate uridyltransferase (GALT), responsible for breaking down galactose, is either defective or missing due to mutations in the GALT gene on chromosome 9. This causes galactose to build up to toxic levels in the blood. This can lead to severe symptoms such as liver enlargement, kidney failure, brain damage, cataracts, poor growth, and intellectual disability. If left untreated, this condition kills up to 75% of sufferers. Galactosemia was first discovered in 1908 by the physician Von Ruess. Classic galactosemia occurs in 1 in 30,000 to 60,000 newborns. Galactosemia type II and type III are less common; type II probably affects fewer than 1 in 100,000 newborns and type III appears to be very rare. Galactosemia primarily affects humans.

Adenosine Deaminase (ADA) Deficiency. Adenosine deaminase (ADA) deficiency is a metabolic disorder that causes immunodeficiency. It is caused by a mutation in the ADA gene on chromosome 20. The gene codes for the enzyme adenosine deaminase (ADA). Adenosine deaminase converts a toxic substance called deoxyadenosine to deoxyinosine, which is not harmful. Without this enzyme, the body is unable to break down deoxyadenosine, which is generated when DNA is broken down. A buildup of deoxyadenosine in cells destroys infection-fighting T and B lymphocytes. Because ADA deficiency affects the immune system, people who have the disorder are more susceptible to all kinds of infections, particularly those of the skin, respiratory system, and gastrointestinal tract. They may also be shorter than normal. Sadly, most babies who are born with the disorder die within a few months. ADA deficiency is extremely rare. It is estimated to occur in approximately 1 in 500,000 newborns worldwide. Approximately 15 percent of people with SCID have ADA deficiency. ADA deficiency primarily affects humans.

Smith-Lemli-Opitz Syndrome (SLOS). Smith-Lemli-Opitz syndrome (SLOS) is a metabolic developmental disorder caused by mutations in the DHCR7 (7-dehydrocholesterol reductase) gene on chromosome 11. This gene codes for an enzyme called 7-dehydrocholesterol reductase that is involved in the production of cholesterol. Cholesterol is an essential component of the cell membrane and tissues of the brain. People who have SLOS are unable to make enough cholesterol. A lack of this enzyme also allows toxic byproducts of cholesterol production to build up in the blood, nervous system, and other tissues. The combination of low cholesterol levels and an accumulation of other substances affects many parts of the body. This condition is characterized by distinctive facial features, small head (microcephaly), intellectual disability or learning problems, and behavioral problems. Many affected children have the characteristic features of autism, a developmental condition that affects communication and social interaction. Malformations of the heart, lungs, kidneys, gastrointestinal tract, and genitalia are also common. Smith-Lemli-Opitz syndrome affects an estimated 1 in 20,000 to 60,000 newborns. SLOS primarily affects humans.

Thalassemia. Thalassemias are a group of inherited blood disorders characterized by defective production of hemoglobin, leading to reduced or absent hemoglobin in red blood cells. This results in anemia, which can cause fatigue, weakness, pale skin, and other serious health complications. Thalassemia is classified into **beta (β) thalassemia** and **alpha (α) thalassemia**, depending on which globin chain is affected. Beta thalassemia is caused by mutations in the **HBB gene** on chromosome 11p15.4, which encodes beta-globin. Mutations that completely prevent beta-globin production are referred to as beta-zero (β0) thalassemia, while mutations that reduce beta-globin production are called beta-plus (β+) thalassemia. The disease is further categorized based on severity into minor (trait), intermedia, and major (Cooley's anemia). Individuals with beta thalassemia minor are usually asymptomatic or have mild anemia, whereas beta thalassemia intermedia causes mild to moderate anemia and may affect growth and bone development. Beta thalassemia major is severe, appearing within the first two years of life, and typically requires regular blood transfusions. Children with this form may develop life-threatening anemia, jaundice, enlarged spleen, liver, and heart, bone deformities, delayed puberty, and complications from iron overload due to transfusions. Alpha thalassemia is caused by reduced or absent alpha-globin due to deletions or mutations in the **HBA1 and HBA2 genes** on chromosome 16p13.3. Severity depends on the number of affected alleles. Loss of all four alpha-globin alleles leads to hemoglobin Bart hydrops fetalis syndrome, a condition usually resulting in stillbirth or early neonatal death. Loss of three alleles causes hemoglobin H (HBH) disease, which presents with mild to moderate anemia, jaundice, hepatosplenomegaly, and sometimes bone abnormalities, with affected individuals typically surviving into adulthood. Loss of two alleles results in alpha thalassemia trait, causing mild anemia and small, pale red blood cells, while loss of a single allele produces silent carriers who usually have no symptoms. Both alpha and beta thalassemias highlight the critical role of hemoglobin in oxygen transport, and disease management ranges from monitoring in mild cases to lifelong blood transfusions and iron chelation therapy in severe forms.

Medium Chain Acyl-CoA Dehydrogenase (MCAD) Deficiency. Medium chain acyl-CoA dehydrogenase (MCAD) deficiency is a rare genetic metabolic disorder of fatty acid oxidation characterised by a deficiency of the enzyme medium chain acyl-CoA dehydrogenase that impairs the body's ability to break down medium-chain fatty acids into acetyl-CoA. Failure to break down these fats can lead to the abnormal accumulation of fatty acids in the liver and the brain. Abnormally low levels of the MCAD enzyme may also hamper or interrupt other processes associated with the metabolism of fatty acids. Signs and symptoms of MCAD deficiency typically appear during infancy or early childhood and can include vomiting, lack of energy (lethargy), and low blood sugar (hypoglycemia). In rare cases, symptoms of this disorder are not recognized early in life, and the condition is not diagnosed until adulthood. People with MCAD deficiency are at risk of serious complications such as seizures, breathing difficulties, liver problems, brain damage, coma, and sudden death. MCAD deficiency is caused by a mutation in the ACADM (acyl-CoA dehydrogenase medium chain) gene (1p31) which encodes the mitochondrial medium-chain acyl-CoA dehydrogenase (MCAD) enzyme. MCAD deficiency has a worldwide birth prevalence of 1 in 15,000, however this is higher in northern European populations. MCAD deficiency primarily affects humans.

Niemann-Pick Diseases (NPD). Niemann-Pick disease (NPD) is a group of lipid storage disorders called sphingolipidoses in which harmful quantities of lipids accumulate in the spleen, liver, lungs, bone marrow and brain. It can occur at any age but mainly affects children. Niemann-Pick disease primarily affects humans. Traditionally, NPD is classified into four subtypes: type A, B, C, and E. Affected organs, symptoms, and treatments vary based on the particular type of Niemann-Pick disease. However, every type is severe and can shorten a person's life expectancy.

Niemann-Pick disease type A and type B are caused by mutations in the sphingomyelin phosphodiesterase-1 (SMPD1) gene that encodes the lysosomal enzyme called acid sphingomyelinase (ASM). The condition is sometimes called acid sphingomyelinase deficiency (ASMD). ASM breaks down the lipid sphingomyelin to ceramide and phosphocholine. If ASM is missing or does not work properly, sphingomyelin builds up inside cells. This causes cell dysfunction and, over time, cell death. NPD type A occurs mainly in infants, who show severe, progressive brain disease. There is no cure, so most children do not live beyond their first few years. NPD type B usually occurs later in childhood and is not associated with primary brain disease. Most people affected with type B survive into adulthood. Niemann-Pick disease (NPD) types A and B affect 1 in 250,000 individuals.

Niemann-Pick type C (NPC) is a rare neurodegenerative lysosomal storage disorder caused by defects in the lysosomal proteins NPC1 or NPC2 and characterized by the accumulation of cholesterol, sphingomyelin, and other lipids in endosomes and lysosomes. It is caused by mutations in either NPC1 (95% of cases) or NPC2 genes (Vanier and Millat, 2003). With these mutations, the body doesn't have the proteins it needs to move and use cholesterol and other lipids in cells. Cholesterol and other lipids build up in the cells of the liver, spleen or lungs. The disease primarily affects the liver, spleen, brain, bone marrow. This causes problems with eye movements, walking, swallowing, hearing and thinking. Symptoms vary widely, can appear at any age and get worse over time. NPD type C affects 1 in 150,000 persons.

Type E is a rare type of Niemann-Pick disease that occurs in adults. Little is known about it.

Roberts Syndrome. Roberts syndrome is a rare genetic disorder characterized by fetal growth restriction, limb reductions, and facial abnormalities. Its prevalence is unknown. Children with Roberts syndrome are born with abnormalities of all four limbs. Infants with a severe form of Roberts syndrome are often stillborn or die shortly after birth. Mildly affected individuals may live into adulthood. People with Roberts syndrome may also have abnormal or missing fingers and toes, joint deformities, and numerous facial abnormalities including an opening in the lip (a cleft lip) with or without an opening in the roof of the mouth (cleft palate), a small chin (micrognathia), ear malformations, wide-set eyes (hypertelorism), outer corners of the eyes that point downward (down-slanting palpebral fissures), small nostrils, and a beaked nose. They may have a small head (microcephaly) or clouding of the clear front covering of the eyes (corneal opacities). In severe cases affected individuals have a sac-like protrusion of the brain (encephalocele) at the front of their head. In addition, people with Roberts syndrome may have heart, kidney, and genital abnormalities. The syndrome is caused by mutations in the ESCO2 (Establishment of Sister Chromatid Cohesion N-Acetyltransferase 2) gene (8p21.1) which codes for functional ESCO2 protein that is important for proper chromosome separation during cell division. Roberts Syndrome is named after John B. Roberts, who first described the disease in 1919. Roberts Syndrome primarily affects humans.

Maple Syrup Urine Disease (MSUD). Maple syrup urine disease (MSUD), also known as branched-chain α-ketoacid dehydrogenase deficiency, is a rare but serious inherited disorder caused by the body's inability to properly process amino acids, leading to a distinctive odor of maple syrup in their urine (sweet-smelling urine). Maple syrup urine disease is caused by mutations in any of three genes – BCKDHA, BCKDHB or DBT encoding branched-chain alpha-ketoacid dehydrogenase (BCKAD) complex enzymes which is responsible for breaking down the 3 essential branched-chain amino acids (BCAAs): leucine, isoleucine and valine. Any amino acids that are not needed are usually broken down and removed from the body. Mutations in any of these three genes result in decreased or no activity of the enzymes. As a result, these amino acids and their toxic byproducts accumulate in the body leads to the serious health problems associated with MSUD. It affects about 1 in every 185,000 babies born worldwide. MSUD primarily affects humans and certain cattle breeds, specifically polled Shorthorn, polled Hereford, and Hereford calves. The signs, symptoms and severity of maple syrup urine disease varies greatly among affected patients and depends on the type of MSUD and amount of residual enzyme activity. Classic MSUD is the most severe and most common form of the disease, characterized by little or no enzyme activity. Infants with this form of MSUD generally show symptoms such as poor feeding, vomiting, drowsiness, increased irritability, lethargy, abnormal movements, delayed development, breathing difficulties, hypertonia, neurological symptoms, intellectual disabilities, depression, loss of bone mass, increased fractures, pancreatic inflammation, chronic headaches, nausea and vomiting. If untreated, maple syrup urine disease can lead to seizures, coma, and death as early as the first two weeks of life.

Autosomal Dominant

Each affected person usually has an affected parent; occurs in every generation. Examples of diseases with autosomal dominant inheritance include Huntington's disease, familial hypercholesterolaemia, polydactyly, brachydactyly, Piebald spotting, woolly hair, neurofibromatosis type 1, neurofibromatosis type 2, Pachyonychia congenita, Marfan syndrome, hereditary nonpolyposis colorectal cancer, hereditary multiple exostoses, tuberous sclerosis (epiloria), Von Willebrand disease, hereditary spherocytosis, adult polycystic kidney disease, myotonic muscular dystrophy, hypertrophic cardiomyopathy (HCM), acute intermittent porphyria, *etc.*

Huntington's Disease (HD). Huntington's disease (HD) is an autosomal dominant neurodegenerative disorder that affects a person's ability to think, talk and move. The frequency of the disease is approximately 1/20,000. HD is caused by a mutation in the Huntingtin (HTT) gene located on chromosome 4. The HTT gene provides instructions for making a defective protein called huntingtin. The disease destroys cells in the basal ganglia, the part of the brain that controls thinking, emotion and movement. Symptoms start between the ages of 30 and 50. The disease gets worse over time and ultimately death. Symptoms include poor memory, depression and/or mood swings, lack of coordination, twitching or other uncontrolled movements, and difficulty walking, speaking, and/or swallowing. In the late stages of the disease, a person will need help doing even simple tasks, like getting dressed. The disease was named for Dr. George Huntington, who first described it in 1872. Huntington's disease primarily affects humans.

Familial Hypercholesterolaemia (FH). Familial hypercholesterolemia (FH) is commonly caused by a mutation in the low-density lipoprotein receptor (LDLR) gene on chromosome 19. LDLR gene encodes a protein called low-density lipoprotein receptor (LDLR), which is involved in passing 'bad' cholesterol, known as LDL (low-density lipoprotein) from the blood into cells for use by, or removal from, the body. Heterozygous individuals generally have cholesterol levels 3 to 5 times higher than normal. Homozygous individuals have cholesterol level 7 to 12 times higher than normal. Having high cholesterol causes a gradual build-up of fatty material in coronary arteries that means arteries gradually become narrower, making it harder for blood to flow to vital organs that puts higher risk of having heart attack or stroke. Mutations in other genes also can cause inherited high cholesterol. Those genes include the PCSK9, LDLRAP1 or APOB (Apolipoprotein B) gene. One in about 200-250 adults have the FH genetic mutation. FH primarily affects humans.

Polydactyly (Extra Digits). The name comes from the Greek *poly* (many) and *dactylos* (finger). Polydactyly, also known as polydactylia, polydactilism, or hyperdactyly, is a condition where a person is born with extra fingers or toes on one or both of their hands and feet. Extra digits may be poorly developed and attached by a small stalk (generally on the little finger side of the hand). Or, they may be well-formed and may even function. The incidence of polydactyly is 1 in 700 pregnancies. Mutations in several genes have been associated with polydactyly, including ZNF141, GLI3, IQCE, GLI1, FAM92A1, KIAA0825 MIPOL1, PITX1, IQCE, and DACH1. Polydactyly can affect various other mammals, including dogs, cats, and camelids like llamas and alpacas.

Brachydactyly (Short Digits). The term brachydactyly (BD) is derived from the Greek *brachy* (short) and dactylos (digit). Brachydactyly is a shortening of the fingers and toes because the bones in the affected digits don't grow as long as they should. It is due to abnormal development of phalanges, metacarpals, or both. The shortness is relative to the length of other long bones and other parts of the body. To date, many different forms of brachydactyly have been identified. Researchers have identified several genes that play a role in the development of brachydactyly including HOXD3 (Homeobox D3), BMPR1B (Bone Morphogenetic Protein Receptor Type 1B), and GDF5 (Growth Differentiation Factor 5). Brachydactyly is inherited as an autosomal dominant trait with incomplete penetrance and variable expressivity. Brachydactyly type A1 (BDA1) is caused by mutations in the Indian hedgehog gene (IHH) on chromosome 2q35-36. Brachydactyly type A2 is caused by a mutation in one of two genes: BMPR1B (4q22-24) or GDF5. Brachydactyly type C is caused by mutation in gene CDMP1 (Cartilage-Derived Morphogenetic Protein-1). Brachydactyly primarily affects humans.

Piebald Spotting (Piebaldism). Piebald spotting (piebaldism) is a rare autosomal dominant genetic condition, typically present at birth, marked by the congenital absence of melanocytes (the cells responsible for producing the pigment melanin)   in certain areas of the skin and hair, leading to white patch of skin or hair. Piebaldism can be caused by mutations in the KIT (KIT Proto-Oncogene, Receptor Tyrosine Kinase) and SNAI2 (Snail Family Transcriptional Repressor 2) (often called SLUG) genes. The KIT gene is responsible for sending the body signals to make certain cells, including melanocytes. When there is a mutation in the KIT gene, melanocytes are altered. This leads to a lack of pigmentation in areas of the skin or hair. When there is a mutation in the SNAI2 or SLUG gene, a protein called snail 2 is negatively affected which is responsible for cell development, including the development of melanocytes. The prevalence of piebaldism is unknown. It is also found in various breeds of dogs, horses, cattle, sheep, and other animals.

Woolly Hair. Woolly hair is an uncommon congenital anomaly of the scalp hair characterized by extremely curly hair having a rough texture involving a localized area of the scalp or covering the entire side. Dominant mutations in the LPAR6 (Lysophosphatidic acid receptor 6), also known as P2RY5 and GPR87 (G Protein-Coupled Receptor 87) gene on chromosome 13q14.2 resulted in 'wooly hair'. Woolly hair primarily affects humans. Autosomal recessive woolly hair-3 (ARWH3) is caused by homozygosity for a missense mutation in the KRT25 (Keratin 25) gene on chromosome 17q21. The KRT25 gene encodes a type I keratin protein, particularly involved in hair structure and potentially linked to conditions like woolly hair. Its prevalence worldwide is unknown.

Neurofibromatosis Type 1 (NF1). Neurofibromatosis type 1 (NF1) is a disorder characterized by the growth of noncancerous tumors called neurofibromas. Neurofibromas usually form on or just underneath the skin, as well as in the brain and peripheral nervous system. But they can also develop in other parts of the body, such as the eye. Complications of NF1 may include vision loss, bone fractures or dysplasia, nerve damage, high blood pressure, tumors that return after treatment, low self-esteem, *etc.* Some NF1 tumors can develop into cancer. The disorder is caused by a mutation in NF1 gene on chromosome 17. The gene codes for a protein called neurofibromin that is a tumor suppressor. When the gene is mutated, it causes a loss of neurofibromin. This allows cells to grow without control. NF1 primarily affects humans, but spontaneous cases of NF1-like symptoms, including neurofibromas and malignant peripheral nerve sheath tumors, have also been documented in various large animal species,

Neurofibromatosis Type 2. Neurofibromatosis type 2 (NF2) is an autosomal dominant disorder characterized by the development of multiple tumors involving the central nervous system (CNS) (Dinh *et al.*, 2020*)*. The incidence of NF2 is about 1 in 25,000 to 40,000 individuals. NF2 is caused by mutations in the NF2 gene located in the long arm of chromosome 22 (22q12.2). The

NF2 gene encodes protein known as merlin, which acts as a tumor suppressor. If the NF2 gene is faulty, it leads to tumours developing in the nervous system. Symptoms may become apparent during childhood, adolescence, early adulthood or later in adult life which may include problems with balance and walking (gait), dizziness, headache, facial weakness, but more typically ringing or buzzing in the ears (tinnitus), and/or progressive hearing loss. NF2 primarily affects humans.

Pachyonychia Congenita (PC). Pachyonychia congenita (PC) is a rare keratinizing skin disorder that primarily affects the skin, nails and mouth. It is caused by mutations in the KRT6A (keratin 6A), KRT6B (keratin 6B), KRT6C (keratin 6C) genes located on the long arm of chromosome 12 (12q13.13), as well as KRT16 (keratin 16), or KRT17 (keratin 17) genes located on the long arm of chromosome 17 (17q21.2). These genes encode tough, fibrous proteins called keratins. Keratins form networks that provide strength and resilience to the tissues that make up the skin, hair, and nails. Mutation in keratin genes alter the structure of keratin proteins. The altered proteins are unable to form strong, stable networks within cells. Without this network, skin cells become fragile and are easily damaged, making the skin less resistant to friction and minor trauma. This disorder does not affect lifespan, but patients do experience constant pain. PC primarily affects humans.

Marfan Syndrome. Marfan syndrome is an inherited disease that affects the body's connective tissue. The syndrome can affect entire body including skeletal system, heart and blood vessels, eyes, skin, lungs, and nervous system. The damage caused by Marfan syndrome can be mild or severe. Marfan syndrome is caused by the mutation of fibrillin-1 (FBN1) gene located on chromosome 15 that encodes protein called fibrillin-1. FBN1 is a large glycoprotein that serves as a structural component of microfibrils and provides force-bearing mechanical support in elastic and nonelastic connective tissue. Marfan syndrome is fairly common, affecting 1 in 10,000 to 20,000 people. Marfan syndrome primarily affects humans, but it also occurs in cattle.

Hereditary Non-polyposis Colorectal Cancer (HNPCC): Hereditary Non-polyposis Colorectal Cancer (HNPCC), also known as Lynch syndrome, is an autosomal dominant trait associated with a greatly increased risk of colorectal, uterine, and other cancers. People with Lynch syndrome may have colon

78

cancer before age 50, endometrial cancer before age 50, abdominal pain or bloating, appetite loss, bloody stools, fatigue, unexplained weight loss, *etc.* Most cases of HNPCC are due to inherited mutations in DNA mismatch repair (MMR) genes and their encoded proteins which correct errors made during DNA replication. Lynch syndrome primarily affects humans.

Hereditary Multiple Osteochondromas (HMO): Hereditary multiple osteochondromas (HMO), previously called hereditary multiple exostoses (HME), also known as diaphyseal aclasis, osteochondromatosis, or simply multiple osteochondromas, is a rare autosomal dominant condition, characterized by the growth of multiple osteochondromas (bony growths) in children and adolescents. Osteochondromas can be associated with a reduction in skeletal growth, bony deformity, restricted joint motion, shortened stature, premature osteoarthritis, short- and long-term pain and compression of peripheral nerves. HME is caused by a mutation in either exostosin glycosyl transferase-1 (EXT1) gene located at the long arm of chromosome 8 (8q24.11), or exostosin glycosyl transferase-2 (EXT2) gene located at the short arm of chromosome 11 (11p11-12) that encode related members of a putative tumor suppressor family. The disease have been documented in various animal species, including horses, dogs, cats, lions, and lizards.

Tuberous Sclerosis (TS). Tuberous sclerosis (TS), also known as tuberous sclerosis complex (TSC), is a rare genetic disorder characterized by the growth of benign tumours in different parts of the body. Mutations in either the TSC1 or TSC2 gene cause tuberous sclerosis complex. The TSC1 and TSC2 genes encodes the proteins hamartin and tuberin, respectively. These proteins are involved in regulating cell growth, and the mutations lead to uncontrolled growth and multiple tumours throughout the body. The tumours most often affect the brain, skin, kidneys, heart, eyes and lungs. But any part of the body can be affected. Symptoms can range from mild to severe, depending on the size or location of the growths. The tumours caused by tuberous sclerosis can result in a range of associated health problems including epilepsy, developmental problems, learning disabilities, behavior problems, eye problems, skin abnormalities, kidney problems, lung problems, heart problems, a build-up of fluid on the brain (hydrocephalus), *etc.* Tuberous sclerosis complex affects 1 in 6,000 to 10,000 people. TSC is primarily a human genetic disorder.

Von Willebrand Disease (VWD). Von Willebrand disease (VWD) is the most common lifelong bleeding disorder caused by the deficiency of a blood clotting protein called von Willebrand factor (VWF). It is estimated to affect 1 in 100 to 10,000 individuals. Symptoms of von Willebrand disease can include frequent or hard-to-stop nosebleeds, excessive bruising, bleeding gums, heavy bleeding after an injury/surgery/childbirth/miscarriage, heavy menstrual bleeding and iron-deficiency anemia. Mutations in the VWF gene located at the short arm of chromosome 12 (12p13.2) cause von Willebrand disease. The gene encodes von Willebrand factor, which is essential for the formation of blood clots. The VWF gene is highly polymorphic, and this polymorphism leads to a big variation in the normal spectrum of presentations and disease severity. VWD affects both humans and several species of animals, including dogs. In dogs, it's the most common inherited bleeding disorder, with over 40 breeds being affected.

Hereditary Spherocytosis (HS). Hereditary spherocytosis (HS) is an inherited blood disorder characterized by hemolytic anemia. It is the most prevalent cause of hemolytic anemia due to abnormal red cell membrane. Symptoms can range from mild to severe and may include pale skin, fatigue, anemia, jaundice, gallstones, stomach pain, shortness of breath, lack of energy, lack of appetite, headaches, palpitations, irritability in children, enlargement of the spleen, short stature, delayed puberty, and/or skeletal abnormalities. There are different types of Hereditary spherocytosis, which are distinguished by severity and genetic cause.

HS is caused by mutations in one of the following genes: ANK1 (8p11.21), SPTA1 (1q21), SPTB (14q23.3), SLC4A1 (17q21.31) and EPB42 (15q15-q21), that encode the red blood cell membrane proteins and result in an overly rigid, misshapen cell. Instead of a flattened disc shape, these cells are spherical. Dysfunctional membrane proteins interfere with the cell's ability to change shape when traveling through the blood vessels. The misshapen red blood cells, called spherocytes, are removed from circulation and taken to the spleen for hemolysis. The shortage of red blood cells in circulation and the abundance of cells in the spleen are responsible for the signs and symptoms of hereditary spherocytosis. Mutations in the ANK1 gene are responsible for approximately half of all cases of hereditary spherocytosis. The other genes associated with hereditary spherocytosis each account for a smaller percentage of cases of this condition. It is most commonly inherited in an autosomal dominant manner, but may be inherited in an autosomal recessive manner. Hereditary spherocytosis occurs in 1 in 2,000 individuals of Northern European ancestry.

Autosomal Dominant Polycystic Kidney Disease (ADPKD). Autosomal dominant polycystic kidney disease (ADPKD), adult polycystic kidney disease, is one of the most common, life-threatening inherited renal disorder that causes cysts to develop in the kidneys and is the most common genetic cause of renal failure worldwide. The most common symptoms are pain in the back and sides, between the ribs and hips, and headaches. The pain can be short term or ongoing, mild or severe. Signs and symptoms of ADPKD often develop between the ages of 30 and 40. Symptoms can include early-onset hypertension, hematuria, headaches, urinary tract infections (UTIs), liver cysts, pancreatic cysts, seminal vesicle cysts, kidney stones, kidney enlargement, cardiovascular abnormalities, bronchiectasis, *etc.* ADPKD affects 1 in 400 to 1,000 people. Mutations in PKD1 (16p13.3) or PKD2 (4q22.1) genes which codes for the protein polycystin cause ADPKD. ADPKD primarily affects humans and dogs, though similar symptoms and pathological features have been observed in other domestic animals like pigs, cattle, and sheep. PKD also occurs in Persian and Himalayan cats. Many of these cats develop kidney failure, while some only develop isolated cysts that do not impair normal kidney function.

Autosomal recessive polycystic kidney disease (ARPKD) is a rarer type of kidney disease. In this type problems usually start much earlier, during childhood.

Myotonic Dystrophy (MD). Myotonic dystrophy (MD) is the second most common form of muscular dystrophy that causes progressive muscle weakness and wasting. It affects about 1 in 8,000 people. Symptoms of myotonic dystrophy can appear anytime between birth and old age and can range from mild to severe. People with the most severe form of the disorder have extreme muscle stiffness (myotonia) and many other symptoms, including cataracts, small testes, premature balding in the front of the scalp (in men), irregular heartbeats, slurred speech, problems with swallowing, bowel problems, behavioural and personality problems, excessive sleepiness or tiredness, diabetes, intellectual disability, *etc.*

Both Myotonic dystrophy type 1 and type 2 are caused by an expansion of DNA tandem repeats, which results in an RNA gain of function mutation (Hahn and Salajegheh, 2016). Myotonic dystrophy type 1 is caused by expansion of cytosine-thymine-guanine (CTG) repeat in the 3'-untranslated region of the DM1 protein kinase (DMPK) gene on chromosome 19q13.3 (Brook *et al.*, 1992). Myotonic dystrophy type 2 results from the expansion of cytosine-cytosine-thymine-guanine (CCTG) tetranucleotide repeat located in the intron of the CCHC-type zinc finger nucleic acid-binding protein (CNB or ZNF9) gene on chromosome 3q21.3 (Ranum *et al.*, 1998). Scientific evidence suggests that excess messenger RNA generated from the abnormal DNA repeats is toxic and interferes with the production of many proteins in cells, which, in turn, causes signs and symptoms in various organs in myotonic dystrophy. There's no cure for DM, but certain treatments and therapies can help manage symptoms and improve quality of life. Myotonic dystrophy primarily affects humans.

Acute Intermittent Porphyria (AIP). Acute intermittent porphyria (AIP; also called Swedish porphyria) is a rare multifactorial disorder characterized by partial deficiency of the heme biosynthetic enzyme porphobilinogen deaminase (PBGD), also called hydroxymethylbilane synthase (HMBS). This enzyme deficiency can result with toxic accumulation of aminolevulinic acid and porphobilinogen in the body. It is caused by mutations in the HMBS gene (11q24.1-q24.2). AIP is an autosomal dominant disorder with low penetrance. The deficiency by itself is not sufficient to produce symptoms of the disease. Additional factors such as hormonal changes associated with puberty, the use of certain drugs, excess alcohol consumption, infections, and fasting or dietary changes are required to trigger the appearance of symptoms. Symptoms include severe abdominal pain, nausea, vomiting, peripheral neuropathy, constipation, a rapid heartbeat and increased blood pressure, pain in the back, arms and legs, urinary retention, cognitive and behavioral abnormalities, and seizures. AIP primarily affects humans.

Hypertrophic Cardiomyopathy (HCM). Hypertrophic cardiomyopathy

(HCM) is a congenital or acquired heart muscle disease characterized by increased left ventricular wall thickness (hypertrophy) causing dynamic left ventricular outflow obstruction, diastolic dysfunction, myocardial ischemia, arrhythmias, autonomic dysfunction, and mitral valve regurgitation. Symptoms include dyspnea, chest pain, fainting, palpitations, heart failure, stroke, and sudden death. Most cases (60 to 70%) of hypertrophic cardiomyopathy are caused by mutations in one of several sarcomere protein genes that encode components of the contractile machinery of the heart; the most commonly involved genes are MYH7, MYBPC3, TNNT2, and TNNI3. Other genes may also be involved in this condition, including some that have not been identified. Incomplete penetrance and variable expression can cause unpredictable manifestations, even within the same family. It has been observed in a wide range of mammalian species, including humans, cats, dogs, and several others.

Single Gene Genetic Disorders in Cattle

Some of the most common single-gene genetic disorders in cattle include Bovine Leukocyte Adhesion Deficiency (BLAD), Complex Vertebral Malformation (CVM), curly calf syndrome, marble bone disease, syndactyly (mule foot), Double muscling (DM) and hypotrichosis.

Bovine Leukocyte Adhesion Deficiency (BLAD). Bovine leukocyte adhesion deficiency (BLAD) is an autosomal recessive immunological disorder affecting young Holstein calves. Animals with BLAD have recurrent and prolonged mucosal and epithelial infections, fever, low appetite, chronic pneumonia, and diarrhea. They also have severe ulcers of oral mucosa, stunted growth, and impaired wound healing. These animals shows signs of immunodeficiency, but the appearance of affected animals varies. The disease is due to a single base substitution of adenine with guanine at position 383 in the CD18 gene, also known as ITGB2 (integrin beta chain beta 2) gene, situated on chromosome 21, which subsequently leads to replacement of aspartic acid with glycine at position 128 in the corresponding protein (D128G). The ITGB2 gene encodes the CD18 protein, which is a key component of leukocyte adhesion and is involved in various immune processes. Mutations in ITGB2 can lead to absent, reduced, or aberrant CD18 protein expression, resulting in impaired immune function.

Curly Calf Syndrome: Curly Calf Syndrome, also known as Arthrogryposis Multiplex (AM), is a recessive lethal genetic defect in. The calves are stillborn and have a twisted or curved spine and legs. That is how it gets the term "curly calf syndrome." AM calves typically have poor muscle development causing them to appear small and thin. The joints are commonly rigid, with the hind legs extended. AM is caused by a mutation in which a significant section of DNA missing. The deletion of this genetic information affects 3 genes – one involved in immune function, one involved in nervous system-muscle connection, and one with unknown function.

Marble Bone Disease: Marble bone disease, also known as osteopetrosis, is an extremely rare genetic disorder in which the bones become extremely dense, hard, and brittle. The disease progresses as long as bone growth continues. It's named "marble bone" due to the bone's dense, stony appearance on X-rays. Osteopetrosis is caused by a defect in the osteoclasts, the cells responsible for bone resorption (breaking down old bone tissue). Marble bone disease affects both humans and certain animal species, including cattle, specifically the Red Angus breed, are known to be susceptible.

Syndactyly (Mule Foot). Syndactyly, also known as mule foot, is an autosomal recessive rare developmental disorder in cattle, where the digits of the feet are partially or completely fused. It is caused by a recessive mutation to the gene LRP4 located on chromosome 15. The LRP4 gene encodes a protein, low-density lipoprotein receptor-related protein 4 (LRP4), which is critical for the formation and maintenance of the neuromuscular junction (NMJ). It is most common in Holsteins and Aberdeen Angus breed. Affected cattle are born with toes on one or more feet fused together, resulting in singular hooves (much like those of a mule or horse). Walking on fused hooves can be painful and stressful to affected cattle.

Double Muscling. Double muscling (DM), or muscular hypertrophy, is a syndrome affecting cattle characterized phenotypically by an increase in muscle mass, reduction in fat and reduction of skeletal weight. The highest frequency of occurrence is found in only two breeds, *i.e.*, the Belgian Blue and the Piedmontese. These animals are characterized by an excellent conformation and an extremely high carcass yield, coinciding with a reduced organ mass. As a consequence, voluntary feed intake is reduced, but feed efficiency is considerably improved. DM animals are more susceptible to respiratory disease, stress and dystocia, requiring extra attention for accommodation and welfare. The syndrome is associated with some production problems such as reduced fertility, dystocia and reduced calf survival. Cattle showing the syndrome, however, have higher meat yield, a higher proportion of expensive cuts of meat, and lean and very tender meat. DM is due to an autosomal recessive mutation of the MSTN (MSTN) gene located on chromosome 2, which results in inactivation of the myostatin gene. Myostatin gene codes for the growth regulating factor myostatin. Myostatin is a myokine that is produced and released by myocytes and acts on muscle cells to inhibit muscle growth.

Hypotrichosis (Hairlessness). Bovine hypotrichosis is an autosomal recessive congenital condition characterized by a partial or complete absence of hair follicles, potentially leading to a lack of hair or a thin, sparse coat. Calves are born with little or no hair; in most cases, calves will grow a short curly coat of hair as they age. Calf generally having black diluted charcoal or chocolate coloured. Hypotrichosis may occur with other congenital defects, including anodontia (missing teeth), hypoplasia of the thyroid or thymus, and corneal opacity. Affected animals are prone to environmental stress (cold and wet) and skin infections are more prevalent. The disease is caused by mutations in various genes. These mutations can be inherited through autosomal recessive or sex-linked inheritance patterns. In some cases, like in Hereford cattle, it's linked to a recessive mutation in the KRT71 gene. Other forms of bovine hypotrichosis can be associated with anodontia (lack of teeth) and are linked to deletions in the ED1 gene or X-linked recessive inheritance.

Single Gene Disorders in Goats

Two well-known examples are Myotonia congenita (fainting goats) and G6-Sulfatase deficiency.

Myotonia Congenita. This is an autosomal dominant disorder where goats experience sudden muscle contractions when startled, sometimes causing them to fall. It's characterized by a delay in muscle relaxation due to a defect in chloride channels. Goats with this condition are often referred to as "fainting goats. Mutations in the CLCN1 gene cause myotonia congenita. The CLCN1 gene provides instructions for making a protein called CLC-1 chloride channel that is essential for the proper function of skeletal muscle cells. These mutations affect the CLC-1 chloride channel. The defective chloride channels triggers prolonged muscle contractions, which are the hallmark of myotonia.

G6-Sulfatase Deficiency or Mucopolysaccharidosis type IIID (MPSIIID). This is an autosomal recessive lethal disorder primarily affecting Nubian goats and their crosses. It's caused by a mutation in the G6-S (N-acetylglucosamine-6-sulfatase) gene, leading to the inability to break down heparin-sulfate glycosaminoglycans, which then accumulate in the body. Affected goats exhibit growth retardation, delayed motor development, and early death. N-acetylglucosamine-6-sulfatase is a naturally occurring enzyme that helps maintain connective tissues throughout the body.

Pedigree Analysis

There are two very useful tools for studying how traits are passed from one generation to the next. One tool is a pedigree analysis, and the other is a Punnett square. The study of the inheritance of genetic traits in human is complicated by the facts that human generation interval is about 20 years, humans produce relatively few offspring compared to almost all other species, and no controlled crosses can be done due to ethical reasons. Human geneticists perform pedigree analysis to study the pattern of the inheritance even when there is limited data. A pedigree chart is a graphical representation of a family's genetic history. It is also called family tree. It shows the members of a family, their inter-relations, and occurrence of phenotypes of a particular gene or individual over several generations (Figure 2.5). Pedigree charts are commonly used for humans, racehorses, and dogs. Pedigrees are used to analyze the pattern of inheritance of simple Mendelian traits throughout a family. Pedigree analysis is the process of determination of the mode of inheritance of a gene. If more than one individual in a family is affected with a disease, then it is assumed that the disease is inherited. By analyzing a pedigree, we can determine genotypes, identify phenotypes, and predict how a trait will be passed on in the future. The information from a pedigree makes it possible to determine how certain alleles are inherited: whether they are dominant, recessive, autosomal, or sex-linked.

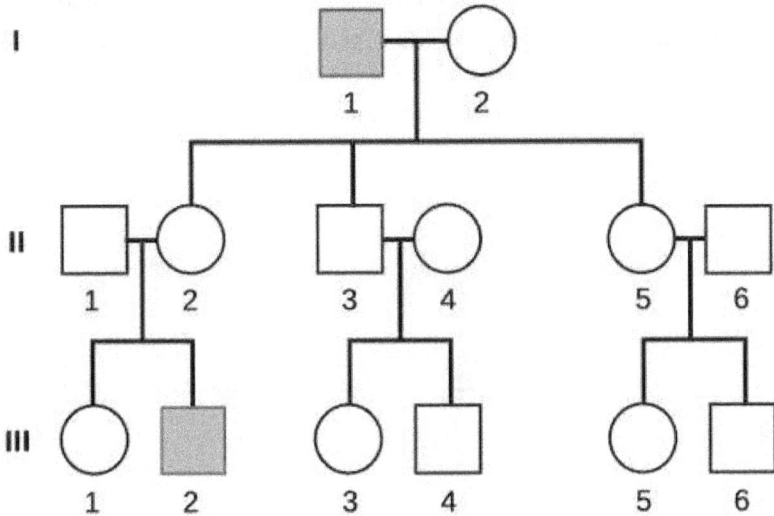

Figure 2.5 Pedigree illustrating a three-generation family and the inheritance of an X-linked recessive trait.

Pedigrees are drawn using standard symbols and formatting to make them easier to understand. Each family member is represented by a symbol:

- ❖ Males are represented by squares.
- ❖ Females are represented by circles.
- ❖ Filled symbols represent affected individuals, and empty ones represent normal individuals.
- ❖ A marriage or mating is represented as a horizontal line (marriage line) between the square and the circle.
- ❖ Offspring symbols appear from left to right according to their order of birth and are indicated by a suspending vertical line drawn perpendicular to the horizontal line. Twin offspring are connected by a triangle.
- ❖ A diagonal line through a symbol indicates that person is deceased.

- ❖ Individuals on the same row belong to the same generation.
- ❖ Each generation is identified by a Roman numeral (I, II, III, and so on). I being the first generation.
- ❖ Each individual within the same generation is identified by an Arabic numeral (1, 2, 3, and so on).
- ❖ Specific combinations of numerals (like II-3) uniquely identify each individual in the pedigree. Many pedigrees do not include numerals.

CHAPTER THREE   MODIFICATIONS
OF MENDELIAN RATIOS

One of Gregor Mendel's greatest contributions to the study of heredity was the concept of dominance. Mendel observed that a heterozygous offspring can show the same phenotype as the homozygous parent, so he concluded that there were some traits that dominated over other inherited traits. While alleles are transmitted from parent to offspring according to Mendelian principles, they sometimes fail to display the clear-cut dominant-recessive relationship observed by Mendel.

Modifications of 3 : 1 Phenotypic Ratio

1. Modifications of Dominance Relationships

With complete dominance, the dominant allele fully masks the effect of the recessive allele, so that the phenotype of the heterozygote is the same as the dominant homozygote. Here one (dominant) allele product totally suppressing the effect of other (recessive) allele. All seven traits that Mendel selected have one allele completely dominant over the other. The 3 : 1 ratio depends on complete dominance of one phenotype over the other. In the absence of complete dominance, the 3 : 1 phenotypic ratio of monohybrid crosses has been found to be modified variously. This may be due to a number of factors.

Incomplete dominance or semi-dominance or partial dominance

When neither allele for a trait is fully dominant, resulting in a blended phenotype in heterozygous individuals, it is said to show incomplete, or semi, or partial, dominance. In heterozygote, both parental phenotypes are mixed together to form a third, different phenotype. In incomplete dominance the phenotype of the heterozygote is somewhere intermediate between the phenotypes of the two homozygotes. In incomplete dominance, one allele produces a functional product, and the other allele either produces no product or produces a nonfunctional product at the molecular level.

This type of gene interaction was discovered by Karl Correns in 1900, while experimenting with four-O'-clock plants (*Mirabilis jalapa*). When he crossed the homozygous red-flowered (*RR*) plant with homozygous white-flowered (*rr*) plant, he noticed a strange phenotype in F_1 hybrids. The flowers of heterozygous plants (*Rr*) were pink, instead of being red or white by virtue of dominance. He concluded that this was due to an intra-allelic interaction in which the dominant allele could express itself partially in heterozygous condition. Since the character appeared to be intermediate between the dominant and recessive phenotypes, the phenomenon was called 'incomplete dominance'. Further, if selfing is done among F_1 hybrids, the phenomenon of incomplete dominance persists and phenotypic and genotypic ratios obtained in F_2 generation are Red 1: Pink 2: White 1. Here flower color is determined by a gene with two alleles, one for red flowers (*R*) and one for white flowers (*r*). The allele for red flower colour (*R*) is incompletely dominant over the allele for white flower colour (*r*),

Image 3.1 Red, white, and pink flowers in snapdragon (source: https://ngb.org/snapdragon-pictures/).

Other examples of incomplete dominance are the flower color of snapdragon (*Antirrhinum majus*) (Image 3.1). One of the genes for flower color in these plants has two alleles, one for red flowers and one for white flowers. A plant that is homozygous for the red allele ($C^R C^R$) will have red flowers, while a plant that is homozygous for the white allele ($C^W C^W$) will have white flowers. On the other hand, the heterozygote ($C^R C^W$) will have pink flowers. Neither the red nor the white allele is dominant, so the phenotype of the offspring is a blend of the two parents.

In snapdragons, broad leaves are also incompletely dominant over narrow leaves. The heterozygotes are intermediate in leaf breadth.

Plumage color in Andalusian chickens presents a good example of incomplete dominance. Crosses between a true-breeding black strain (homozygous $C^B C^B$) and a true-breeding white/splashed strain (homozygous $C^W C^W$) give F_1 birds ($C^B C^W$) with bluish-grey plumage, called Andalusian blues by chicken breeders. In Andalusian × Andalusian crosses the two alleles will segregate in the offspring and produce black, Andalusian blue, and white fowl in a ratio of 1 : 2 : 1 (Figure 3.1). The American Poultry Association (APA) recognizes the blue variety (heterozygous form) in Americana, Andalusian, Cochin, Jersey Giant, Langshan, Orpington, Plymouth Rock, Sumatra, and Wyandotte.

Parents	$C^B C^B$ Black	×	$C^W C^W$ White
Gametes	C^B		C^W
F_1	$C^B C^W$ Andalusian blues		

F_2		Male Gametes		
			½ C^B	½ C^W
	Female Gametes	½ C^B	¼ $C^B C^B$ Black	¼ $C^B C^W$ Andalusian blues
		½ C^W	¼ $C^B C^W$ Andalusian blues	¼ $C^W C^W$ White

Figure 3.1 Incomplete dominance in Andalusian fowls.

The frizzle phenotype of chicken is most commonly found in Bantam, Cochins, Silkies, and Serama breeds. In 1936, researchers Walter Landauer and Elizabeth Upham observed that frizzle chickens have feathers that curled outward and upward rather than lying flat against their bodies. The allele responsible for the frizzle phenotype is known as the frizzle (F) allele. It is an autosomal incompletely dominant allele. The F allele codes for the protein keratin, an important component of feathers that causes the feathers to all curl outward and upward rather than lying flat against the body. If a chicken inherits only one copy of the *F* allele (*Ff*) it will express the frizzle trait. Having two copies of the *F* allele (*FF*), results in health problems associated with the extreme expression of the trait, now called "frazzled". The feathers fall out, reducing insulation, which can be fatal. If a chicken inherits two normal feather alleles (*ff*), it will have normal feathering without the frizzle trait. Along with producing defective feathers, the frizzle allele also led to abnormal body temperatures, higher metabolic and blood flow rates, and greater digestive capacity. Furthermore, chickens who had this allele also laid fewer eggs than their wild-type counterparts.

The crest phenotype in chicken is characterised by a tuft of elongated feathers atop the head. It is most commonly seen in Sultans, Polish, Houdans, and Silkies, though the shape of the crests vary wildly from breed to breed. This is due to the feather length, width, and shape that changes by breed. A similar phenotype is also seen in several wild bird species. Crest in domestic chicken is determined by an autosomal gene variant (Cr) showing incomplete dominance. It is caused by a 195 bp duplication in the intron of the HOXC10 gene. This duplication leads to ectopic expression of HOXC10 and other HOXC genes in the cranial skin, causing the development of the characteristic tuft of elongated feathers. Generally, the heterozygous forms of crested have smaller shaped crests, while the homozygous form has larger and more developed crests. Males will often have more prominent crest shapes than females in both forms.

The naked neck trait in chickens is controlled by an autosomal incompletely dominant allele, Na, located near the middle of chromosome 3. It exhibits incomplete dominance, meaning that the heterozygous state (Na/na) shows a phenotype that is intermediate between the homozygous wild-type (na/na) and homozygous naked neck (Na/Na) phenotypes. Homozygous naked neck individuals (Na/Na) have a significantly reduced feather coverage, with a nearly bare neck and abdomen, and a lack of feathers around the keel bone. Heterozygous individuals (Na/na) have an isolated tuft of feathers on the ventral side of the neck above the crop, and a moderate reduction in overall feather coverage compared to the wild-type. Homozygous wild-type individuals (na/na) have normal feather coverage, with feathers present on the neck and other areas of the body. Heat stress causes behavioral, physical, and physiological changes in poultry, with severe financial impacts. The Na allele is associated with increased tolerance for heat as it allows for better heat dissipation. The naked neck (Na) allele is not only responsible for defeathering the neck region, but also restricts the feathering areas around the body by 20-30% in the heterozygous (Na/na) and up to 40% in the homozygous (Na/Na) genotype (Horst, 1988). Na is also associated with a small increase in meat yield and lower body fat content. Homozygous individuals (Na/Na) also results in high embryo death and reduced hatchability and diminishes floating and flying capacity. The gene is located on Chromosome 3. The naked neck mutation (Na) was originated from a large insertion (~ 180 bp) of a protein-coding gene – GDF7 (Growth Differentiation Factor 7).

Table 3.1 Examples of incomplete dominance inheritance

Species/breed	Homozygous Phenotype	Heterozygote (Intermediate)	Homozygous Phenotype
Sheep	Black fleece	Gray fleece	White fleece
Dogs	Long tail	Medium-length tail	Short tail
Rabbits	Long fur	Medium fur	Short fur
Humans	Curly hair	Wavy hair	Straight hair
	Close set eyes	Average	Distant
	Large eyes	Medium eyes	Small eyes
	Long mouth	Medium-sized mouth	Small mouth
	Large nose	Medium-sized nose	Small nose
	Darker eyebrow	Same color as the hair	Lighter than the hair
Fruit color of eggplants	Deep purple	Light violet	White
Roses	Red colour	Pink colour	White colour

In chickens, the terms "muff" and "beard" refer to specific feather formations on the face. A "muff" describes the feathers on the sides of the face, while a "beard" refers to the feathers directly under the beak. These are often controlled by the same gene, known as the muffs and beard (Mb) gene. Muffs and beard (Mb) is known as an autosomal incomplete dominant trait. There are three genotypes for the Mb locus regulating the Mb trait: (1) Mb/Mb homozygous (carrying two copies of the gene and exhibiting a full muff/beard), (2) Mb/mb heterozygous (carrying one copy of the gene and exhibiting a partial muff and beard), and (3) mb/mb homozygous (wild-type, absence of muff and beard).

Sheep exhibit incomplete dominance in the trait for eye color. When a brown-eyed sheep is crossed with a green-eyed sheep, blue-eyed offspring are produced.

In some cats the gene for tail length shows incomplete dominance. Cats with long tails and cats with no tails are homozygous for their respective alleles. Cats with one long tail allele and one no tail allele have short tails.

There are many examples of incomplete dominance inheritance in many different species. A few of these are listed in Table 3.1.

Incomplete dominance is not support for the blending theory of inheritance, because alleles maintain their integrity in the heterozygote and segregate during gamete formation.

Codominance

Codominance means that both alleles are equally dominant (simultaneously expressed) and the heterozygote shows the phenotypic characteristics of both homozygotes. Since the characters expressed by both the alleles exist simultaneously in equal amount, there is no appearance of intermediate phenotype in heterozygotes. In codominance, both alleles produce a functional product at the molecular level. A well-known case is type AB in the ABO blood group system in humans. The L^A allele and the L^B allele are equal in their dominance and will be expressed equally if they are paired together into the genotype $L^A L^B$. Neither the L^A allele nor the L^B allele is dominant over each other, so each type is expressed equally in the phenotype giving the human an AB blood type.

The human MN blood group system is another example of codominance. MN system has two alleles denoted as L^M and L^N. In MN system three blood types occur: M, N, and MN. They are determined by the genotypes $L^M L^M$, $L^N L^N$, and $L^M L^N$, respectively. The heterozygote in this case has both the M and N antigens and shows the phenotypes of both homozygotes. A mating between a (homozygous) M-type person ($L^M L^M$) and a (homozygous) N-type person ($L^N L^N$) would result in all heterozygous ($L^M L^N$) offspring. Mating between heterozygotes ($L^M L^N \times L^M L^N$) would result in a ratio of 1 M-type ($L^M L^M$) : 2 MN-type ($L^M L^N$) : 1 N-type ($L^N L^N$). A phenotypic ratio of 1 : 2 : 1 has thus replaced the 3 : 1 ratio, because the alleles are both expressed in the heterozygote; that is, the alleles are codominant. The genes responsible for the MN blood group antigens, GYPA and GYPB, are located on the long arm of chromosome 4, specifically in the region 4q28.2-q13.1. These two genes encode glycophorin A and B, respectively, which are proteins found on the surface of red blood cells. In terms of transfusion compatibility, this system is of less clinical importance than the ABO system.

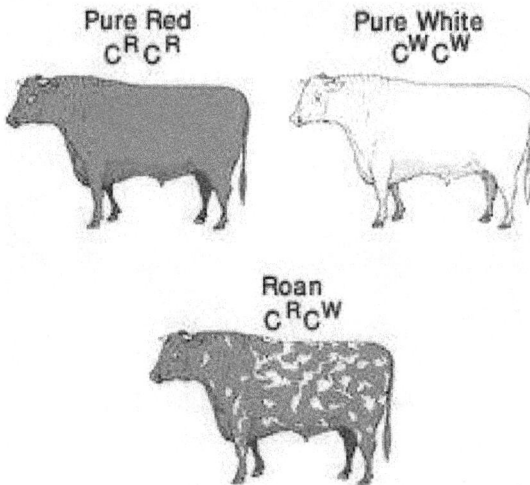

Pure Red
$C^R C^R$

Pure White
$C^W C^W$

Roan
$C^R C^W$

Image 3.2 Cross between red bull and white cow in Shorthorn cattle.

Roan is a pattern common in Shorthorn cattle and Belgian Blue cattle. The roan coat color pattern is controlled by the mast cell growth factor (MGF) gene, also known as the KITLG gene or Steel locus. This gene is located on bovine chromosome 5 (BTA5). In Shorthorn cattle, the coat colors (red, white or roan) are controlled by a single pair of alleles. When a red bull ($C^R C^R$) is crossed with a white cow ($C^W C^W$), it will produce "red roan" calf ($C^R C^W$) (Image 3.2). Roan means there are red hairs and white hairs intermixed in the coat of the calf. The individual hairs are not a blend of red and white color. The red hairs and white hairs of a roan coat are present in distinct regions.

In Belgian Blue cattle breed, coat color variation is mainly under the influence of a single autosomal locus, the roan locus, characterized by a pair of codominant alleles: C^B (black) and C^W (white). Heterozygous ($C^B C^W$) animals have a mixture of black hairs and white hairs, yielding the "blue roan" phenotype typical of the breed.

Image 3.3 Flowers with mixed patches of red and white colour in Camellia plants (source: https://c7.alamy.com/comp/2H7053N/white-camellia-japonica-striped-with-red-2H7053N.jpg).

Codominance is also seen in plants. A classic example of codominance is inheritance of flower colour in Camellia plants. If white flowered Camellia plants are crossed with red flowered Camellia plants, F_1 generation produce flowers with mixed patches of red and white colour (Image 3.3). The alleles for both red and white petal colour are codominant and both the colours are expressed simultaneously in equal amounts.

Codominance can also be observed in Rhododendrons whose flower color is determined by two alleles those are codominant. The C^R allele causes red flowers and the C^W allele causes white flowers. In heterozygote ($C^R C^W$) there is a simultaneous expression of red and white alleles displaying flowers with both red and white petals.

In some chicken breeds, the black and white alleles for feather color are codominant. Homozygotes for the white allele are white, for the black allele are black. But the heterozygotes express both alleles known as erminette. Erminette chickens are speckled with both black feathers and white feathers.

An example of codominance occurs in cats. There are four alleles, wild type (black, C), burmese (cb), siamese (cs), and albino (c). Albino has no pigment, and the others have different amounts of pigment. Wild type is dominant to the rest, and albino is recessive to the rest. A cat with two burmese alleles is lighter than wild type and darker than a cat with two siamese alleles. The dominance hierarchy of these alleles is C > cb > cs > c. A cat with a burmese allele paired with a siamese allele is called a Tonkinese (cbcs) cat. As both the burmese and siamese alleles produce a catalyst involved in the pigment biosynthetic pathway, the two alleles are codominant to each other.

The inheritance pattern of sickle-cell anemia in humans exhibits codominance at the hemoglobin level. If an individual has just one copy of the mutated gene, they are said to be a sickle cell trait, carrier for sickle cell disease. They have two types of hemoglobin, normal (Hb^A) and sickle (Hb^S).

Overdominance (also called Super dominance or Heterodominance)

Overdominance, also called superdominance or heterodominance, is a rare condition in which the **heterozygote genotype (Aa)** shows **greater fitness, vigor, or advantage** than either of the homozygotes (AA or aa). Overdominance can also be described as heterozygote advantage regulated by a single gene.

The classic example of overdominance is the occurrence of the Mendelian inherited sickle-cell anaemia in humans that is a recessive trait at phenotypic level, but at molecular level it is an example of codominance. Individuals who are homozygous for the sickle-cell allele (*ss*) develop the disease. Individuals who are heterozygous (*Ss*) for the sickle cell anaemia express both normal and sickle hemoglobin, so they have a mixture of normal and sickle red blood cells and in most situations are phenotypically normal. Heterozygous (*Ss*) individuals are partial resistance to falciparum malaria because mixed normal and sickle red blood cells "collapse" around the parasites and filter them out of the blood. Thus, people who carry the sickle-cell allele are more likely to recover from malarial infection. This mutation negatively affects human survival by causing sickle-cell anemia, but positively affects human survival by protecting against malaria.

A similar phenomenon is observed in carriers of the cystic fibrosis allele, where heterozygotes (Cc) show resistance to typhoid fever compared to normal homozygotes (CC), while homozygous mutants (cc) suffer from cystic fibrosis.

In animals, a well-known example of overdominance is **warfarin resistance in rats**. Warfarin is a common rodenticide that interferes with vitamin K metabolism, preventing blood clotting. Rats that are **homozygous susceptible (ss)** die when exposed to warfarin, while **homozygous resistant (SS)** rats survive the poison but often suffer from vitamin K deficiency, which impairs normal clotting and reduces overall fitness. In contrast, **heterozygotes (Ss)** are resistant to warfarin while still maintaining normal vitamin K metabolism. This gives them a clear survival advantage over either homozygote, making it a classic case of overdominance in animals.

2. Lethal Alleles

A large proportion of genes in an individual's genome are essential for survival. They are called essential genes. An allele which produces effects severe enough to cause death of an organism is called a lethal allele. It occasionally arises by mutations in essential genes. Lethal alleles may be dominant, incompletely dominant, recessive, semilethal, or conditional and they can be sex-linked or autosomal. If the lethal allele is dominant, then both homozygous dominant and heterozygous individuals will die. If it is a recessive lethal, then only the homozygous individuals will die. Death from different lethal alleles may occur at any time from fertilization of egg to advanced age. Some lethal alleles are expressed as deaths *in utero* (homozygosity for yellow allele in mice), where they either go unnoticed, or are noticed as spontaneous abortions. Other lethal alleles, such as those responsible for Duchenne muscular dystrophy, PKU, and cystic fibrosis, exert their effects in childhood. The time of death can even be in adulthood, as in Huntington's disease. Over 100 chicken mutant loci have been shown to have lethal effects.

Many genes have alleles that affect the rate of mortality but are not lethal. In general, these are termed deleterious, or detrimental, alleles. They result in various levels of mortality, ranging from a few percent more than the wild type genotype to nearly lethal (almost 100% mortality). The total of all the deleterious and lethal genes that are present in individuals is called genetic load, a kind of genetic burden that the population has to carry.

Dominant lethal alleles

Dominant lethal alleles kills organisms both in homozygous and heterozygous states. Dominant lethal alleles are rarely detected due to their rapid elimination from populations. While dominant lethal alleles are lethal in both homozygous and heterozygous individuals, their lethality can only be transmitted if it manifests after reproductive age. Dominant lethal alleles are much rarer than lethal recessives, because they are always expressed. A good example of a disease caused by a dominant lethal allele is the rare – Huntington's disease that causes neurodegeneration and ultimately death in humans. Because the onset of Huntington's disease is usually between the ages of 30 and 50, individuals carrying the allele can pass it on to their offspring. This allows the allele to be maintained in the population. Dominant traits can also be maintained in the population through recurrent mutations or if the penetrance of the gene is less than 100%. Examples of human diseases caused by dominant lethal alleles include myotonic muscular dystrophy, hereditary nonpolyposis colorectal cancer, tuberous sclerosis (epiloia), *etc.*

Incompletely dominant lethal alleles

Incompletely dominant lethal alleles are separate concepts in genetics. In incompletely dominant lethal alleles, the heterozygous condition leads to a distinct (intermediate/blended/recognizable) phenotype (though abnormal but viable) and homozygous state causes death. Examples include Manx gene in cat, achondroplasia, brachydactyly, bull-dog dwarfism, the creeper gene in Japanese Bantams, and the ear-tuft (Et) allele in Araucanas. They are incompletely dominant lethal alleles.

Creeper gene in Japanese Bantams. In chickens, a condition referred to as "creeper" exists whereby the bird has very short legs and wings and appears to be creeping when it walks. Japanese bantams bred for exhibition must have short legs, according to the Standard of Perfection. Unfortunately this trait comes with an incompletely dominant lethal allele called creeper (*C*). This is also found in other short-legged breeds such as Scots Dumpies. Specifically, it's caused by a 25-kb deletion on chromosome 7 that includes the Indian hedgehog (IHH) and non-homologous end-joining factor 1 (NHEJ1) genes. The creeper allele in homozygous state (*CC*) causes embryos to die within 3 days of incubation, resulting in a reduced hatch rate. The heterozygous condition (*Cc*) results in creeper birds. Wings and legs of creeper birds are considerably shortened and deformed giving it a squatty appearance. When two creepers were mated, a ratio of 2 creepers to 1 normal instead of 3 to 1 appeared (Figure 3.2). This is a characteristic ratio for all crosses involving lethals. In this particular case, the *CC* class is missing.

Parents	*Cc* Creeper	×	*Cc* Creeper
Gametes	½ *C*, ½ *c*		½ *C*, ½ *c*
Progeny	1 *CC* (Dies), 2 *Cc* (Creeper), 1 *cc* (Normal)		

Figure 3.2 Incompletely dominant lethal allele in chicken.

The ear-tuft (Et) allele in Araucanas chickens. South American Araucana chickens are known for their distinctive features like tufted ears, rumplessness (no tail), and beautiful, unique blue eggs. The ear-tuft (Et) allele in Araucana chickens is an autosomal incompletely dominant lethal allele located on chromosome 15 with reduced penetrance in heterozygotes. This allele is responsible for the distinctive ear tufts (feathered skin projections near the ear canal on each side of the head) seen in some Araucana chickens. The homozygotes (Et/Et) is lethal and often die during 17-19 days of incubation or shortly after hatching. All ear-tufted chickens must be heterozygous (Et/et), with varying expressivity.. These chicks have deformed ears, mouths, or throats. The recessive homozygotes (et/et) are normal (non-tufted).

Bull-dog dwarfism. Bull-dog dwarfism, also known as Bulldog Calf Syndrome or Congenital Bovine Chondrodysplasia, is a serious developmental bone defect occurring due to disturbed endochondral osteogenesis in cattle, particularly Dexter cattle. This leads to severe disproportionate dwarfism, a short vertebral column, and a large head. The skull is vaulted and much rounder than normal, the nose flattened, upper lip split, the lower jaw protruding, and the swollen tongue thrust out. It is caused by mutations in the aggrecan (ACAN) gene, leading to dwarfism in heterozygotes. Affected fetuses are naturally aborted around 4-7 month of gestation in homozygotes. The aggrecan (ACAN) gene is located on chromosome 15q26.1. It encodes aggrecan, a major proteoglycan found in the extracellular matrix of cartilage, particularly in the articular and growth plate cartilage. Mutations in ACAN can be associated with skeletal dysplasia and spinal degeneration.

Manx syndrome. "Manx syndrome" or "Manxness" is a genetic condition predominantly affecting Manx cats, characterized by the absence or shortening of the tail. It is produced by an allele that is lethal in the homozygous state. It is caused by a mutation in the Brachyury gene, located on chromosome 18. The Manx and normal alleles are denoted by M and m, respectively. A single dose of the Manx allele causes abnormal development of the coccygeal and sacral vertebrae, resulting in the absent or shortened tail in the Mm heterozygote. In MM homozygote, the double dose of the gene produces such an extreme developmental abnormality that the embryo does not survive. Animals that suffer this defect include Holstein cattle, Land race, large white pigs and cat.

Achondroplasia. Achondroplasia, commonly known as short-limbed dwarfism, is a genetic condition that affects bone growth, leading to abnormally short limbs and other skeletal abnormalities. It is the most common form of short-limbed dwarfism, occurring in about one in every 40,000 births. The presence of a single copy of the Achondroplasia allele (Aa) causes the disorder. At maturity, affected people have abnormally short arms and legs relative to other body parts but are able to reproduce. The homozygous dominant genotype results in a lethal phenotype. These fetuses are stillborn or die shortly after birth. The recessive homozygotes (aa) are normal. Achondroplasia is caused by a mutation in the fibroblast growth factor receptor (FGFR3) gene located on chromosome 4. Achondroplasia affects humans and various animal species, including dogs, horses, cats, chickens, and pigs.

Brachydactyly. In Brachydactyly the fingers and toes are unusually short in heterozygous condition (Bb). But, this condition is lethal during early years to homozygous dominant individuals (BB) due to major skeletal defects. It is categorized into five types. Brachydactyly type A1 is caused by mutations in Indian hedgehog (IHH) gene, type A2 by mutations in Bone morphogenetic protein receptor type-1B (BMPR1B) gene, type A3 by mutations in Homeobox D13 (HOXD13) gene, type B by mutations in Receptor Tyrosine Kinase Like Orphan Receptor 2 (ROR2) gene, and type C by mutations in cartilage-derived morphogenetic protein-1 (CDMP1) gene. Brachydactyly types D and E can also be caused by mutations in HOXD13 gene. IHH gene encode signaling molecules that play a role in regulating embryonic morphogenesis (Marigo *et al.*, 1995). When a single mutated copy of the allele is present, the phenotype has just few deformations of skeletal bones. If an organism inherits two mutated copies of IHH allele no protein essential for skeletal bones formation is produced and development of embryo cannot be continued – the embryo dies. The recessive homozygotes (bb) are normal.

Recessive lethal alleles

Recessive lethal alleles confer no detectable effect in the heterozygote at all, only homozygous recessive individuals express the lethal phenotype and die. The majority of lethal genes are recessive. Recessive lethal alleles are more interesting, the offspring can survive with one copy of the gene (heterozygous). It has been estimated that we each carry a small number of recessive lethal in our genomes. The lethal effect is expressed in the homozygous progeny of a mating between two individuals carrying the same recessive lethal in the heterozygous condition.

Lethal genes were first discovered by French geneticist Lucien Cuénot in 1905 while studying the inheritance of coat colour in mice. He found that when yellow mice were mated *inter se*, they never bred true. After more experimentation, it was eventually realized that all yellow mice were heterozygote and those zygotes homozygous for yellow died at an early stage of gestation.

Yellow fur in mice. Yellow fur in mice is dominant to black; *yy* mice are black,

Yy mice are yellow. When two yellow mice (*Yy*) are mated the progeny would be expected to be in the proportion shown in Table 3.2. However, *YY* is lethal and any mice with this genotype die *in utero*. Hence live-born progeny from this cross are in the ratio given in Table 3.3.

Table 3.2 Expected inheritance of yellow coat color in a cross of *Yy* × *Yy* mice

Genotype	Phenotype	Ratio
yy	Black fur	1
Yy	Yellow fur	2
YY	Yellow fur	1

The 3 : 1 phenotypic ratio has been distorted to a 2 : 1 ratio. The yellow allele (*Y*) has a dominant affect with regard to coat color, but acts as a recessive allele with respect to the lethality phenotype, since only homozygote die. In mice, yellow fur is primarily caused by mutations in the Agouti gene (At) located on chromosome 2.

Table 3.3 Actual inheritance of yellow coat color in a cross of *Yy* × *Yy* mice

Genotype	Phenotype	Ratio
yy	Black fur	1
Yy	Yellow fur	2

White plants in snapdragon. In 1907, Edwin Baur reported a recessive lethal gene in snapdragon (*Antirrhinum majus*) (Baur, 1907). In snapdragon there are three kinds of plants; green plants with chlorophyll (CC), yellowish green plants with carotenoids referred to as pale green, golden or aurea plants (Cc), and white plants without any chlorophyll (cc). When two aurea plants are crossed the F_2 progeny has identical phenotypic and genotypic ratio of 1 : 2 : 1 (Green : Aurea : White). Since the white plants lack chlorophyll pigment, they will not survive. So, the F_2 ratio is modified into 1 : 2. In this case the homozygous recessive genotype (cc) is lethal.

Cuénot and Baur discovered these first recessive lethal genes because they altered Mendelian inheritance ratios. Examples of human diseases caused by recessive lethal alleles include cystic fibrosis, sickle-cell anemia, tay-sachs disease, thalassemia, phenylketonuria, spinal muscular atrophy, adenosine deaminase deficiency, maple syrup urine disease, *etc*.

Faded shaker in chicken. The faded shaker (fs) is an autosomal recessive lethal allele located on chromosome 16 in chickens causing a congenital tremor, dilution of down and feather melanin, and reduced weight of the cerebellum. The homozygotes (fs/fs) show incomplete penetrance and mortality can occur from 18 days of incubation to three months of age, with most deaths occurring before hatching.

Blood ring in chicken egg: A blood ring in a chicken egg during incubation indicates that the embryo started developing but then died early in the incubation process. It appears as a circular blood vessel formation inside the egg, often visible when candling. The vitelline arteries, which are essential for transporting nutrients from the yolk sac to the embryo, are absent in affected embryos. This is often due to a genetic cause, specifically an autosomal recessive lethal gene (called blr) that is expressed between 48 and 66 h of incubation. The gene prevents proper blood vessel formation in the early embryo, leading to its death.

Cornish lethal: The "Cornish lethal" refers to a lethal mutation in White Cornish and Cornish-Rock crosses, which are widely used in broiler chicken production. It is usually caused by recessive lethal alleles. It results in severely shortened leg and wing bones, causing the embryo to die around day 17-19 of incubation or shortly after hatching, or be unable to hatch.

The complex vertebral malformation (CVM). The complex vertebral malformation (CVM) is an autosomal recessive lethal congenital disorder primarily affecting Holstein cattle, causing skeletal malformations and often resulting in fetal resorption, abortion, or stillbirths. Affected calves have anomalies in the vertebral column, including hemi vertebrae, fused and misshapen vertebrae and ribs, scoliosis, and vertebral syncostosis. It is caused by a single base substitution (missense mutation) from guanine to thymine at nucleotide position 559 of the bovine solute carrier family 35 member 3 (SLC35A3) gene which causes an amino acid substitution at position 180 (valine to phenylalanine) in the corresponding protein. The SLC35A3 gene codes for a nucleotide-sugar transporter that plays an important role in mechanisms controlling the formation of vertebrae from the unsegmented paraxial mesoderm. The mutation inhibits the function of the transporter. Therefore, the defective transporter molecule leads to vertebral malformations. It causes intra-uterine mortality through the entire gestation period leading to repeat breeding and involuntary culling of cows and thereby economic losses.

Atresia ani. Atresia ani, or imperforate anus, is a congenital anorectal malformation (ARM) which occurs as a result of an autosomal recessive gene (Bademkiran *et al.*, 2009) where a normal anal opening is absent at birth. This condition is quite common in pigs, sheep and a time in cattle. Affected animals may survive up to 10 days and are identified by peculiar features like depression, anorexia, colic marked abnormal abdominal distension and in most cases with no defecation. Consequently, death of the affected animal occurs during the first few days of life.

Bovine spinal muscular atrophy. Spinal muscular atrophy is a progressive lethal autosomal recessive disorder. It is reported mainly in advanced backcrosses between American Brown Swiss and European Brown Cattle but it is also observed in Holstein-Friesian calves. The gene AFG3L2 (AFG3 Like Matrix AAA Peptidase Subunit 2) respective for this disease was located on distal part of chromosome 24 (24q24). AFG3L2 encodes a subunit of the mitochondrial m-AAA protease that plays a key role in proteostasis of inner mitochondrial membrane proteins, and which is essential for axonal and neuron development. The condition is characterised by severe muscular atrophy, progressive quadriparesis and sternal recumbancy. The preliminary signs observed at 3rd to 4th weeks of age like symmetric weakness of the rear legs, locomotive difficulties and slight dyspnoea. Animals usually look alert and have a good appetite and normal suckling reflex. Death occur after 2-4 weeks, usually as a consequent respiratory failure due to atrophy of the respiratory muscles.

The absence of legs in cattle. In cattle, a lethal allele (l) in homozygous condition results calves to be born without legs (amputated)" with malformed head and internal abnormalities. Affected calves are born at term but most calves are stillborn or die shortly after birth, probably due to respiratory failure.

Tay-Sachs disease in humans. Tay-Sachs disease in humans is an autosomal recessive lethal disorder. Individuals with one copy of the mutated allele are carriers and have a normal phenotype, while those with two copies develop the disease and die.

Semilethal or sublethal genes

Lethal alleles which kill only some individuals of the population but not all are known as semi lethal (lethality between 50% and 100%). Hemophilia is a hereditary disease caused by deficiencies in clotting factors, which results in impaired blood clotting and coagulation. Normally, clotting factors help forming a temporary scab after a blood vessel is injured to prevent bleeding, but hemophiliacs cannot heal properly after injuries because of their low levels of blood clotting factors. Therefore, affected individuals bleed for a longer period of time until clotting occurs. This means that normally minor wounds can be fatal in a person with hemophilia. The alleles responsible for hemophilia are thus called semilethal or sublethal, because they cause the death of only some of the individuals or organisms with the affected genotype.

Conditional lethals

Whereas certain alleles would be lethal in virtually any environment, others are viable in one environment but lethal in another and are known as conditional lethals. For example, many of the phenotypes favored and selected by agricultural breeder would almost entirely be eliminated in nature as a result of competition with the members of the natural population. Modern grain varieties, and chicken strains provide good examples; only careful nurturing by the farmer has maintained such phenotypes for our benefit. Alleles of a number of genes in the fruit fly *Drosophila melanogaster* affect viability and may be lethal at elevated temperatures (say, greater than 28^0C) but have little or no effect at lower temperatures. Such genes show their lethal effects in extreme environmental situations and are known as conditional lethals.

The favism is an example of a sex-linked conditional lethal, conferring to carrier glucose-6-phosphate dehydrogenase deficiency that causes the organism to develop hemolytic anemia after eating of fava beans (*Vicia faba*).

Essential Genes and Lethal Alleles

Essential genes are those genes of an organism that are thought to be critical for its survival. Alleles of certain genes result in the lack of production of necessary functional gene product and this gives rise to a lethal phenotype. The existence of lethal alleles of a gene indicates that the normal product of the gene is essential for the function of the organism; the gene is called an essential gene. Essential genes are genes which, when mutated, can result in a lethal phenotype.

Pleiotropy

Greek word *pleion* means more and *tropos* means character. The phenomenon in which a single gene affects two or more distinct traits is called pleiotropy. An allele may correspond to the dominant phenotype for one trait and the recessive phenotype for another. Pleiotropy was first noticed by geneticist Gregor Mendel. Mendel noticed that plant flower color (white or purple) was always related to the color of the leaf axil (area on a plant stem consisting of the angle between the leaf and upper part of the stem) and seed coat. The term "pleiotropy" was first coined by German geneticist Ludwig Plate in 1910. The *Y* allele in mice is pleotropic in affecting both coat color and viability. Another example of multiple effects is the gene affecting seed shape in garden peas; this gene also affects starch grain morphology. Many genetic diseases in humans are caused by genes that have pleiotropic effects, *e.g.*, sickle-cell anemia, achondroplasia, phenylketonuria, *etc*. Sickle-cell anemia causes enlarged spleen, muscle pain, low red blood cell count, resistance to malaria, and early death. All of this is caused by a single mutation in one of the hemoglobin genes. Phenylketonuria is a human disease that causes mental retardation and reduced hair and skin pigmentation but is caused by one gene defect. A human genetic disorder called Marfan syndrome is caused by a mutation in one gene, yet it affects many aspects of growth and development, including height, vision, and heart function. In fact, all genes may be pleiotropic, with their various effects simply not yet recognized. The study of pleiotropic genes is important to genetics as it helps us to understand how certain traits are linked in genetic diseases.

Gene Interactions and Modified Dihybrid Ratio

As noted earlier, the 3 : 1 phenotypic monohybrid ratio can be distorted by factors such as incomplete dominance, codominance, or lethal effects of certain alleles. These also affect the 9 : 3 : 3 : 1 ratio, but other factors can also modify this ratio.

Gene interaction is the determination of a single trait by two alleles of a single gene, or genes at more than one locus where the effect of one gene depends on the effect of a different gene located elsewhere in the genome. For the determination of single trait of an organism, two alleles of a single gene may interact in various ways, such as, complete dominance, incomplete dominance or codominance. These kinds of genetic interactions between the two alleles of a single gene are referred to as inter-allelic or allelic or intra-genic genetic interactions. In addition to intra-genic genetic interactions two or more nonallelic genes may also interact with one another for the determination of single phenotypic trait of an organism. This kind of genetic interaction is referred as nonallelic or inter-genic genetic interactions. These are of two kinds: epistasis and nonepistatic inter-genic genetic interaction.

Epistasis

Epistasis is a Greek word meaning 'standing over'. It was first used in 1909 by William Bateson to describe a masking effect. In many cases, contrast to Mendelian genetics, more than one gene can affect a single trait. Different genes affecting the same trait interact so that the phenotype expressed is a function of the particular combination of alleles present at different loci. Epistasis is the interaction between nonallelic genes in which one gene masks or suppresses the expression of another gene. Genes whose phenotype is expressed is said to be epistatic/inhibiting, and genes whose phenotype are masked, altered or suppressed is said to be hypostatic. Epistasis can occur in both directions between two gene pairs. This is usually seen as a distortion of the 9 : 3 : 3 : 1 phenotypic ratio with a reduction in the number of different phenotypes observed. Any type of gene interaction that results in the F_2

dihybrid ratio of 9 : 3 : 3 : 1 being modified into some other ratio is called epistasis. Epistasis is an interaction at the phenotypic level. The genes that are involved in a specific epistatic interaction may still show independent assortment at the genotypic level.

Difference between Dominance and Epistasis

Epistasis is the interaction between different (two or more) genes in which one gene masks the effect of other gene at different loci. Dominance is the interaction between different alleles of the same gene in which one allele masks the effect of other allele at the same loci.

Kinds of Epistasis

When in dihybrid crosses, the epistasis occurs between two genes, less than four phenotypes appear in F_2. Such bigenic (two genes) epistasis may be of following types:

Dominant epistasis (12 : 3 : 1)

It is also known as simple epistasis. When the dominant allele of one gene masks the expression of both (dominant and recessive) alleles of the second gene. For example, if the $A-$ genotype renders the $B-$ and bb genotypes indistinguishable, then the dihybrid ratio is 12 : 3 : 1, because the $A- B-$ and $A- bb$ genotypes are expressed as the same phenotype.

9/16	$W- B-$	White	12
3/16	$W- bb$	White	
3/16	$ww B-$	Black	3
1/16	$ww bb$	Brown	1

An example of dominant epistasis is that of white, black, and brown color in sheep. White is a dominant allele in sheep. Hence, both WW and Ww whites exist. Sheep that are ww are colored, and may then express the effect of alleles at the B locus. $B-$ sheep are black, and bb sheep are brown. Heterozygous sheep ($Ww Bb$) are white, and, when mated together, will give an F_2 progeny as shown above. The dihybrid ratio is therefore 12 white : 3 black : 1 brown.

9/16	W– G–	White	12
3/16	W– gg	White	
3/16	ww G–	Yellow	3
1/16	ww gg	Green	1

Another example of dominant epistasis is that of white, yellow and green fruit color in the summer squash (*Cucurbita pepo*). White is a dominant allele in squash. Hence, both *WW* and *Ww* whites exist. Squash that are *ww* are colored, and may then express the effect of alleles at the *G* locus. *G*– squash are yellow, and *gg* squash are green. When the white fruit with genotype *WWgg* is crossed with yellow fruit with genotype *wwGG*, the F_1 plants have white fruit and are heterozygous (*WwGg*). When F_1 heterozygous plants are crossed they give rise to F_2 with the phenotypic ratio of 12 white : 3 yellow : 1 green as shown above.

Recessive epistasis (9 : 3 : 4)

When homozygosity for the recessive allele of one gene masks the expression of both (dominant and recessive) alleles of the second gene, it is known as recessive epistasis. This type of gene interaction is also known as supplementary epistasis. For example, if the *aa B*– and *aa bb* individuals have the same phenotype, then the 9 : 3 : 4 ratio results.

In the mouse, the agouti (hair made with bands of black pigment and yellow pigment) pattern results from the presence of a dominant allele *A*, and in *aa* animals the coat color is black. A second dominant allele, *C*, is necessary for the formation of the hair pigments of any kind, and *cc* animals are albino (white fur). In a cross of *AA CC* (agouti) × *aa cc* (albino), the F_1 progeny are *Aa Cc* and agouti. Crosses between F_1 males and females produce F_2 progeny in the following proportions:

9/16	A– C–	Agouti	9
3/16	A– cc	Albino	3
3/16	aa C–	Black	3
1/16	aa cc	Albino	1

The dihybrid ratio is therefore 9 agouti : 3 black : 4 albino.

In Labradors retrievers, the dominant *B* allele determines black coat color, and *bb* dogs are brown. However, a second gene is also involved in this case, the dominant *E* allele has no effect on black or brown coat color, but dogs carrying *ee* are yellow, regardless of the genotype at the *B* gene.

9/16	*B– E–*	Black	9
3/16	*B– ee*	Yellow	3
3/16	*bb E–*	Brown	3
1/16	*bb ee*	Yellow	1

A good example of recessive epistasis is found for grain colour in maize. There are three different colours of grain in maize, *viz.*, purple, red and white. The dominant allele *P* causes purple kernel color, while the homozygous recessive genotype *pp* causes red kernels. At a different locus, dominant allele *R* permits kernel color, while the recessive allele *r* in homozygous condition inhibits color and will produce plants with white grains. A cross between purple (*RRPP*) and white (*rrpp*) grain colour strains of maize produced plants with purple grain colour in F_1. Inter-mating of these F_1 plants produced progeny with purple, red and white grains in F_2 in the ratio of 9 : 3 : 4 as shown below:

9/16	*P– R–*	Purple	9
3/16	*P– rr*	White	3
3/16	*pp R–*	Red	3
1/16	*pp rr*	White	1

Dominant and recessive epistasis or Dominant inhibitory epistasis (13 : 3)

If the dominant allele of one gene and the recessive allele in homozygous condition of another gene produce the same phenotype, the 9 : 3 : 3 : 1 ratio is modified into a 13 : 3 ratio. In such case, the genotype *A– B–, A– bb* and *aa bb* produce same phenotype. It is also known as inhibitory gene interaction.

Experiments reveal that the gene for white plumage of white leghorns is dominant over the gene for coloured plumage of coloured varieties. But the gene for white plumage of white Wyandotte's or white Plymouth Rock or white Silkie chickens is recessive to the gene for coloured plumage of coloured varieties. Therefore, the gene which produces white plumage in white leghorns is different from the gene for white plumage in white Wyandotte's.

Dominant and recessive epistasis is illustrated by the difference between white Leghorn chickens (genotype *CC II*) and white Wyandotte or white Plymouth Rock or white Silkie chickens (genotype *cc ii*). *C* (color) locus determines whether the birds will be colored or white. The dominant allele C allows melanin production, while its recessive allele c in homozygous condition prevents melanin production, such that birds with $C-$ genotype are colored, while cc are white. But the dominant I allele inhibits the action of feather coloration gene C. The F_1 generation of a dihybrid cross between these breeds has the genotype $Cc\ Ii$, which is expressed as white feathers because of inhibitory effects of the I allele. In the F_2 generation, only the $C-\ ii$ genotype has colored feathers, so there is a 13 : 3 ratio of white : colored as shown below.

9/16	$C-I-$	White	9
3/16	$C-ii$	Colored	3
3/16	$cc\ I-$	White	3
1/16	$cc\ ii$	White	1

Another example of dominant and recessive epistasis is anthocyanin pigmentation in rice. A dominant allele, I controls the green color while P controls the purple coloration in rice. The allele I is dominant over i and is epistatic to both P and p, meaning that whenever I is present, it masks the expression of the P locus. As a result, plants with genotypes $I-\ P-$ or $I-\ pp$ appear green, since the dominant allele I suppresses the production of purple pigment. Only in the absence of I (that is, in ii) does the effect of the P locus become visible: $ii\ P-$ plants are purple. A cross between the green ($I-\ pp$) and purple ($ii\ P-$) plants produced green plants ($I-\ P-$) in the F_1 generation; but in F_2 generation, 13 were green and 3 purple (13 : 3 ratio) as shown below.

9/16	$I-P-$	Green	9
3/16	$I-pp$	Green	3
3/16	$Ii\ P-$	Purple	3
1/16	$ii\ pp$	Green	1

Duplicate recessive epistasis, or complementary gene action, or complementation (9 : 7)

If the (homozygous) recessive alleles in either or both of the two genes result in the same mutant phenotype, the 9 : 3 : 3 : 1 ratio is modified into a 9 : 7 ratio. In such case, the genotype $aa\ B-$, $A-\ bb$, and $aa\ bb$ produce the same phenotype. Both dominant alleles when present together only ($A-\ B-$) then they can complement each other and produce a different phenotype.

There is no widely cited example of duplicate recessive epistasis. The standard example is flower color in sweet peas. In sweet pea (*Lathyrus odoratus*) two varieties of white flowering plants were noted. Each variety bred true and produced white flowers in successive generations. According to Bateson and Punnett, when two such white flowering varieties of sweet pea, with *CCpp* and *ccPP* genotypes, were crossed, the offspring had purple flowers in F_1 generation; but in F_2 generation, 9 were purple and 7 white (9 : 7 ratio). This is a modification of 9 : 3 : 3 : 1 ratio, where only one character *i.e.*, flower-colour is involved. It is clear in the above example that for the production of the purple flower-colour both dominant genes (*C* and *P*) are necessary to remain present. In the absence of either of the genes (*C* or *P*), the flowers are white. Thus, genes *C* and *P* were complementary to each other.

9/16	C– P–	Coloured flowers	9
3/16	C–pp	White flowers	3
3/16	cc P–	White flowers	3
1/16	cc pp	White flowers	1

If a homozygous dominant pea plant with coloured flowers (*CCPP*) is crossed with homozygous recessive with white flowers (*ccpp*), the F_1 generation have coloured flowers (*CcPp*). However, on selfing the F_1 plants, the normal Mendelian dihybrid ratio (9 : 3 : 3 : 1) is not obtained, instead the interactions of complementary nature between the two genes *C* and *P* genes gives a modified phenotypic ratio (9 : 7).

Rex rabbits have a dense, velvet-like fur (rex fur), which is recessive to normal fur. Normal fur is the wild type and is characterized by longer, heavier guard hairs and a shorter, softer undercoat. In contrast, rex fur shows an approximately **43% reduction in guard hair length** and **20% reduction in undercoat length** compared to normal fur, along with a decrease in guard hair diameter and an altered guard hair shape (Diribarne *et al.*, 2011). When Rex rabbits of different breeds are mated with one another, the rex fur does not appear in the F_1. All the F_1 progeny have normal fur. Normal fur develops only when at least one dominant allele is present at both loci. This means that both genes must contribute to the expression of the wild-type fur structure. In contrast, rex fur appears whenever either or both locus is homozygous recessive. This interaction leads to the characteristic 9 : 7 ratio of normal to rex fur observed in the F_2 generation.

9/16	R– N–	Normal fur	9
3/16	R– nn	Rex fur	3
3/16	rr N–	Rex fur	3
1/16	rr nn	Rex fur	1

When both dominant alleles are present simultaneously, they are referred to as complementary genes and result in a distinct phenotype.

Duplicate dominant epistasis or duplicate epistasis or duplicate gene interaction (15 : 1)

If the dominant allele in either or both of the two gene loci results in the same phenotype without any cumulative effect (*i.e.*, independently), the 9 : 3 : 3 : 1 ratio is modified into a 15 : 1 ratio. In such case, the genotypes *A– B–, A– bb,* and *aa B–* produce same phenotype.

Both Hereford and Simmental cattle have a characteristic white facial color (white-faced). The white face pattern appears to be dominant over self-color (no white face). Pure-breeding Hereford cattle can be represented as *HH ss,* and Simmental as *hh SS,* where *H* represents the dominant allele for the Hereford face pattern and *S* the dominant allele for the Simmental pattern. Their F_1 could then be *Hh Ss* with a white face. When F_1 cattle (*Hh Ss*) are mated together, the F_2 phenotypic ratio will be 15 white face : 1 self-color (no white face) as shown below:

9/16	H– S–	White face	9
3/16	H– ss	White face	3
3/16	hh S–	White face	3
1/16	hh ss	Self-color	1

Presence of feathered shanks in chickens is another example of duplicate dominant epistasis. The trait is controlled by two dominant alleles (*F* and *S*). They are interacting in such a way that presence of at least one copy of any of these two dominant alleles results in feathered shanks. The homozygous recessive genotype (*ffss*) exhibits "unfeathered shanks". In a cross between a pure-breeding Black Langhans (feathered shanks) and pure-breeding Buff Rocks (unfeathered shanks), the F_1 generation all have feathered shanks. When the F_1 generation is crossed, the F_2 generation contains chickens with feathered shanks to unfeathered shanks in a ratio of 15 feathered shank : 1 unfeathered shank as shown below:

9/16	$F-S-$	Feathered shank	9
3/16	$F-ss$	Feathered shank	3
3/16	$ffS-$	Feathered shank	3
1/16	$ffss$	Unfeathered shank	1

Seed capsule of shepherd's purse plant (*Capsella* spp) exhibits duplicate gene interaction. Plant seed capsules from the shepherd's purse species can be either triangular or oval. There are two non-allelic dominant genes (*A* and *B*) that govern the shape of seed of shepherd's purse. To produce triangular seed, either both of them (*A* and *B*) or at least one of them (*A* or *B*) should be present either in homozygous or heterozygous condition. When both the genes are present in a homozygous recessive state, the seed capsule adopts an ovoid shape.

9/16	$A-B-$	Triangular shape	9
3/16	$A-bb$	Triangular shape	3
3/16	$aaB-$	Triangular shape	3
1/16	$aabb$	Ovoid shape	1

Another good example of duplicate dominant epistasis is awn character in rice. Development of awn in rice is controlled by two dominant duplicate genes (*A* and *B*). Presence of any of these two alleles can produce awn. Dominant allele *A* is epistatic to alleles *B* and *b*. Similarly, dominant allele *B* is epistatic to alleles *A* and *a*. The presence of either or both of these dominant alleles is responsible for the awn character in rice. The awn less condition develops only when both these genes are in homozygous recessive state (*aabb*). A cross between awned and awn less strains produced awned plants in F_1. Inter-mating of F_1 plants produced awned and awn less plants with 15 : 1 ratio in F_2 generation.

9/16	$A-B-$	Awned	9
3/16	$A-bb$	Awned	3
3/16	$aaB-$	Awned	3
1/16	$aabb$	Awn less	1

Similar gene action is found for nodulation in peanut and non-floating character in rice.

Duplicate genes with cumulative effect, or duplicate interaction or polymeric gene interaction (9 : 6 : 1)

This dihybrid ratio is observed when homozygosity for the recessive allele of either of two genes but not both results in the same phenotype, but the phenotype of the double heterozygote is distinct. The 9 : 6 : 1 ratio results from the fact that both single recessive have the same phenotype.

9/16	R– S–	Red	9
3/16	R– ss	Sandy	3
3/16	rr S–	Sandy	3
1/16	rr ss	White	1

For example, red coat color in Duroc-Jersey pigs requires the presence of two dominant alleles R and S. Pigs of genotype $R–$ ss and rr $S–$ have sandy colored coats, and rr ss pigs are white. The F_2 dihybrid ratio will be 9 red : 6 sandy : 1 white.

A well-known example of polymeric gene interaction is fruit shape in summer squash. There are three types of fruit shape in this plant, *viz.*, disc, spherical and long. The disc shape is controlled by two dominant genes (*A* and *B*), the spherical shape is produced by either dominant allele (*A* or *B*) and long shaped fruits develop in double recessive (*aa bb*) plants. A cross between disc shape (*AA BB*) and long shape (*aa bb*) strains produced disc shape fruits in F_1. Inter-mating of F_1 plants produced plants with disc, spherical and long shape fruits in 9 : 6 : 1 ratio in F_2. Here plants with $A–$ $B–$ (9/16) genotypes produce disc shape fruits, those with $A–$ $bb–$ (3/16) and aa $B–$ (3/16) genotypes produce spherical fruits, and plants with aa bb (1/16) genotype produce long fruits. Thus, in F_2, normal dihybrid phenotypic ratio 9 : 3 : 3 : 1 is modified to 9 : 6 : 1 ratio. Similar gene action is also found in barley for awn length.

9/16	A– B–	Disc shape fruits	9
3/16	A– bb–	Spherical fruits	3
3/16	Aa B–	Spherical fruits	3
1/16	Aa bb	Long fruits	1

The various epistatic ratios can be summarized in Table 3.4.

Table 3.4 Examples of epistatic F_2 phenotypic ratios from a cross of *Aa Bb* × *Aa Bb*

Kinds of interaction	A- B-	A- bb	aa B-	aa bb
Dominant epistasis; *A* epistatic to *B* and *b*.	12		3	1
Recessive epistasis; *aa* epistatic to *B* and *b*.	9	3	4	
Duplicate interaction	9	6		1
Duplicate recessive epistasis; *aa* epistatic to *B* and *b*; *bb* epistatic to *A* and *a*	9	7		
Duplicate dominant epistasis; *A* epistasis to *B* and *b*; *B* epistatic to *A* and *a*	15			1
Dominant and recessive epistasis; *A* epistatic to *B* and *b*; *bb* epistatic to *A* and *a*	13 (12 + 1)	3		

Non-Epistatic Inter Genic Genetic Interaction

In all the examples of Mendelian dihybrid crosses, the two genes have acted independently in terms of phenotype. For example, the allelic pair for round/wrinkled peas had no effect on the allelic pair for long/short stem. If, however, the two allelic pairs affect the same phenotypic characteristic, there is a chance for gene product interaction to produce new phenotypes. This is a kind of interactions between nonallelic genes that control the same general phenotypic attribute and the F_2 phenotypic ratio 9 : 3 : 3 : 1 remain unaltered. A classic example of such gene interactions is comb shape in chickens. The comb is a fleshy growth on the top of the chicken's head. There are many different breeds of domestic chicken. Each breed possesses a characteristic type of comb. The Wyandottes and Hamburgs have 'rose' comb; Brahmas, Cornish or Indian games have 'pea' comb; and Leghorns, Plymouth Rocks and Rhode Island Reds have 'single' comb (Image 3.4). In the first decade of the twentieth century, British geneticists William Bateson and R.C. Punnett conducted research showing that the comb shape in chickens was caused by the interaction between two different genes on two different chromosomes. Crosses made between true-breeding rose-combed and single-combed birds showed that rose was completely dominant over single. When the F_1 rose-combed birds were bred together, there was a clear segregation into 3 rose : 1 single in the F_2. Similarly, pea comb was found to be completely dominant over single, with 3 pea : 1 single ratio in the F_2. When true-breeding rose and pea varieties were crossed, the F_1 progeny was found with a different type of comb called walnut comb. It is a medium sized, solid comb, with a bumpy surface like a walnut shell. When the F_1 walnut-combed birds were bred together, in F_2 all four comb types occurred in a ratio of 9 walnut : 3 rose : 3 pea : 1 single (Figure 3.3).

Parents	Rose comb *RR pp*		×	Pea comb *rr PP*
F_1	*Rr Pp* All walnut comb			
F_2	9/16	*R– P–*		Walnut
	3/16	*R– pp*		Rose
	3/16	*rr P–*		Pea
	1/16	*rr pp*		Single

Figure 3.3 Inheritance of comb type in chickens.

Through continued research, Bateson and Punnett deduced that the walnut comb depends on the presence of two dominant alleles, R and P, both located at two independently assorting gene loci. In $R–$ *pp* birds, a rose comb results, in *rr* $P–$ birds, a pea comb results, and in *rr pp* birds, a single comb results. Thus, it is the interaction of two dominant alleles that produces a new phenotype, each of which individually produces a different phenotype.

William Bateson used comb type of chickens to show that Mendelian genetics apply to animals as well as plants.

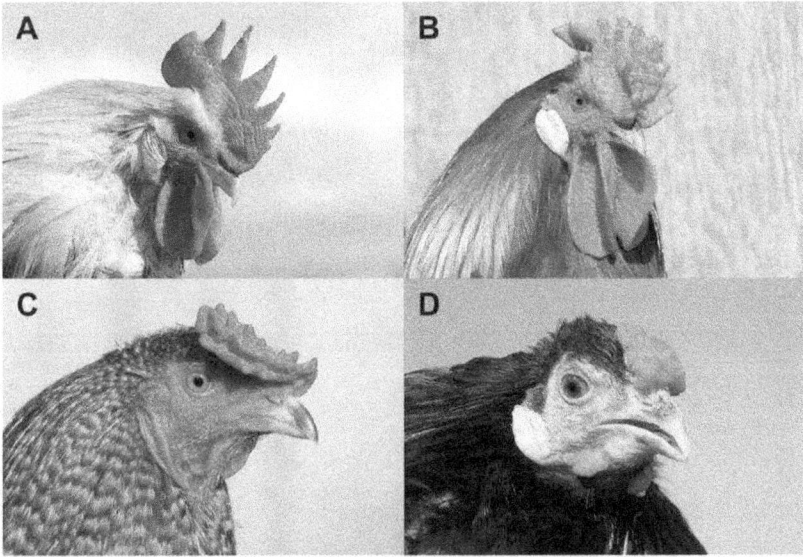

Image 3.4 Four comb phenotypes in chickens (A) Single-combed male, (B) Rose-combed male, (C) Pea-combed male, and (D) walnut-combed male (source: https://www.researchgate.net/profile/Ranran-Liu/publication/228116327/figure/fig7/AS:341389037588495@1458404897971/Four-comb-phenotypes-in-chickens-explained-by-segregation-at-the-Rose-comb-and-Pea-comb.png).

Modifier Genes

Modifier genes are genes that do not directly produce a trait but instead alter or influence the expression of other genes. Unlike major genes, which directly determine a phenotype, modifier genes act in the background to adjust the **intensity, distribution, or visibility** of a trait. They may enhance, reduce, or fine-tune the effect of another gene without completely masking it, which makes them different from epistatic genes. Instead of masking another gene's effect, a modifier gene alters the degree of expression, often producing quantitative differences. Their effects may include **reduced penetrance, modification of dominance, variation in expressivity,** and **phenotypic**

pleiotropy. Thus, modifier genes are crucial in explaining why individuals with the same major genotype can still exhibit different phenotypic outcomes. In mice, coat color is controlled by the *B* gene. The *B* allele produces black coat color and is dominant to the *b* allele that produces a brown coat color. The intensity of the color, either black or brown is controlled by another gene, the *D* gene. At this gene, the *D* allele controls full color whereas the *d* allele produces a dilute or faded expression of the color at the *B* gene. The *D* gene does not mask the effect of the *B* gene, rather it modifies its expression. If a cross is made among mice that are *Bb Dd* the following result will be seen:

9/16 *B– D–*   (black)

3/16 *B– dd*   (dilute black)

3/16 *bb D–*   (brown)

1/16 *bb dd*   (dilute brown)

CHAPTER FOUR CHROMOSOMAL BASIS OF INHERITANCE

Cells are the basic units of life. The chemical reactions characteristic to life occur in and on cells. Cell division is at the heart of the reproduction of cells and organisms because cells arise only from preexisting cells.

Cell Cycle

Cell cycle was described by Howard and Pele in 1953. The cell cycle is the sequence of events occurring in an ordered fashion which results in cell growth and cell division. It is the period from the beginning of one division to the beginning of the next. Though it differs in some aspects between animal cells, plant cells and fungi, the overall procedure is similar between them. The primary function of the cell cycle is to duplicate the genome precisely and divide it equally between two daughter cells. The cell cycle is divided into two basic phases (Image 4.1):

1. Interphase
2. M Phase (Mitosis phase)

The M phase represents the phase when the actual cell division or mitosis occurs. Mitosis is a relatively small part of the total cell cycle; the majority of the time, a cell is in interphase. Interphase represents the phase between two successive M phases. In interphase the chromatin is dispersed throughout the nucleus and individual chromosomes are not usually visible. Interphase is the time of high metabolic activity during which the cell is preparing for division by undergoing both cell growth and DNA replication in an orderly manner. The purpose of interphase is cell growth. The interphase is divided into three phases (Image 4.1):

1. G_1 phase (1st gap/growth phase)
2. S phase (Synthesis phase)
3. G_2 phase (2nd gap/growth phase)

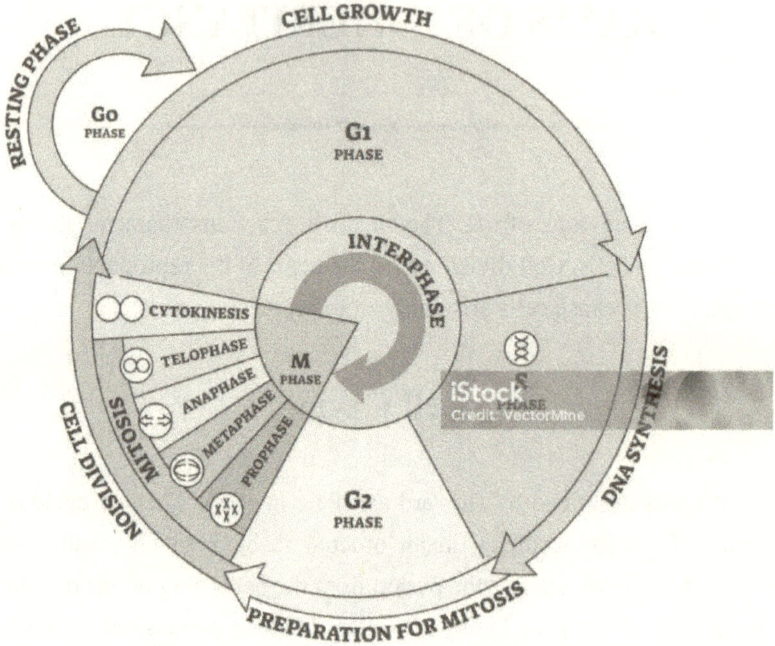

Image 4.1 Cell cycle (source: https://www.istockphoto.com/vector/cell-cycle-growth-mitosis-synthesis-and-division-stages-outline-diagram-gm2160379608-580918294).

A cell enters the G_1 phase after the M phase of the previous cycle, and thus, it is termed as the first gap phase or the first growth phase. During this phase the cell is metabolically active, grows in size, performs normal function, and increases the production of RNA, protein, ATP, raw materials (pentose sugar, phosphoric acid and nitrogenous bases) for DNA duplication, cell organelles, and so many things in order to prepare for later stages but does not replicate its DNA. The most important event of the G_1 phase is the transcription of all three types of RNAs which then undergo translation to form proteins and enzymes necessary for other events in the cell cycle. It is often the longest phase in the cell cycle. In rapidly dividing human cells, the G_1 phase typically lasts about 11 hours.

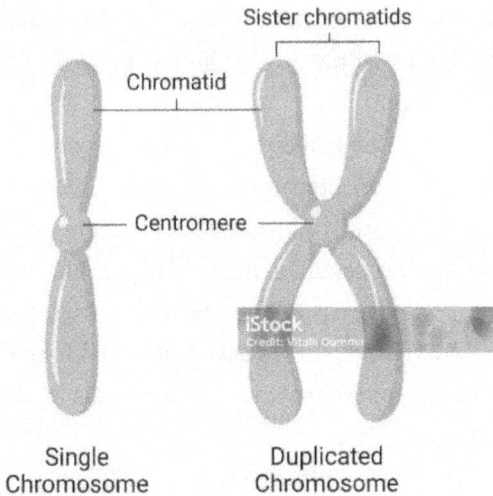

Image 4.2 Sister chromatids (source:
https://www.istockphoto.com/vector/vector-illustration-of-chromosome-
structure-gm1216602702-354821856).

S phase marks the period during which DNA replication and synthesis of histone proteins take place. The phase is moderately long, occupying about 30% of the total cell cycle time. Inside the nucleus, chromosomes begin to replicate, producing two sister chromatids (Image 4.2). Sister chromatids are two identical copies of a single replicated chromosome that are attached to each other by the replicated but unseparated centromere by specific proteins. During mitotic anaphase, when the centromere separates and the chromatids are pulled apart, each chromatid is then referred to as a daughter chromosome. Chromosome replication is the process by which living cells duplicate their DNA to produce identical copies of each chromosome. Replication of chromosomes in their entirety occurs during the next few minutes after replication of the DNA has been completed. Once DNA replication is complete, the newly formed DNA molecules associate with proteins to assemble fully replicated chromosomes. The two chromatids of the same chromosome are called sister chromatids, while chromatids of homologous chromosomes are called non-sister chromatids. During this time the amount of DNA per cell doubles. If the initial amount of DNA is denoted as 2C then it increases to 4C. However, there is no increase in the chromosome number; if the cell had diploid or 2N number of chromosomes at G_1, even after S phase the number of chromosomes remains the same, *i.e.*, 2N. In animal cells, during the S phase, DNA replication begins in the nucleus and the centrioles duplicates in the cytoplasm.

G_2 phase is the final, and usually the shortest phase in which the cell undergoes a period of rapid growth to prepare for M phase. During this phase, cell organelles are replicated and cell synthesizes tubulin protein required for spindle formation, protein required for plasma membrane formation, ATP molecules required for movement of chromosomes from equator to pole (30 ATP/chromosome), RNA and other materials required for cell division. DNA repair is a crucial step in the G_2 phase as it repairs breaks that might be present in the DNA strand after replication.

The cell cycle is commonly described in terms of these three interphase periods followed by mitosis, M. The order of events is therefore $G_1 \rightarrow S \rightarrow G_2 \rightarrow M$. After M phase, the proliferating cells enter the G_1 phase of the next cell cycle. When an organism is growing, this cycle repeats itself many times, eventually resulting in an individual with billions of cells. The time it takes to complete one cell cycle is the generation time. The length of time required for a complete cell cycle varies with cell type. In higher eukaryotes, the majority of cells require 18 to 24 hours. The relative duration of the different stages in the cell cycle also varies considerably, depending upon the organism, cell type, temperature, and other factors. Mitosis, requiring $1/2$ hour to 2 hours, is usually the shortest period. The longest phase is G_1. In a given organism, variation in the length of the cell cycle depends primarily on the length of G_1, with the duration of S, G_2, and M being approximately the same in all cell types. For example, cancer cells and early fetal cells of humans spend minutes in G_1, while some differentiated adult cells (such as nerve cells) spend years in G_1. The cell cycle is regulated by various stimulatory and inhibitory factors that decide whether the cell needs to divide or grow.

Some cells may exit the cell cycle by entering a quiescent state or resting phase called G_0 phase from the G_1 phase for extended periods, where it neither divides nor grows but still fully functional and perform specialized functions. Some cells such as neurons stop dividing completely and are permanently in G_0 phase, whereas others can re-enter the G_1 phase and continue the cell cycle if necessary. Most cells in the body are in G_0. The withdrawal of growth factors from animal cells at a critical period during G_1 causes them to cease growth and exit from the cell cycle.

Cell Cycle Checkpoints

Each step of the cell cycle is monitored by internal controls called checkpoints. A checkpoint is a stage in the eukaryotic cell cycle at which the progression of a cell to the next stage in the cycle can be halted until conditions are favorable. It is a control mechanism where certain conditions must be met before the cell can progress to the next step. There are three major checkpoints in the cell cycle: one near the end of G_1, a second at the G_2/M transition, and the third near the end of metaphase (Image 4.3).

The G_1 Checkpoint – Size and Nutrient Verification

This checkpoint, also called the restriction checkpoint, takes place near the end of G_1. It is the main decision point for a cell at which it must choose whether or not to divide. It determines whether all conditions are favorable for cell division to proceed. The cell verifies that it is large enough to divide, that its DNA is intact, and if there is enough access to nutrients and stimulating growth factors. A cell that does not meet all the requirements will not be allowed to progress into the S phase. The cell can halt the cycle and attempt to remedy the problematic condition, or signaled for apoptosis or moved to the G_0 and await further signals when conditions improve.

The G_2 Checkpoint – DNA Quality Control

The G_2 checkpoint is the second checkpoint in the cell cycle which is present at the transition between G_2 and M phase. After the second growth phase, the cell checks that the DNA was completely and correctly replicated during the S phase. If it passes it enters the M phase. If the checkpoint mechanisms detect problems with the DNA, the cell cycle is halted, and the cell attempts to either complete DNA replication or repair the damaged DNA. If the cell is unable to repair the DNA, it undergoes apoptosis. This prevents it from passing the damaged DNA on to the daughter cells.

M Checkpoint – Nuclear Division Setup Check

This checkpoint takes place near the end of the metaphase stage of karyokinesis. It is also called the spindle checkpoint. The checkpoint ensures that all the replicated chromosomes of cells entering the anaphase are firmly attached to at least two spindle fibers from opposite poles of the cell. If not, the cell pauses mitosis until all sister chromatids have been attached in the right way.

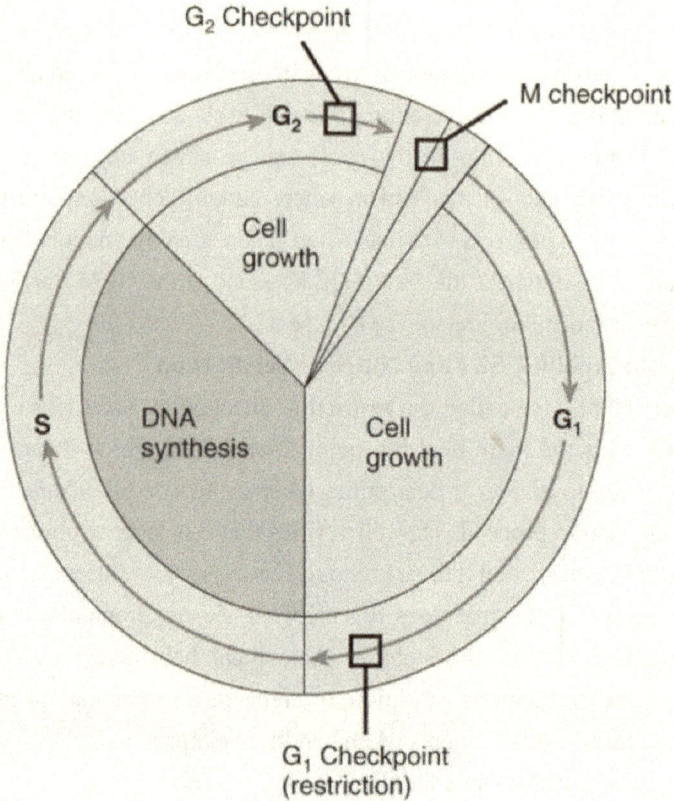

Image 4.3 Cell cycle checkpoints (source: https://www.researchgate.net/figure/Schematic-view-of-different-phases-of-the-cell-cycle-cell-cycle-checkpoints-and-key_fig1_335057938).

In addition to the internally controlled checkpoints, there are two groups of intracellular molecules that regulate the cell cycle. Positive regulator molecules allow the cell cycle to advance to the next stage. Negative regulator molecules monitor cellular conditions and can halt the cycle until specific requirements are met. Regulator molecules may act individually, or they can influence the activity or production of other regulatory proteins.

Mitosis

Growth of multicellular organisms is basically a process of increasing the number of cells. This involves two processes; the distribution of copies of the genetic information from the parent cell to the two daughter cells and the cytoplasmic division. In this way, a body grows or wounds are repaired. Each of the new cells will include all the genetic information possessed by all of the other living cells of the body.

Mitosis is the process of cell duplication, or reproduction, during which one cell gives rise to two genetically identical daughter cells (Image 4.4). It is the most crucial and dramatic phase of the cell cycle. The most important event of this phase is the karyokinesis where the chromosomes separate into two distinct cells. Mitosis begins with the condensation of chromosomes which then separate and move towards opposite poles. A cell entering the M phase has a 4C concentration of DNA and ends with two cells, each containing a 2C concentration of DNA. Mitosis occurs in both haploid and diploid cells. Mitosis is usually accompanied by cytokinesis, the process in which the cell itself divides to yield two daughter cells. Mitosis without cytokinesis results in multinucleate cells. It is a continuous process, but for purposes of discussion it is usually divided into four distinct stages characterized by the appearance and orientation of the homologous pairs of chromosomes.

Prophase. As a cell leaves G_2 it enters prophase of mitosis. Prophase is marked by the initiation of condensation of chromosomal material. Chromatin in the nucleus begins to condense and becomes visible in the light microscope as chromosomes. Centrioles have separated and taken positions on the opposite poles of the cell. The spindle fibers that are aggregates of microtubules begin to form at opposite poles of the cell, extending from the centrioles. At the end of prophase, the nucleoli disappear and the nuclear membrane (envelope) begins to break down.

Metaphase. Metaphase is the second and the longest stage of cell division. Metaphase begins when the nuclear envelope has completely disappeared, leaving the chromosomes free in the cytoplasm. The nuclear spindle becomes prominent. The chromosomes (sister chromatids) lined up at the middle of the cell with their centromeres aligned at the exact center, or equator, of the cell. This arrangement of chromosomes along a plane midway between the poles is called the metaphase/equatorial plate. Hence, the metaphase is characterized by all the chromosomes coming to lie at the equator. Spindle fibers connect the centromere of each sister chromatid to the poles of the cell. Condensation of chromosomes is completed at this stage and their size, length, and centromere location are easiest to study. Chromosomes are usually examined during mitotic metaphase. During mitosis, each chromatid becomes condensed approximately ten thousand-fold reaching maximal condensation at metaphase.

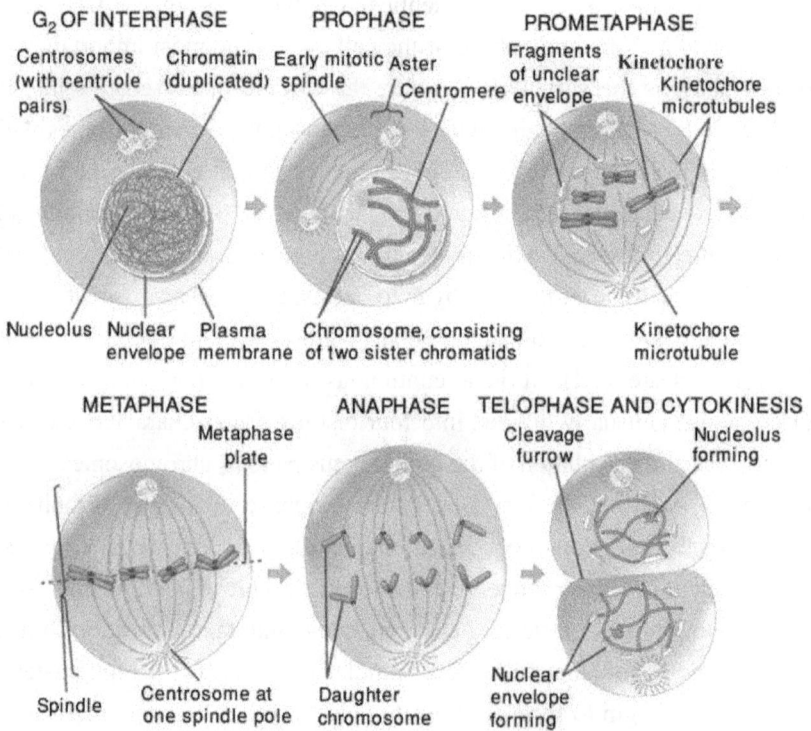

Image 4.4 A diagrammatic representation of mitosis in an animal cell.

Anaphase. Anaphase is the shortest period of mitosis. The centromere of each chromosome split and pulled apart by the spindle fibers, causing the sister chromatids to separate, creating two daughter chromosomes. One of the daughter chromosomes is pulled to one side of the cell, while the other is pulled to the opposite pole.

Telophase. Telophase is the final stage of mitosis where the two sets of daughter chromosomes reach opposite poles of the cell where they begin to decondense and lose their individuality. The individual chromosomes can no longer be seen and chromatin material tends to collect in a mass in the two poles. The nuclear envelope reforms around the clusters of chromosomes at each end of the cell to form two distinct daughter nuclei, the spindle fibers break down and dissolve, and the nucleolus, Golgi complex and endoplasmic reticulum (ER) reform. At this point, nuclear division is completed: the cell has two nuclei. Cytokinesis begins and two daughter cells are formed. The daughter cells have chromosome number identical to the original cell.

Cytokinesis

Cytokinesis is the final stage of cell division in plant and animal cells. Cytokinesis is the process that occurs immediately after the telophase in which the cytoplasmic content of the cell is divided into two halves to form two distinct and complete cells. Cytokinesis begins at the very end of anaphase and continues throughout telophase of both mitosis and meiosis. Cytokinesis ensures that each daughter cell receives all the cytosol and cellular organelles it needs to begin its new life.

Plant and animal cells differ greatly in their external structure. Animal cells simply have a cell membrane encircling the cell whereas plant cells are significantly more rigid and require more structure and water storage so they possess a cell wall. Animal and plant cells complete cytokinesis differently (Image 4.5). Cytokinesis in animal cells is accomplished via a cleavage furrow, an indentation of the cell (plasma) membrane of the dividing cell. A contractile ring composed of actin filaments and myosin motor proteins located immediately under the plasma membrane around the equator of the dividing cell pulls the cleavage furrow deeper into the cell that eventually pinches the cell into two halves, each part containing its own nucleus and cytoplasmic organelles.

However, in plant cells, during cytokinesis small vesicles are formed between the daughter cells. These small vesicles combine to form a cell plate at the middle of the dividing cell which separates the cytoplasm and cell organelles into equal halves. The cell plate extends to two sides and joins with the plasma membrane. The fusion of growing cell plates with the existing plasma membrane occurs and produces two daughter cells, each with its own plasma membrane. Between the two membranes of the cell plate the new cell is formed. Cellulose gets deposited on the cell plate and the cell wall is formed. Cytokinesis, like the rest of the cell cycle, is also regulated by several factors that are responsible for the initiation of division as well as the termination.

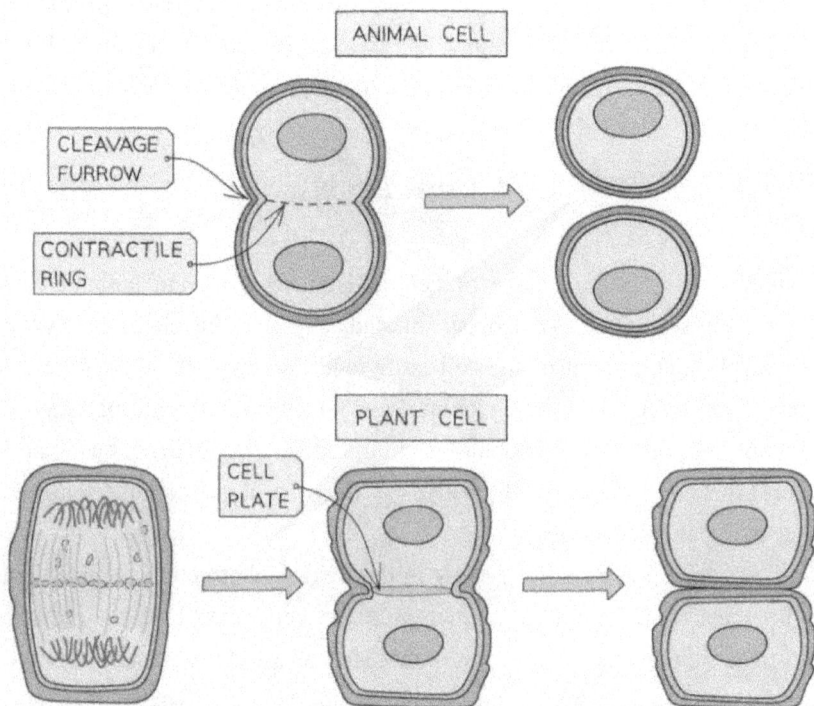

Image 4.5 Diagram illustrating cytokinesis in plant and animal cell (source: https://www.savemyexams.com/a-level/biology/aqa/17/revision-notes/2-cell-structure/2-3-cell-division-in-eukaryotic--prokaryotic-cells/2-3-4-cytokinesis/).

Genetic Significance of Mitosis

Mitosis maintains a constant amount of genetic material from generation to generation. Mitosis produces two daughter nuclei that contain identical chromosome numbers and that are genetically and cytologically identical to one another and to the parent nucleus from which they arose. The growth of multicellular organisms is due to mitosis. Cell growth results in disturbing the ratio between the nucleus and the cytoplasm. It therefore becomes essential for the cell to divide to restore the nucleo-cytoplasmic ratio. A very significant contribution of mitosis is cell repair. The cells of the upper layer of the epidermis, cells of the lining of the gut, and blood cells are being constantly replaced.

Meiosis

If a species is to retain the original number of chromosomes and produce offspring with chromosomes from both male and female parent, then the parents must have a mechanism to produce cells with half the number of chromosomes. Meiosis is a basic and important process of cell division that takes place in germ cells of sexually reproducing eukaryotic organisms in order to halve the number of chromosomes and produce haploid gametes for fertilization. A diploid cell (never a haploid cell) entering in meiosis has a 4C concentration of DNA that undergoes two rounds of nuclear division to produce four specialized haploid cells, each containing a 1C concentration of DNA (Image 4.6). Two nuclear divisions of meiosis are called meiosis I and meiosis II. Meiosis I is the reduction division, where the homologous chromosomes segregate to opposite poles of the spindle. The second meiotic division (meiosis II) is the same as mitosis, where sister chromatids are segregated. As with mitosis, both meiotic divisions can be divided into four stages, based on the position of chromosomes and other characteristics.

The first meiotic division (meiosis I)
Meiosis I is divided into prophase I, metaphase I, anaphase I, and telophase I. These are the same phases that are used to describe mitosis, but behavior of the chromosomes in meiosis I is very different from that in mitosis. The interphase period before meiosis I is similar to that in mitosis.

Prophase I. The most complex and lengthy phase of the whole process of meiosis is prophase I. It is much longer in meiosis than in mitosis. Based on chromosomal behavior prophase I is divided into five substages, *i.e.*, leptotene, zygotene, pachytene, diplotene and diakinesis.

Leptotene (thin-thread). The spindle appears. Nuclear envelopes disappear. The chromosomes start to condense and the compaction continues throughout leptotene. The chromosomes become gradually visible under the light microscope.

Zygotene (joined-thread). Zygotene is the second stage of meiotic prophase-I. The chromosomes continue condensing during zygotene into distinct threadlike structures. Each chromosome now appears thicker as the sister chromatids are closely aligned. Homologous chromosome of each pair come together and line up side by side and gene by gene with each other. This process, termed synapsis begins. One member of a pair is the maternal homologue, the other is the paternal homologue. In synapsis, the genes on the chromatids of the homologous chromosomes are aligned precisely with each other. The process of synapsis is incredibly precise. It can start at either end of the chromosome and move towards the center or vice versa. In some cases, synapsis can even initiate at multiple points on the chromosome. When the two homologous chromosomes, consisting of four chromatids, are paired, the structure is called a bivalent, or a tetrad. The number of bivalents in a cell is equal to half the number of chromosomes that were initially present in the cell. Homologous chromosomes pair up during meiosis to ensure correct chromosome segregation and facilitate crossing over, which is essential for generating genetic diversity in offspring.

Pachytene (thick-thread). Throughout this period, the bivalents continue to shorten and thicken. Synapsis is complete and crossing over takes place. Synapsis is accompanied by the formation of complex structure called synaptonemal complex that allows chiasma to form. The complex is made of protein and DNA, and found between the synapsed homologs.

Diplotene (double-thread). The beginning of diplotene is recognized by the dissolution of the synaptonemal complex. The synapsed chromosomes begin to separate. The process begins at the centromere and the bivalent are held together by chiasmata.

Diakinesis (moving apart). Homologous chromosomes seem to repel each other and the segments not connected by chiasmata move apart. The homologous chromosomes in a bivalent remain connected by at least one chiasma and do not separate until anaphase I. During this phase the chromosomes are fully condensed and the meiotic spindle is assembled to prepare the homologous chromosomes for separation. By the end of diakinesis, the nucleolus disappears and the nuclear envelope also breaks down. Diakinesis represents transition to metaphase.

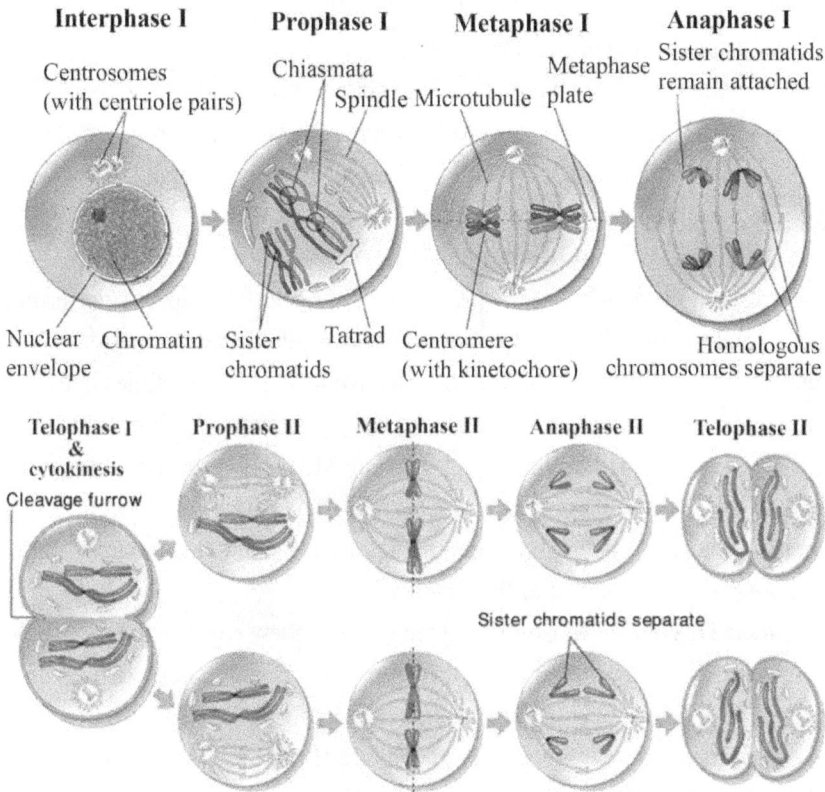

Image 4.6 A diagrammatic representation of meiosis.

Metaphase I. By the beginning of metaphase I, the nuclear envelope has completely break down. The bivalents or tetrads become aligned on the equatorial plate. The two centromeres of a homologous chromosome pair attach to the microtubules from the opposite poles of the spindle.

Anaphase I. Spindle fibers pull the homologous chromosomes towards opposite poles. As the homologs move apart, the chiasmata completely terminalize. Each chromosome still has two sister chromatids. At this stage, each of the separated chromosomes containing two sister chromatids is called a dyad.

Telophase I and Cytokinesis. The dyads complete their migration to opposite poles of the cell, and the spindle disappears. The nuclear membrane may or may not reform, depending on the species, but in any case, cytokinesis does occur, resulting in two new cells, each with the haploid number of chromosomes, which are still in the form of sister chromatids. A degree of chromosome decondensation is observed.

Thus, meiosis I, which begins with a diploid (2N) cell that contains one maternally derived and one paternally derived set of chromosomes, ends with two daughter cells, each of which is haploid (N) and contains one mixed-parental set of replicated chromosomes.

The second meiotic division (meiosis II)

Each of the two daughter cells produced as a result of meiosis I go through the second meiotic division (meiosis II) that results in the formation of four cells. Meiosis II is very similar to mitosis. Meiosis II is sometimes called the equational division because the chromosome number remains the same in each cell before and after the second division. Meiosis II proceeds shortly after telophase I. There is no replication of chromosome between the two divisions; the chromosomes present at the beginning of the second division are identical to those present at the end of the first division.

Prophase II. The chromosomes are seen to recondense within the two nuclei. Prophase II is generally similar to mitotic prophase. The nuclear membrane disappears by the end of prophase II. A spindle apparatus forms and the chromosomes progress toward the metaphase II plate.

Metaphase II. The chromosomes align at the equator in mitosis-like fashion. The microtubules from opposite poles of the spindle get attached to the kinetochores of sister chromatids of each chromosome.

Anaphase II. The centromeres of sister chromatids finally separate and sister chromatids of each pair are pulled to opposite poles by the spindle fibers. Once the centromere has separated at anaphase II, each chromatid is referred to as individual chromosome.

Telophase II and Cytokinesis. Chromatids (now individual chromosome) arrive to the opposite poles of the cell. New nucleus reconstructed around them, and cytokinesis takes place. After telophase II, the chromosomes become more extended and again are invisible under the light microscope.

The end result of meiosis is the formation of four genetically different haploid (N) cells from one original diploid (2N) cell. No daughter cells formed during meiosis are genetically identical to either mother or father.

Genetic Significance of Meiosis

Chromosome number of each species is conserved across generations in sexually reproducing organisms through a cycle of meiosis and fertilization. Meiosis generates unique haploid gametes. During sexual reproduction, fusion of the unique haploid gametes produces truly unique diploid offspring. If not reduced via meiosis prior to fertilization, the chromosome number would double in each generation.

Meiosis increases the genetic variation in the population of organisms from one generation to the next through genetic recombination. **Genetic recombination** refers to the process of recombining genes to produce new combination of alleles that differ from those of either parent. Genetic recombination happens as a result of independent assortment of genes located on non-homologous chromosomes, crossing over of genes located on homologous chromosomes, and the random union of these genes at fertilization. Genetic recombination is the primary mechanism that produces genetic variation in sexually reproducing organisms within a species, which is important for survival of the species.

Interchromosomal recombination (non-homologous recombination). Meiosis permits the different maternal and paternal chromosomes to randomly assort into each gamete. In metaphase I of meiosis, each maternally derived and paternally derived chromosome has an equal chance of aligning on one or the other side of the equatorial metaphase plate. When paired homologous chromosomes separate at anaphase I, one member of each pair moves to the opposite poles of the cell. As a result, each nucleus generated by meiosis will contain a random combination of maternally and paternally inherited chromosomes. Hence we have recombination due to independent assortment of non-homologous chromosomes. The general formula states that the number of possible chromosome combinations in the nuclei resulting from meiosis is 2^n, where n is the number of chromosome pairs. Each human can produce over 8.38 million (2^{23}) different gametes by random shuffling of maternal and paternal chromosomes in metaphase I of meiosis. Each couple can produce over 70 trillion (one of 8.38 million possible chromosome combinations for each sperm × one of 8.38 million possible chromosome combinations for each ovum) different zygotes during fertilization. This figure does not take account diversity created by crossing over.

Intrachromosomal recombination (homologous recombination, also known as general recombination). Intrachromosomal recombination occurs by crossing over. The frequency of recombination in gametes is further increased due to crossing over. Crossing over between maternal and paternal chromatid pairs during prophase I produce recombinant chromosomes with some maternally and paternally inherited alleles by the next generation. It occurs in eukaryotes, bacteria and viruses naturally. On average 2 or 3 crossovers takes place in each human chromosome.

Key Differences between Mitosis and Meiosis

Differences between mitosis and meiosis are given below on the basis of various criteria:

Criteria	Mitosis	Meiosis
Function	Cellular reproduction and general growth and repair of the body.	Sexual reproduction.
Occurs in	Somatic cells of all organisms. The cells undergoing mitosis may be haploid or diploid.	Meiosis occurs in germ cells only. The cells undergoing meiosis are always diploid.
Genetically	The daughter nuclei or cells formed after mitosis are genetically similar to the parent one.	The daughter nuclei or cells formed after meiosis are neither genetically similar to the parent one nor to one another.

Crossing over	No crossing over.	Crossing over occurs.
Synapsis	No.	Yes.
Number of divisions	1	2
Number of daughter cells produced	Two diploid cells.	Four haploid cells.
Chromosome number	Remains the same.	Reduced by half.
Steps	The steps of mitosis are Interphase, Prophase, Metaphase, Anaphase and Telophase.	The steps of meiosis are Interphase, Prophase I, Metaphase I, Anaphase I, Telophase I, Prophase II, Metaphase II, Anaphase II, and Telophase II.
Cytokinesis	Occurs in Telophase.	Occurs in Telophase I and Telophase II.
Centromeres split	The centromeres split during anaphase.	The centromeres do not split during anaphase I, but during anaphase II.

Nondisjunction

Disjunction is the normal separation of homologous chromosomes (in meiosis I) or sister chromatids (in meiosis II or mitosis) to opposite poles at anaphase of nuclear division. Nondisjunction is the failure of homologous chromosomes or sister chromatids to segregate during mitosis or meiosis. Nondisjunction occurs spontaneously. Nondisjunction results in an uneven distribution of chromosomes during cell division. If nondisjunction occurs during meiosis I, the homologous chromosomes do not separate. If nondisjunction happens in meiosis II, sister chromatids do not separate. Nondisjunction of one or more chromosomes during meiosis I or meiosis II is typically responsible for generating gametes with abnormal number of chromosomes (Image 4.7).

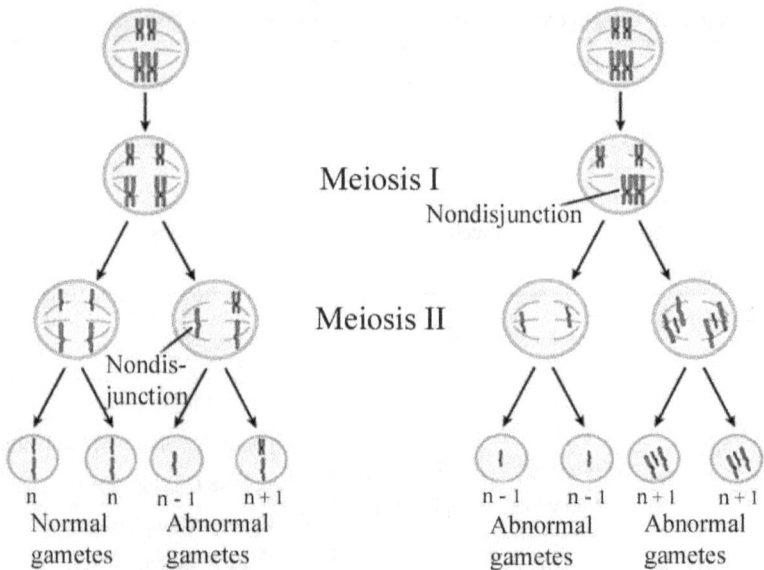

Image 4.7 Nondisjunction in meiosis I and meiosis II (adapted from: http://opentextbc.ca/biology/wp-content/uploads/sites/96/2015/02/Figure_07_03_02.jpg).

Offspring results from fertilization of a normal gamete with one after nondisjunction will have an abnormal chromosome number or aneuploidy. It is also a common cause of early spontaneous abortions. Nondisjunction can involve autosomes or the sex chromosomes. Nondisjunction can also produce gametes with extra or missing sex chromosomes. Nondisjunction can occur in mitosis, giving rise to somatic cells with unusual chromosome complements. Nondisjunction of chromosomes occurs when the centromere fails to function properly. Nondisjunction also occurs when problems with the meiotic spindle cause errors in daughter cells. Spermatogenesis rarely results in nondisjunction, but during oogenesis it is much more probable. Examples of genetic disorders caused by nondisjunction include Down syndrome (trisomy 21), Edwards syndrome (trisomy 18), Patau syndrome (trisomy 13), Turners syndrome (XO), Klinefelter syndrome (XXY), Triple X syndrome (XXX), XYY syndrome (XYY), *etc.*

Anaphase Lag

Anaphase lag is a type of chromosomal error that occurs during cell division, either in mitosis or meiosis, when a chromosome or chromatid **moves too slowly (lags behind)** than the others during anaphase and fails to reach the spindle pole before cytokinesis is completed. The lagging chromosome is excluded from the newly formed daughter nuclei, where it may be lost or form a micronucleus, resulting in one normal cell and one cell with monosomy. This error can arise when one homologous chromosome in meiosis I or one sister chromatid in meiosis II or mitosis is delayed or arrested in its movement. Anaphase lag is often caused by spindle fibre defects or improper attachment of the chromosome to the spindle apparatus. Unlike nondisjunction, in which chromosomes migrate together to the same pole, anaphase lag results from the failure of a single chromosome to be incorporated into the nucleus. It is a frequent cause of aneuploidy and mosaicism, and may occur spontaneously or under the influence of external factors such as spindle poisons or radiation.

Homologous and Nonhomologous (Heterologous) Chromosomes

The pairs of autosomes are called homologous chromosomes (Image 4.8). Each member of a pair is called a homolog. Typically, each homolog is inherited from each parent. Each chromosome in the homologous pair is identical in size, shape, gene location, location of the centromere, location of bands and constriction, and physicochemical properties, and are called homomorphic but they are not identical in terms of the specific alleles they carry. Homologous chromosomes have all of the same genes arranged in the same order, but the alleles may be different. Homologous chromosomes pair during meiosis.

Chromosomes that contain different genes and that do not pair during meiosis are called nonhomologous chromosomes.

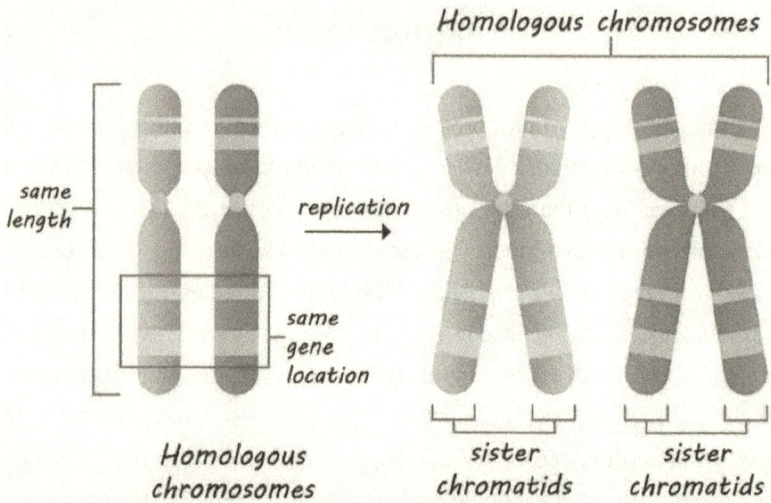

Image 4.8 Homologous and nonhomologous chromosomes (source: https://stock.adobe.com id=405568045).

Eukaryotic chromosomes are transmitted during cell division by mitosis and during reproduction by meiosis. Eukaryotes have multiple linear chromosomes in a number characteristic of the species. Most have two versions of each chromosome, and so are diploid (2N). Diploid cells are produced by haploid (N) gametes that fuse to form a zygote. The zygote then undergoes development, forming a new individual.

Meiosis and Mendel's Principles

Behavior of chromosomes during meiosis provides the physical/biological/cellular basis for the Mendel's principles of segregation and independent assortment. Physical separation of homologous chromosomes in anaphase I is the physical basis of Mendel's principle of segregation. Random combination of maternally and paternally inherited chromosomes carrying their respective alleles in progeny nuclei is the physical basis of Mendel's principle of independent assortment.

140

Locations of Meiosis in the Life Cycle

Meiosis in animals is found only in ovaries and testes, and even in these tissues is restricted to cells that are destined to form gametes (germ cells).

In plants, the gametophyte is the haploid part of the life cycle and sporophyte is the diploid part of the life cycle. In a fungus, for example *Neurospora crassa* (bread mold), the haploid state is the major part of the life cycle.

The Chromosome Theory of Inheritance

In 1902 Walter Sutton (an American who at that time a graduate student) and Theodor Boveri (a German biologist) independently recognized that the transmission of chromosomes from one generation to the next parallels inheritance of Mendelian factors: genes are in pairs (so are chromosomes); the alleles of a gene segregate equally into gametes (so do the members of a pair of homologous chromosomes); different genes act independently (so do different chromosome pairs). To explain this correlation, they proposed the chromosome theory of inheritance. This theory states that i) Mendelian factors or genes are located on chromosomes, and ii) it is the chromosomes that undergo segregation and independent assortment. The chromosome theory of inheritance is one of the foundations of genetics and explains the physical reality of Mendel's principles of inheritance.

Meiotic Drive

According to Mendel's law of segregation each member of a pair of alleles will be equally represented in an organism's gametes so that on average half of an organism's offspring inherit one of the alleles and the other half the other allele. There are, however, some curious cases in which Mendel's laws are broken, and one of the alleles is consistently found in more than half offspring. The term 'meiotic drive' refers to the biological process that results in the unequal transmission of alleles, haplotypes, or chromosomes from a parental genome to gametes during meiosis. Meiotic drive is a violation of the Mendelian inheritance. In the absence of meiotic drive, the two copies of each gene in a diploid organism are transmitted to offspring with equal probability, whereas meiotic drive results in over representation of the driving allele among the surviving products of meiosis. It is also called transmission ratio distortion or segregation distortion. Meiotic drive systems are known in plants, animals, and fungi and can occur in females, males, and during haploid gametogenesis. Female meiosis may be particularly susceptible to meiotic drive since only one of the four products of meiosis develops into the mature gamete, while the other products of meiosis become the polar bodies of the egg. The segregation distorter gene of *Drosophila melanogaster* is an example. Sex chromosome meiotic drive manifests as unequal sex ratio.

CHAPTER FIVE   MULTIPLE ALLELES

Although an organism can carry only two alleles of one gene, many alleles of a single gene can be present in population. When more than two alleles of a single gene exist in a species, they are referred to as multiple alleles and the set of alleles itself is called multiple allelic series. When a trait is governed by multiple alleles it is called multiple allelism. It is a type of non-Mendelian inheritance pattern.

Any change in the nucleotide sequence of a gene is a different allele. Each nucleotide contains a base: adenine (A), thymine (T), guanine (G), or cytosine (C). So, a gene of n nucleotides can theoretically mutate at any of the positions to any of the three other nucleotides. The number of possible single-nucleotide differences in a gene of length n is therefore $3n$; each of these DNA sequences, if it exists in the population, is an allele. When $n = 5,000$, for example, there are potentially 15000 alleles (not counting any of the possibilities with more than one nucleotide substitution). So, there are millions of possible alleles for any gene. Most of the potential alleles may not actually exist at any one time, but many of them may be present in any population. Indeed, for most genes in most populations, the 'normal' or 'wild type' allele is not a single-nucleotide sequence but rather a set of different nucleotide sequences, each capable of carrying out the normal function of the gene. About 70% of human genes are monoallelic, about 30% are di-allelic and a few are multiple-allelic. The number of possible genotypes in a multiple allelic series depends on the number of alleles involved. In general, the number of genotypes possible in a population of a diploid organism with n different alleles are, $n(n + 1)/2$ of which n are homozygotes and the remaining are heterozygotes. For the ABO blood group system with three alleles ($n = 3$), the number of genotypes is $3 \times 4/2 = 6$. If there were four alleles, there would be $4 \times 5/2 = 10$ possible genotypes. The ten genotypes resulting from four alleles are composed of 4 homozygotes and $4 \times 3/2 = 6$ heterozygotes.

Characteristics of Multiple Alleles

1. Multiple alleles exist at the population level and a population of organisms may contain any number of alleles.
2. A person or cell does not have more than two alleles for a single gene.
3. A gamete may contain only one allele of each gene.
4. Multiple alleles are situated at the same locus of homologous chromosomes and affecting same character.
5. No crossing-over takes place between the members of the multiple alleles.
6. Multiple alleles never show complementation with each other. The cross between two mutant alleles will always produce mutant type phenotype.
7. The wild-type allele is mostly dominant over the mutant alleles.
8. The mutant allele may be dominant, recessive or have an intermediate phenotypic effect.

Difference between Multiple Alleles and Multiple Genes

Multiple alleles refers to different versions of one gene and occupy the same locus in the chromosome. Multiple genes are located in different loci and have cumulative effects which control a single trait.

Symbolism for Multiple Alleles

The members of the multiple allelic series are conventionally represented by the same letter or symbol with appropriate superscripts or subscripts to represent the different alleles.

Examples of Multiple Alleles

Multiple alleles are common in all organisms, including animals, plants, and humans. Here are some common examples of multiple alleles in each group.

Coat Color (*C* gene) in Rabbits

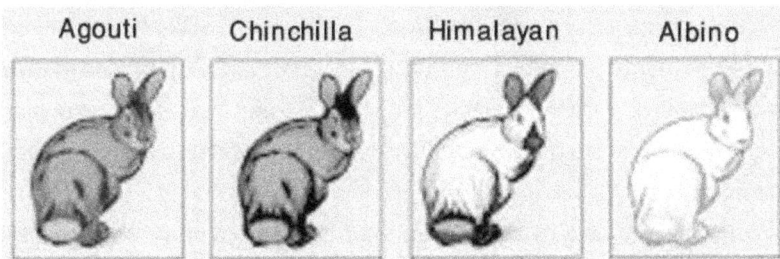

Image 5.1 Coat Color Phenotypes in Rabbits (source: https://zipfslaw.org/wp-content/uploads/2017/10/figure_12_02_05.jpg).

Table 5.1 *C* gene in rabbits

Coat color phenotype	Genotype
Full color	CC or Cc^{ch} or Cc^h or Cc
Chinchilla	$c^{ch}c^{ch}$ or $c^{ch}c^h$ or $c^{ch}c$
Himalayan	$c^h c^h$ or $c^h c$
Albino	cc

In rabbits, multiple alleles of one gene are responsible for a number of different coat/fur color phenotypes. The alleles in this series are C (full or agouti color), c^{ch} (Chinchilla, a light grayish color), c^h (Himalayan, albino with black paws, nose, ears and tail), and c (albino or white, lacks of pigmentation). When homozygous each produces a distinct fur pattern (Image 5.1). In heterozygotes there is a clear pattern of dominance. Agouti is dominant over all the other alleles, Chinchilla is dominant over Himalayan and albino, while Himalayan is dominant only over albino which fails to produce any pigment and hence is recessive to all the others. The dominance hierarchy of these alleles is $C > c^{ch} > c^h > c$. The phenotypes and corresponding genotypes for alleles at *C* locus in rabbits are summarized in Table 5.1.

ABO Blood Group System in Humans

Blood is a complex, living tissue that contains many cell types and proteins. The term blood group/type is generally based on the presence or absence of agglutinogens (a type of antigen) on the surface of erythrocytes (RBC). These are identified by characteristic agglutination reactions with specific antibodies. Blood group system refers to all of the genes, alleles, and possible genotypes and phenotypes that exist for a particular set of blood type antigens. As of 2019, the International Society of Blood Transfusion (ISBT) have recognized a total of 41 human blood group systems (Storry *et al*, 2019). Knowledge of blood types in different species is important as transfusion of incompatible blood can result in severe hemolytic transfusion reactions and even death, in some instances.

One commonly known genetic variation resulting from multiple alleles is the ABO blood group system in humans. The ABO blood group antigens are the most well-known and the most immunogenic of all the blood group antigens. The ABO blood group system is widely credited to have been discovered by the Austrian scientist Dr. Karl Landsteiner, who identified two types of blood antigens, A and B, and naturally occurring anti-A antibody and anti-B antibody in human in 1900 at the University of Vienna in the process of trying to learn why blood transfusions sometimes cause death and at other times save a patient. Landsteiner discovered the ABO blood group system by mixing the red cells and serum of each of his staff. He divided human blood into three groups: A, B and O. Two years later (1902), two students of Dr. Landsteiner; Alfred von DeCastello and Adriano Sturli discovered the fourth blood group AB. So, human blood is divided into four blood groups: A, B, AB and O. In 1930, Landsteiner was awarded the Nobel Prize in Physiology or Medicine for his discovery of blood types. Ludwik Hirszfeld and Emil von Dungern discovered the inheritance of ABO blood groups in 1910. There are two different types of antigens, type "A" and type "B". Each type has different properties. Human ABO blood type classification system uses the presence or absence of these antigens. The A blood type has only the A antigen and the B blood type has only the B antigen. The AB blood type has both A and B antigens, and the O blood type has neither A nor B antigen. "O blood type" was named after the German word "*Ohne*", which means "without", or "zero", or "null".

There was initially some confusion over how a person's blood type was determined, but the puzzle was solved in 1924 by Bernstein's "three allele model". As an autosomal trait, the ABO blood type is determined by the polymorphism of the ABO gene, located at the 9q34.2 locus (Ellinghaus *et al.*, 2020). It contains 7 exons that span more than 18 kb of genomic DNA. The ABO gene has three main allelic forms: L^A, L^B, and L^O. The L^A and L^B alleles each encode a glycosyltransferase that catalyzes the final step in the synthesis of the A and B antigen, respectively. The L^A allele encodes a glycosyltransferase that makes the A antigen (*i.e.*, adds N-acetyl galactosamine to the glycoprotein H antigen on the surface of red blood cells). The L^B allele codes for a different glycosyltransferase that makes the B antigen (*i.e.*, adds D-galactose instead of N-acetyl galactosamine). The L^O allele encodes an inactive glycosyltransferase that leaves the ABO antigen precursor (*i.e.*, H antigen) unmodified. Therefore, neither A nor B antigen is produced. Both L^A and L^B are dominant to L^O, but alleles L^A and L^B are codominant. Everyone receives one of the three alleles from each parent, giving rise to six possible genotypes and four possible blood types (Table 5.2). There are many variant ABO alleles that encode a number of variant ABO phenotypes, but they do not encode specific antigens other than the A and B antigens.

Table 5.2 ABO blood groups in humans

Blood types (phenotype)	Genotype
O	$L^O L^O$
A	$L^A L^A$ or $L^A L^O$
B	$L^B L^B$ or $L^B L^O$
AB	$L^A L^B$

Blood plasma is packed with proteins called antibodies. The body produces a wide variety of antibodies that will recognize and attack foreign molecules that may enter from the outside world. A person's immune system does not produce any antibodies that will bind to molecules that are part of his or her own body. The immune system produces antibodies against whichever ABO blood group antigens are not found on the individual's RBCs. Thus, individuals of type A will have anti-B antibodies; type B will have anti-A antibodies. Blood group O is the most common, and individuals with this blood type will have both anti-A and anti-B antibodies in their serum. Blood group AB is the least common, and these individuals will have neither anti-A nor anti-B in their serum. L^A allele is somewhat more common around the world than L^B allele but less common than the L^O allele, which is the most common allele and L^B allele is the least common allele worldwide.

Table 5.3 Characteristics of human blood groups and the types of transfusions

Blood groups	Antigen in red blood cells	Antibodies in plasma	Can give blood to groups	Can receive blood from groups
O	None	Anti-A and anti-B	O, A, B, AB	O
A	A	Anti-B	A, AB	A, O
B	B	Anti-A	B, AB	B, O
AB	A and B	None	AB	AB, O, A, B

Blood group antigens play a role in recognizing foreign cells in the blood stream. In order to transfuse blood successfully from one individual to another, the blood groups of donors and recipients must be carefully matched. If the donor blood cells have antigens that are different from those of the recipient, antibodies in the recipient's blood recognize the donor blood as foreign. This triggers an immune response resulting in blood clotting. If the donor blood cells have antigens that are the same as those of the recipient, the recipient's body will not see them as foreign and will not mount an immune response. It is important that individuals be given compatible blood types for transfusions. For example, if a person with blood type B is given blood of type A, the recipient's immune system will recognize the type A cells as foreign and mount an immune response. His or her anti-A antibodies will bind to the antigens on the type A blood cells and initiate a cascade of events that will cause the blood to agglutinate or clump together. Since clumped cells cannot move through the fine capillaries, agglutination may lead to organ failure and possibly death. The characteristics of blood groups and the types of transfusions can be summarized in Table 5.3.

There are two special blood types when it comes to blood transfusions. People with type O blood are universal donors, because its red cells have no antigens that can trigger an immune response. On the other hand, people with type AB blood are universal recipients, because its plasma have no antibodies to react with transfused red cells and can therefore accept A, B, AB, or O blood.

Blood typing (the determination of an individual's blood group) and the analysis of the blood group inheritance are sometimes used in legal medicine in cases of disputed paternity or maternity, or in cases of an inadvertent baby switch in a hospital. In such cases genetic data cannot prove the identity of the parent. Genetic analysis on the basis of blood group can only be used to show that an individual is not the parent of a particular child; for example, a child of phenotype AB (genotype $L^A L^B$) could not be the child of a parent of phenotype O (genotype $L^O L^O$). However, despite their obvious clinical importance, the physiological functions of ABO blood group antigens remain a mystery. People with the common blood type O express neither the A nor B antigen, and they are perfectly healthy.

Frequency of ABO blood groups in man

O group ~ 48%

A group ~ 37%

B group ~ 11%

AB group ~ 4%

There are 11 major blood group systems in cattle and Buffalo. These systems are A, B, C, F, J, L, M, R, S, T and Z. The B system, part of the Bovine Erythrocyte Antigen (BEA) complex, is recognized as one of the most polymorphic systems in veterinary medicine. The system follows a codominant inheritance pattern. B group has over 60 different antigens, making it difficult to closely match donor and recipient. There are five major blood group systems in goats and sheep: A, B, C, M and J. Similar to cattle, B system is highly polymorphic. There are over 30 blood group systems in horses, of which only 8 are major systems: A, C, D, K, P, Q, U, and T. Chickens possess 13 known blood group systems: A, B, C, D, E, H, I, J, K, L, N, P, and R. B system is the major blood group system in chickens. The system is highly polymorphic with codominant inheritance of alleles. There are more than 13 blood types found in dogs. The Dog Erythrocyte Antigen (DEA) is the standard classification. The seven major blood types found in dogs are DEA 1.1, 1.2, 1.3, DEA 4, DEA 3 and 5, and DEA 7. The most commonly found blood type is DEA 1.1. Cats have three main blood types: A, B, and AB. Type A is the most common.

Bombay Blood Phenotype (hh)

In Bombay (now Mumbai) India, an individual was discovered to have an interesting blood type that reacted to other blood types in a way never seen before. Thus, this resulted in the discovery of the Bombay blood type. The serum contained antibodies that attacked all red blood cells of normal ABO phenotypes (*i.e.*, groups O, A, B, and AB). The Bombay (O_h) blood group phenotype is defined in routine blood grouping by the complete absence of A, B and H antigens on red blood cells and in body fluids, and by the presence of anti-A, anti-B and anti-H antibodies in serum (Daniels, 2013.). This is considered to be a significantly rare blood type of the ABO system, with prevalence estimated to be 1 in 1,000,000 in European regions, or up to 1 in 10,000 in Indian regions (Anso and Naegeli, 2023; Ekanem *et al.*, 2020).

Individuals of "Bombay phenotype" lack the H antigen. There is no ill effect with being H deficient. Because the H antigen is the precursor of the ABO blood group antigens, they cannot make the ABO blood group antigens on the surface of their red blood cells, whatever alleles they may have the ABO gene. This can be misleading in paternity cases.

The clinical significance of the Bombay blood group includes the inability to transfuse with other blood groups, such as A, B, AB, and especially the O type (Ekanem *et al.*, 2020). This can lead to fatal hemolytic transfusion reactions, with a classic triad of symptoms including fever, flank pain, and red-brown urine, along with the development of extravascular hemolysis (Patel *et al.*, 2018).

There is a reasonable similarity between the Bombay blood phenotype and the O group. This is due to the absence of the A and B antigens (Qadir *et al.*, 2023). Bombay individuals show agglutination with O cells due to the presence of anti-H antibody, in contrast to the O blood group (Panch *et al.*, 2019). Therefore, reverse grouping is extremely important to establish the correct ABO type and to resolve any ABO blood discrepancy (Qadir *et al.*, 2023).

Individuals with the Bombay blood group can only receive blood transfusions from Bombay donors due to the presence of powerful anti-H antibodies in their serum (Qadir *et al.*, 2023). In order to avoid complications during a blood transfusion, it is very important to detect Bombay phenotype individuals.

Bombay phenotype was discovered by Dr. Y. M. Bhende in 1952. It is transmitted as an autosomal recessive trait. Most people have *HH* or *Hh* genotypes but Bombay phenotypes have *hh* genotype. The Bombay phenotype is the result from homozygosity of mutations in the FUT1 (fucosyltransferase 1) gene (H gene) located on chromosome 19q13.3. This gene encodes α(1,2)fucosyltransferase (H enzyme) that is involved in the creation of a precursor of the H antigen, which is in turn the precursor of A and B antigens.

The Rhesus (Rh) Blood Group System

In addition to the ABO blood group antigens, there is another blood group antigen located on red blood cell surfaces, known as the Rhesus factor or Rh factor or Rh antigen. It was named after Rhesus monkeys, since they were first detected in the RBC of Rhesus monkey and later in man. The Rh antigen refers to a group of closely related antigens. Clinically, the Rh blood group system is the second most import blood group system in humans due to hemolytic disease of newborn and was discovered in 1940 by Karl Landsteiner and Alexander Wiener, when the two biologists carried out immunization experiments on rabbit using blood of rhesus monkey. They found that the antibodies produced after immunization, agglutinate the blood of monkey and also the blood of large number of humans. Genetically this blood group system may be the most complex of all of the human blood group systems. The complexity of the Rh blood group antigens begins with the highly polymorphic gene that encodes them. It is highly polymorphic because it contains more than forty-four different antigens. The most clinically important antigen in the Rh system is the D-antigen. The commonly used terms Rh factor, Rh-positive (Rh$^+$) and Rh-negative (Rh$^-$) refer to the *D* antigen only. Despite its actual genetic complexity, the inheritance of this trait usually can be predicted by a simple conceptual model in which there are two alleles, *D* and *d*. Individuals who are homozygous dominant (*DD*) or heterozygous (*Dd*) are Rh$^+$. Those who are homozygous recessive (*dd*) are Rh$^-$ (*i.e.*, they do not have the key Rh antigens). A person who is Rh$^-$ will produce antibodies against Rh$^+$

blood cells (anti-D antibody) if exposed to them. Exposure can be due to instances such as a blood transfusion with Rh$^+$ blood or a pregnancy where the Rh$^-$ mother has an Rh$^+$ child. A person who is Rh$^+$ can receive blood from someone who is Rh$^+$ or Rh$^-$ without any negative consequences. Rh-negative individuals must always be transfused with Rh-negative blood to avoid immunization and subsequent dangerous reactions. Besides its role in blood transfusion, the Rh blood group system – specifically, the *D* antigen is used to determine the risk of *erythroblastosis fetalis*, or hemolytic disease of the newborn, a disorder in which maternal antibodies destroy the child's red blood cells during late pregnancy and shortly after birth. The foetus suffers from haemolytic anaemia and severe jaundice which results in stillbirth or neonatal death. The gene responsible for the Rh antigen is located on the short arm of chromosome 1 (1p34-p36) in man. Rh antigens are highly immunogenic.

Parents	♂ *Dd*	×	♀ *dd*
Gametes	½ *D*, ½ *d*		*d*
Progeny	½ *Dd* (Rh-positive), ½ *dd* (Rh-negative)		

Figure 5.1 A cross between a Rh-positive father and a Rh-negative mother.

A Rh-positive father (*DD* or *Dd*) and a Rh-negative mother (*dd*) may have a Rh$^+$ child (Figure 5.1). The first time an Rh$^-$ woman becomes pregnant; usually there are no incompatibility difficulties for her Rh$^+$ fetus. However, the second and subsequent births are likely to have life-threatening problems for Rh$^+$ fetuses. The risk increases with each birth. In order to understand why first born are normally safe and later children are not, it is necessary to understand some of the placenta's functions. Placenta is an organ that connects the fetus to the wall of the uterus via an umbilical cord. Nutrients and the mother's antibodies regularly transfer across the placental boundary into the fetus, but her red blood cells usually do not (except in the case of an accidental rupture). Normally, anti-Rh$^+$ antibodies do not exist in the first-time mother unless she has previously come in contact with Rh$^+$ blood. Therefore, her antibodies are not likely to agglutinate the red blood cells (RBC) of her Rh$^+$ fetus. Placental ruptures do occur normally at birth so that some fetal blood gets into the mother's system, stimulating the formation of antibodies to Rh$^+$ blood antigens (anti-Rh$^+$ antibodies). As little as one drop of fetal blood stimulates the production of large amounts of antibodies. When the next pregnancy occurs, a transfer of antibodies from the mother's system once again takes place across the placental boundary into the fetus. The anti-Rh$^+$ antibodies that she now produces react with the fetal blood, causing many of its RBC to burst or agglutinate. As a result, the newborn baby may have a life-threatening anemia because of lack of oxygen in the blood. The baby also usually is jaundiced, fevered, quite swollen, and has an enlarged liver and spleen. This condition is called erythroblastosis fetalis (hemolytic disease of the newborn). In extreme cases so many red cells are destroyed that the fetus dies before birth; more frequently it is borne alive but dies after birth. Keep in mind that only the Rh$^+$ children (*Dd*) are likely to have medical complications. When both the mother and her fetus are Rh$^-$ (*dd*), the birth will be normal. The standard treatment in severe cases is immediate massive transfusions of Rh$^-$ blood into the baby with the simultaneous draining of the existing blood to flush out Rh$^+$ antibodies from the mother. This is usually done immediately following birth, but it can be done to a fetus prior to birth. Because the baby's own Rh$^+$ red blood cells have been replaced with Rh$^-$ ones, the mother's anti-Rh$^+$ antibodies don't agglutinate any additional red cells. Later, the Rh$^-$ blood will be replaced naturally as the baby gradually produces its own Rh$^+$ blood. Any residual anti-Rh$^+$ antibodies from the mother will leave gradually as well because the baby does not produce them.

Mother-fetus incompatibility problems can result with the ABO system also. However, they are very rare – less than 0.1% of births are affected and usually the symptoms are not as severe. It most commonly occurs when the mother is type O and her fetus is A, or B. The symptoms in newborn babies are usually jaundice, mild anemia, and elevated bilirubin levels. These problems in a baby are usually treated successfully without blood transfusions.

About 85% people have Rh antigen (Rh$^+$) and about 15% people lack this antigen (Rh$^-$) in the blood.

Wing Type in Drosophila

In Drosophila wings are normally long. Wide variation in wing size of Drosophila is observed. There are five types of wings in Drosophila, *viz.*, wild (long), nicked, notched, strap and vestigial wings. The nicked wings are notched at the margin, strap wings are very narrow and vestigial wings are miniature in size. These wing sizes are observed when the individuals are in homozygous condition. The variation in wing size is due to multiple alleles of the same gene. The wild type is dominant over all other types. Crosses between other wing types exhibit intermediate expression in F_1 and $1 : 2 : 1$ segregation in F_2.

Eye Colour in Drosophila

In drosophila, eye color is controlled by more than a dozen alleles. The wild type or normal colour of the eye is red. Mutation changed this red eye colour to cherry, eosin, creamy, buff, tinged, honey, ecru, pearl, ivory, blood, wine, apricot, coral, white, *etc*. First white eyed mutant was discovered and later on other colours of eye were reported. The wild-type red colour is dominant over all other mutant alleles and the white colour is recessive to all the alleles. The cross between individuals of other eye colour (mutants) exhibits intermediate expression in F_1 and true expression only in homozygous condition.

Human Leukocyte Antigen (HLA) Gene Family

Human leukocyte antigen (HLA) is a substance that is located on the surface of white blood cells. The HLA gene family is the human version of the major histocompatibility complex (MHC) gene family that occurs in many species. It provides instructions for making a group of related proteins known as the human leukocyte antigen complex. The HLA complex helps the immune system to distinguish the body's own proteins from proteins made by foreign invaders such as viruses and bacteria. In humans, the MHC consists of about 220 genes located close together on chromosome 6. It encodes cell surface molecules specialized to present antigenic peptides to the T-cell receptor (TCR) on T cells. The MHC is the primary determinant of human tissue type, which determines whether organs can be transplanted between people without rejection by the immune system. Genes in this complex are categorized into three basic groups: class I, class II, and class III.

Humans have three main MHC class I genes, known as HLA-A, HLA-B, and HLA-C. The proteins produced from these genes are present on the surface of almost all cells. On the cell surface, these proteins are bound to peptides. MHC class I proteins display these peptides to the immune system. If the immune system recognizes the peptides as foreign (such as viral or bacterial peptides), it responds by triggering the infected cell to self-destruct.

There are six main MHC class II genes in humans: HLA-DPA1, HLA-DPB1, HLA-DQA1, HLA-DQB1, HLA-DRA, and HLA-DRB1. MHC class II genes provide instructions for making proteins that are present almost exclusively on the surface of certain immune system cells. Like MHC class I proteins, these proteins display peptides to the immune system.

The proteins produced from MHC class III genes have somewhat different functions; they are involved in inflammation and other immune system activities. The functions of some MHC genes are unknown.

HLA genes have many possible variations, allowing each person's immune system to react to a wide range of foreign invaders. Some HLA genes have hundreds of identified alleles. As a result of this diversity, an extraordinary number of genotypes generally occur at this gene in a population. Because there are so many different genotypes, donors and recipients generally do not match. Because relatives are more likely to share HLA types, often close relatives are examined for HLA to find a suitable organ donor. Even with a suitable donor for the major HLA types, it is still necessary to use immunosuppressant drugs to reduce the organ recipient's immune response to foreign antigens from other unmatched genes and thus achieve a successful transplant.

Genes for Hemoglobin (*HB*) Variants in Man

Hemoglobin is made of heme, alpha globins, and beta globins. Hemoglobin within red blood cells binds to oxygen molecules in the lungs. These cells then travel through the bloodstream and deliver oxygen to tissues throughout the body. At least 9 different genes direct the production of heme. Changes in these genes may lead to disorders of heme production, a group of conditions separate from the thalassemias. In adults, hemoglobin composed of two subunits of beta-globin and two subunits of alpha-globin. Hemoglobinopathies covers a heterogeneous group of red blood cell disorders caused by mutations in the alpha-globin (*HBA*) or beta-globin (*HBB*) genes and are the most common single-gene genetic disorders in humans. In humans, there are four alpha globin genes located on chromosome 16 used to make alpha-globin. There are two beta globin genes located on chromosome 11 used to make beta-globin. There are more than 1,700 hemoglobin (*HB*) variants have been reported. Approximately 300 variants are in the *HBA* gene, and more than 900 variants are in *HBB* gene. The most common variants in *HBA* gene are *HBZ, HBA2, HBA1, HBM* and *HBQ1*, and in *HBB* gene are hemoglobin E (*HBE*), sickle hemoglobin (*HBS*), and hemoglobin C (*HBC*).

Self-Incompatibility Alleles in Plants

Self-incompatibility (SI) is defined as the inability of a fertile hermaphrodite seed plant to produce zygotes after self-pollination. It prevents self-fertilization in flowering plants which promotes out-crossing in plants. The most common example of multiple alleles in plants is the series of self-incompatibility alleles. Such alleles were often observed in plants belonging to Solanaceae family (potatoes, tobacco, *etc.*) and members of the mustard family, including turnips, rape, cabbage, broccoli, and cauliflower. In these species, self-incompatibility is often governed by a single gene S which is extremely polymorphic; that is there is an abundance of multiple alleles in the population *viz.*, S^1, S^2, S^3, S^4 and so on. The S gene controls the specificity of the self-incompatibility interactions between pollen and pistil. Matching of the S-allele carried by the pollen with one of the two S-alleles carried by the pistil results in the growth arrest of the pollen tube in the style. Only pollen tubes that carry a different S-allele from the pistil will be able to grow down the style to the ovary to achieve fertilization. Crosses between individuals having self-incompatibility alleles will lead to three types of situations as given below:

Fully Sterile: When both male and female have similar alleles, *viz.*, S^1S^2 × S^1S^2 the cross will be incompatible and there will be no seed setting.

Partially Fertile: Such crosses are obtained when male and female plants differ for one allele, *viz.*, S^1S^2 × S^1S^3. This cross will produce S^1S^3 and S^2S^3 fertile progeny. In other words, half of the progeny will be fertile.

Fully Fertile: The fully fertile crosses are obtained when male and female plants differ in respect of both alleles, *viz.*, S^1S^2 × S^3S^4. This cross will produce four fertile genotypes, *viz.*, S^1S^3, S^1S^4, S^2S^3 and S^2S^4. Thus, plants which have self-incompatibility alleles are always heterozygous for this gene.

Now cases of digenic and trigenic self-incompatibility have also been reported.

Test for Allelism

We have two tests for allelism: first, the monohybrid 3 : 1 ratio reveals a dominant and a recessive allele of the same gene, and second, lack of complementation.

Two recessive mutations are considered alleles of different genes if a cross between the homozygous recessive results in dominant (wild type) phenotype in the F_1 progeny. Such alleles are said to complement each other (complementation). On the other hand, two recessive mutations are considered alleles of the same gene if a cross between the homozygous recessive results in mutant phenotype. Such alleles are said to fail to complement (noncomplementation).

Eye color in Drosophila is a good model to demonstrate the complementation test. Five genes (white, ruby, vermillion, garnet and carnation) controlling eye color of *Drosophila* reside on the X chromosome. The dominant wild type allele for each gene produces the deep red eyes. The mutant alleles produce a different color. If mutants from any of these five genes are crossed, the F_1 would express deep red eye color (wild type phenotype).

Five different alleles (buff, coral, apricot, white and cherry) are also known to exist for the white gene, each representing a mutation at a different position in the gene. If mutant flies for any of the five white alleles are crossed, the F_1 offspring would have a mutant eye color. Therefore, complementation between two alleles of the same gene does not occur.

Isoalleles

Within a series of multiple alleles, each of which may again have slightly different forms, that give almost similar phenotypes like multiple allele. Such alleles expressing themselves within the same phenotypic range are called iso-alleles. If the phenotype is wild, these are normal or wild iso-alleles, and if the phenotype is mutant, they will be called mutant iso-alleles. No crossing over occurs in such iso-alleles. For example, A allele of ABO blood group has A^1,

A^2 and A^3 iso-alleles. They are so called as they are to be distinguished in certain heterozygote combination or where the internal environment changes.

Pseudoalleles

Alleles, which have separate gene loci, but often inherit together due to close linkage, govern different expressions of the same character and have very rare chance of crossing over are called pseudo alleles. They appear to act as the same genes; hence they got the name pseudoalleles. *e.g.,* Lozenge eye in Drosophila. A cluster of pseudoalleles is known as pseudoallelic series.

CHAPTER SIX   GENETIC LINKAGE

Enormous numbers of genes are contained within the genetic material of each individual. These genes are arranged on the chromosomes. Each chromosome must contain many genes, because the number of genes of an organism exceed the number of its chromosomes. Chromosomes are inherited as units, so, all the genes located on the same chromosomes, tend to be inherited together. Genes on the same chromosome which are transmitted as a group are said to be linked. The closer two genes are to each other on a chromosome, the greater the probability that they will be inherited together. Linkage is a phenomenon where two or more genes on the same chromosome are inherited together in the same combination for more than two generations. Linked genes are those that reside on the same chromosome and tend to be inherited together as a group. Linked genes do not assort independently. Only genes situated on the same chromosome can show linkage. Genes located on non-homologous chromosomes are unlinked and assort independently at the first meiotic division. Genes on a single chromosome comprise a linkage group. The number of linkage group is the same as the haploid number of chromosomes of the species. In *Drosophila melanogaster*, there are 4 linkage groups corresponding to the 4 pairs of chromosomes. Maize has 10 linkage groups.

Discovery of Genetic Linkage

Linkage in the Sweet Pea

The first experiment to demonstrate linkage was carried out in 1905 by the British geneticists William Bateson, Edith Rebecca Saunders, and Reginald Crundall Punnett in the sweet pea (*Lathyrus odoratus*). They studied the simultaneous inheritance of flower color, dominant purple (P) versus recessive red (p), and pollen shape, dominant long pollen grains (L) versus recessive round pollen grains (*l*). They crossed homozygous pea plants with purple flowers and long pollen grains (*PP LL*) with homozygous plants with red flowers and round pollen grains (*pp ll*), self-fertilized the F₁ double heterozygote (*Pp Ll*), and counted the numbers of different types of F₂ plants. But instead of the expected 9 : 3 : 3 : 1 ratio under independent assortment, they obtained the results given in Table 6.1.

Table 6.1 F₂ progeny of a dihybrid cross in sweet peas

Phenotype	Genotype	Number of Progeny	
		Observed	Expected from 9 : 3 : 3 : 1
Purple, long	*P– L–*	4831	3911
Purple, round	*P– ll*	390	1303
Red, long	*pp L–*	393	1303
Red, round	*pp ll*	1338	435
Total		6952	6952

Comparing the observed numbers to those expected under 9 : 3 : 3 : 1 ratio, they found many more phenotypes like the original parents and many fewer nonparental types. This does not appear to be explainable as a modified Mendelian ratio. Because the parental phenotypes reappeared more frequently than expected, the three researchers hypothesized that there was a coupling, or connection, between the parental alleles for flower color and pollen grain shape (Bateson *et al.*, 1905), and that this coupling resulted in the observed deviation from independent assortment. In other words, due to physical connection between the dominant alleles *P* and *L*, and between the recessive alleles *p* and *l*, the F₁ had actually produced more *PL* and *pl* gametes than would be produced by Mendelian independent assortment.

But why are certain alleles linked? Bateson, Saunders, and Punnett weren't sure. In fact, it was not until the later work of geneticist Thomas Hunt Morgan that this coupling, or linkage, could be fully explained.

Morgan's Linkage Experiments with Drosophila

In 1911, Thomas Hunt Morgan found a similar deviation from Mendel's second law while studying two autosomal genes in Drosophila (*Drosophila melanogaster*). One gene affected eye color (recessive purple, *pr*, and dominant wild type red, *pr⁺*, alleles), and the other gene affected wing length (recessive vestigial, *vg*, and dominant normal, *vg⁺*, alleles). Morgan crossed *prpr vgvg* (purple vestigial) flies with *pr⁺pr⁺ vg⁺vg⁺* (red normal) and then testcrossed the double heterozygous F_1 flies: *pr⁺pr vg⁺vg × prpr vgvg*. Results are given in Table 6.2. Obviously, these numbers deviate drastically from the Mendelian prediction of 1 : 1 : 1 : 1 ratio. The numbers of the parental phenotypes were many times those of the nonparental phenotypes.

Table 6.2 Results of Morgan's first testcross

Phenotype	Genotype	Observed	Expected
Red, normal	*pr⁺pr vg⁺vg*	1339	709.75
Red, vestigial	*pr⁺pr vgvg*	151	709.75
Purple, normal	*prpr vg⁺vg*	154	709.75
Purple, vestigial	*prpr vgvg*	1195	709.75
Total		2839	2839

In a second experiment, Morgan crossed two different genotypes, each of which was homozygous for a wild type dominant allele at one gene and homozygous for a recessive allele at another gene (*pr⁺pr⁺vgvg × prprvg⁺vg⁺*). The F_1 double heterozygous flies were then crossed to the tester line *prpr vgvg* (purple vestigial). The results of this cross were quite different from the first cross (Table 6.3).

Again, these results were not even close to 1 : 1 : 1 : 1 Mendelian test cross ratio. In both cases, the numbers of parental phenotypes were many times those of the nonparental phenotypes. Morgan's group conducted a large number of other crosses of this type and the conclusions were always the same. In each case the parental phenotypic classes were the most frequent while the recombinant classes occurred much less frequently. Approximately equal numbers of each of the two parental classes were obtained, and similar results were obtained for the recombinant classes. Morgan suggested that the genes governing both phenotypes are located on the same pair of homologous chromosomes. As a result, alleles at these genes tend to remain associated between generations because they are physically linked to each other. Thus, when *pr* and *vg* are introduced from one parent, they are physically located on the same chromosome, whereas *pr⁺* and *vg⁺* are on the homologous chromosome from the other parent. This hypothesis also explains repulsions. In that case, one parental chromosome carries *pr* and *vg⁺* and the other carries *pr⁺* and *vg*.

Table 6.3 Results of Morgan's second testcross

Phenotype	Genotype	Observed	Expected
Red, normal	*pr⁺pr vg⁺vg*	157	583.75
Red, vestigial	*pr⁺pr vgvg*	965	583.75
Purple, normal	*prpr vg⁺vg*	1067	583.75
Purple, vestigial	*prpr vgvg*	146	583.75
Total		2335	2335

T.H. Morgan concluded that during segregation of alleles at meiosis, certain genes tend to remain together because they lie near each other on the same chromosome. The closer genes are located to each other on the chromosome, the greater their tendency to remain linked. Thomas Hunt Morgan, who studied fruit flies, provided the first strong confirmation of the chromosome theory.

Significance of Linkage

Linkage plays an important role in determining the scope of hybridization and selection programmes. The main effect of linkage is that it reduces the chances of new gene combination in the offspring. It keeps parental genes/traits together and preserve valuable traits of new variety. Linkage can be valuable in chromosome mapping.

Linkage Symbolism

Geneticists use a notation for linked genes as:

$$\frac{\text{pr} \quad \text{vg}}{\text{pr}^+ \quad \text{vg}^+}$$

where, each line represents a chromosome; the alleles above are on one chromosome, and those below are on the other chromosome.

We can simplify the genotypic designation of linked gene, by drawing a single line, with the genes on each side being on the same chromosome; now the symbol is:

$$\frac{\text{pr} \quad \text{vg}}{\text{pr}^+ \quad \text{vg}^+}$$

But this is still inconvenient for typing and writing, so let's tip the line to give us pr vg/pr⁺ vg⁺, still keeping the genes of one chromosome on one side of the line and those of its homolog on the other. The linked genes in a chromosome are always written in the same order for consistency. The rule that genes are always written in the same order permits geneticist to use a shorter notation in which the wild type allele is written with a plus sign alone. In this notation the genotype pr vg/pr⁺ vg⁺ becomes pr vg/+ +.

Arrangement of Linked Genes

Genes can be linked in two different phases:

Coupling state/phase or cis-configuration. Coupling refers to the linkage of two dominant or two recessive alleles, *i.e.*, the double heterozygote carries the dominant alleles of the two genes on one homolog and the recessive alleles on the other homolog.

Coupling conformation $\dfrac{\text{pr} \quad \text{vg}}{\text{pr}^+ \quad \text{vg}^+}$ or pr vg/pr⁺ vg⁺ or pr vg/+ +

Repulsion conformation $\dfrac{\text{pr} \quad \text{vg}^+}{\text{pr}^+ \quad \text{vg}}$ or pr vg⁺/pr⁺ vg or pr +/+ vg

Repulsion state/phase or trans-configuration. Repulsion indicates that dominant alleles are linked with recessive alleles.

Gametes with either two wild type or two mutant alleles are coupling gametes, and those with one wild type and one mutant allele are repulsion gametes.

163

Syntenic Gene

Syntenic genes are genes that are physically located on the same chromosome, whether or not they exhibit linkage. Genes far apart on the same chromosome would not show evidence of linkage in genetic crosses. Therefore, all linked genes are syntenic, but not all syntenic genes are linked.

Kinds of Linkage

The phenomenon of linkage is of following two kinds:

1. Complete linkage. Genes that are linked and never crossover. Here 100% of the progeny with parental combination. In fact, two genes that are completely linked can only be differentiated as separate genes when a mutation occurs in one of them. There is no other way to identify genes with complete linkage from single genes that show multiple phenotypes.

2. Incomplete linkage. Genes that are linked but can crossover. Here less than 100% but greater than 50% of the progeny with parental combination. Incomplete linkage occurs when two genes show linkage with a recombination level greater than 0% and less than 50%. In incomplete linkage, all expected types of gametes are formed, but the recombinant gametes occur less often than the parental gametes.

Factors Affecting the Strength of Linkage

1. Distance between the genes: As the distance between genes increases, strength of linkage decreases.
2. Age: With increase in age, chances of crossing over decreases which results in the increase of linkage.
3. Temperature: Rise in temperature causes the chances of chiasmata formation. It decreases the strength of linkage.
4. X-rays: Strength of linkage decreases if genes are exposed to X-rays.

Crossing Over (Homologous Recombination)

The term *crossing over* was coined by T.H. Morgan. While linkage explains why parental gene combinations tend to remain together, it does not account for the origin of non-parental combinations. To explain this, Morgan postulated that during meiosis, homologous chromosomes undergo a physical exchange of segments, a process now known as crossing over, intrachromosomal recombination, or homologous recombination. Linkage is disrupted by crossing over during meiosis I, where portions of genetic material from one parental chromosome are exchanged with corresponding portions from the other, without any addition or deletion of base pairs. The probability of crossing over occurring between two loci increases with the distance between them on the chromosome.

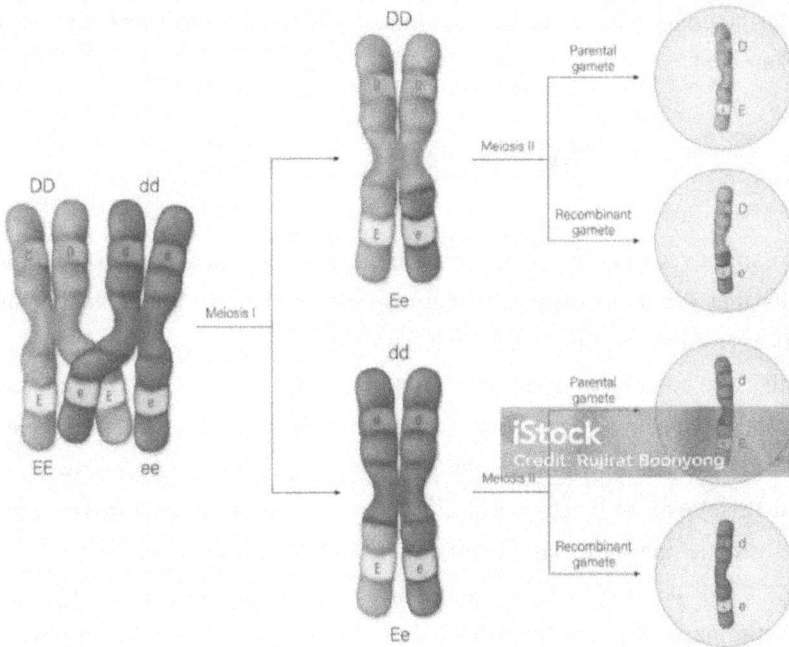

Image 6.1 Crossing over between two non-sister chromatids of the homologous chromosome (source: https://www.istockphoto.com/ Image ID 557894906.).

Crossing over takes place at the pachytene stage of prophase I in meiosis,

165

involving only two of the four chromatids in a homologous pair (bivalent). At this stage, the four chromatids are closely synapsed, allowing non-sister chromatids to exchange equal and corresponding segments of genetic material. This reciprocal exchange is mediated by the enzyme recombinase, which breaks and rejoins chromatids to produce new genetic combinations. The process is observed physically as a *chiasma* (plural *chiasmata*), a cross-shaped structure formed between non-sister chromatids.

When no crossing over occurs, alleles on each homologous chromosome remain in their original combinations. However, when crossing over does occur, the outermost alleles in two chromatids are recombined, producing genetic variation. Two key features of crossing over in eukaryotes are that it (1) results in a reciprocal exchange of genetic material between homologous chromosomes and (2) allows both products of a single exchange to be recovered in different progeny. Through this mechanism, crossing over contributes to genetic recombination and variation in both prokaryotes and eukaryotes.

Mitotic Recombination

Crossing over usually occurs during meiosis but it can also occur in diploid cells that reproduce asexually during mitosis. Mitotic recombination is a process in which a diploid cell undergoing mitosis gives rise to daughter cells with allele combinations different from that in the parental cell. As in meiotic recombination, mitotic recombination generally involves genetic exchange between chromatids of homologous chromosomes, but occurs less frequently than meiotic recombination. Mitotic recombination can give rise to tissues that are genetic mosaics. The frequency of recombinants correlates with the physical distance between the genes and may be used for the construction of genetic maps. In some fungi, like *Aspergillus niger*, this is the only way to construct genetic maps by recombination. It is presumed that mitotic crossing over occurs during mitotic prophase but the exact stage is not known. It is not known whether crossing over occurs during interphase before mitosis or during the transition from interphase to prophase or during the prophase.

Types of Crossing Over (Based on Cell Type)

1. Somatic or Mitotic Crossing Over
   - ❖ Occurs in the **chromosomes of somatic cells** during **mitotic cell division**.
   - ❖ A **very rare phenomenon**.
   - ❖ Has **no genetic significance**, as it does not contribute to gamete formation or variation.
2. Germinal or Meiotic Crossing Over
   - ❖ Occurs during **meiosis**, specifically in **Prophase I (pachytene stage)**.
   - ❖ **Genetically significant**, as it produces **new gene combinations** (recombinants) and increases **variation**.

Kinds of Crossing Over (Based on Number of Chiasmata)

1. Single Crossing Over
   - ❖ Crossing over occurs at **one point** on the homologous chromosome pair.
   - ❖ Results in:
        Two non-crossover chromatids (parental type)
        Two crossover chromatids (recombinants).
2. Double Crossing Over
   - ❖ Crossing over occurs at **two points** in the same chromosome pair.
   - ❖ Two possibilities exist:
a. Reciprocal Chiasma
   - ❖ The **same two chromatids** participate in both crossover events.
   - ❖ Produces **only two crossover chromatids** (the other two remain non-crossover).
b. Complementary Chiasma

- ❖ The chromatids involved in the **second crossover** are **different** from those in the first.
- ❖ Produces **four crossover chromatids**, as **all four chromatids of the bivalent** are involved.

3. Multiple Crossing Over
- ❖ Crossing over occurs at **three, four, or more points** in the same chromosome pair.
- ❖ Depending on the number of exchanges, it may be classified as:
  - Triple crossing over
  - Quadruple crossing over
  - Multiple crossing over

Mechanism/Process of Crossing Over

Crossing over is a complex meiotic event that ensures **exchange of genetic material** between homologous chromosomes. It occurs during **Prophase I of meiosis** and proceeds through the following stages:

Synapsis of Homologous Chromosomes (Zygotene Stage): In the zygotene stage, synapsis of homologous chromosomes occurs. Synapsis is the pairing of two homologous chromosomes, one maternal and one paternal, which are attracted to each other because they possess the same genes at identical loci. As they come closer together, they become connected by the synaptonemal complex. Although homologous chromosomes are initially separated at the onset of meiosis, the precise way in which they locate and pair with each other remains one of the unsolved mysteries of meiotic division.

Tetrad Formation and Actual Crossing Over (Pachytene Stage): In the pachytene stage, tetrad formation occurs and actual crossing over begins. Each homologous pair, now in the form of a tetrad, consists of four chromatids. At specific points, recombination nodules appear, marking the sites of crossing over. The two non-sister chromatids undergo breakage at corresponding sites through the action of the enzyme endonuclease. The broken chromosome segments are then exchanged between chromatids, and the enzyme ligase facilitates the fusion of these exchanged segments. This exchange results in the formation of a chiasma, where chromatids visibly cross each other. Crossing over may occur at one or multiple points depending on chromosome length, and in general, the greater the physical distance between genes, the higher the probability of a chiasma forming between them. Each species is also characterized by a typical number of chiasmata per chromosome.

Terminalization (Disjunction, Diplotene → Diakinesis): After the exchange, the process of terminalization or disjunction begins. During diplotene, homologous chromatids begin to repel one another and partial separation, or desynapsis, takes place. By diakinesis, chromosomes detach from the nuclear envelope, and each bivalent can be seen to consist of four chromatids—sister chromatids remain joined at their centromeres, while non-sister chromatids that have undergone crossing over remain linked at chiasmata. Separation of chromatids proceeds progressively from the centromere towards the chiasmata, and the chiasmata themselves shift toward the ends of the chromosomes in a zipper-like fashion. This process is called terminalization and marks the final step of crossing over.

Factors Affecting Crossing Over between the Genes

The frequency of crossing over is roughly proportional to the physical distance between genes, but a number of other genetic, environmental and physiological factors are known to affect the frequency of crossing over. The following are some of them:

Age of the organism. The age of the organism has an influence on the frequency of crossing over. In female **Drosophila**, it has been observed that the rate of crossing over generally decreases with advancement of age. This indicates that younger females exhibit higher recombination frequencies compared to older ones, suggesting that physiological and cellular changes associated with aging affect the efficiency of meiotic processes.Sex of the organism. The rate of crossing over also differs according to sex.

Temperature. The rate of crossing over in **Drosophila** is influenced by temperature. Studies have shown that recombination frequency increases when the temperature is either above or below the optimum level of about 22°C. This means that both higher and lower temperature conditions tend to enhance crossing over compared to the normal physiological temperature, indicating that environmental factors such as heat or cold stress can significantly affect the process of meiotic recombination.Radiations. Irradiation with X-rays and gamma rays was found to enhance the frequency of crossing over in Drosophila females.

Nutrition. The presence of metallic ions in the diet has been found to affect the rate of crossing over in **Drosophila**. Specifically, ions such as calcium and magnesium, when present in the food, cause a reduction in recombination frequency. This suggests that the availability of certain minerals can influence the cellular processes involved in meiosis, thereby altering the normal pattern of genetic recombination.Chemicals. Certain chemicals which induce chromosomal aberrations are also known to induce or suppress crossing over. Ethyl methane sulphonate is known to induce somatic crossing over. Colchicine also prevents cross over by preventing pairing between chromosomes. High dose of selenium also reduces crossing over frequency. Treatment with mutagenic chemicals like alkylating agents was found to increase the frequency of crossing over in Drosophila.

Physiological Condition: Health and metabolic state of the organism can influence recombination rates.

Interference. The frequency of crossing over is not uniform along the length of a chromosome. It is generally low in regions near the **centromere** and also towards the **telomeres**. These areas are often more tightly packed with heterochromatin, which restricts the formation of chiasmata and reduces the chances of recombination. In contrast, crossing over tends to occur more frequently in the interstitial or middle regions of chromosomes where euchromatin is abundant and genes are more actively expressed.

Structural chromosomal changes. Structural changes in chromosomes, such as **inversions** and **translocations**, can significantly affect the frequency of crossing over. These chromosomal rearrangements reduce recombination in the regions where they occur because the altered alignment of homologous chromosomes during meiosis interferes with the proper formation of chiasmata. As a result, crossing over is suppressed in the affected chromosomes, which can influence the inheritance patterns of genes located within or near the rearranged segments.

Significance of Crossing Over

Crossing over is universal in occurrence, occurs in plants, animals, bacteria, viruses and molds. The importance of crossing over is in the genetic diversity. Crossing over results in reshuffling of alleles on different chromosomes and breaks up allele combinations in linkage groups. Each chromosome that goes into a gamete is a combination of maternal and paternal chromosomes. The importance is that it creates vast numbers of possible combinations of alleles in the gametes and increases genetic variation in organisms, and, therefore, useful in breeding. Crossing over clearly provides direct evidence for linear arrangement of linked genes in chromosomes. Crossing over is a useful tool for locating genes in the chromosomes. The frequency of crossing over is the basis for constructing chromosome/linkage maps.

Parentals and Recombinants

Parentals (Parental Types). Gametes/progeny that have the same combinations of alleles as one or the other of the parents are called parentals (nonrecombinants).

Nonparentals (Recombinant Types). Progeny/gametes in which the new combinations of alleles are shown are called nonparentals or recombinants.

Consider a double heterozygote *Aa Bb* where the two genes involved are on different chromosomes (Figure 6.1). Assume that this individual has inherited the dominant alleles *A* and *B* from one parent, and recessive alleles *a* and *b* from the other. In meiosis, four genetically different gametes, *AB, Ab, aB*, and *ab* will be produced in equal proportions. Those gametes carrying *AB* or *ab* are referred to as parentals and those carrying *aB* and *Ab* as recombinants. The recombination frequency is obviously 50%. This is the maximum recombination frequency that can be obtained between any two genes, and genes that show this recombination frequency are said to be unlinked and assort independently of each other. Unlinked genes may be on different chromosomes, or so far apart on the same chromosome that they are often separated by recombination. When genetic analysis indicates that two genes show significantly less than 50 percent genetic recombination, the two genes are considered to be linked. The process by which the linkage status of genes is analyzed is known as linkage analysis.

P	AA BB		×		aa bb
Gametes	AB				ab
F$_1$			Aa Bb		
Gametes	AB, ab Parental type gametes			Ab, aB Recombinant type gametes	

Figure 6.1 Detection of recombination in diploid organisms.

Rate of Recombination

Rate of recombination is the proportion of recombinant offspring observed in a testcross, or in other crosses. In order to calculate the recombination frequency, we use the following formula:

$$\text{Recombination frequency} = \frac{\text{number of recombinant progeny}}{\text{total number of progeny}} \times 100\%$$

For example, in Table 6.2, 305 (151 + 154) of 2839 progeny were recombinant, making the rate of recombination 305/2839 = 0.107 or 10.7%. The rate of recombination estimates the linkage of genes and ranges from 0 (for very tightly linked loci) to 50% (for very loosely linked loci, or those on different chromosomes). If the two genes are very close to each other, the number of recombinant offspring or gametes should be close to zero. The maximum frequency of recombination between any two genes in a cross is 50 percent; this happens when the genes are in nonhomologous chromosomes and assort independently, or when the genes are sufficiently far apart in the same chromosome that at least one crossover is formed between them in every meiosis.

The Sign of Meiotic Recombination

There are two types of meiotic recombination:

Intrachromosomal recombination: Crossing over produces intrachromosomal recombination. The sign of intrachromosomal recombination is a recombinant frequency of less than 50 percent. If two genes are inherited together more than 50% of the time, this is evidence that they are linked on the same chromosome.

Interchromosomal recombination: Mendelian independent assortment brings about interchromosomal recombination and results in a recombinant frequency of 50 percent. The physical linkage of parental gene combinations prevents the independent assortment of two genes.

Interference and Coincidence

The crossing over percentage is highly modified by Interference and Coincidence. Crossing over takes place at chiasma. In most higher organisms it has been found that one chiasma formation reduces the probability of another chiasma formation in an immediately adjacent region of the chromosome, probably because of physical inability of the chromatids to bend back upon themselves within certain minimum distances. The term interference (i) was coined by Muller which refers to the tendency of one crossover to reduce the chance of another crossover in its adjacent region. The net result of this interference is the observation of fewer double crossover types than would be expected according to map distances. The strength of interference varies in different segments of the chromosome. Interference in quantified by first calculating a term called the **coefficient of coincidence** (c.o.c.) which was also coined by Muller to explain strength or degree of interference. The **coefficient of coincidence** is the ratio of observed double crossovers to the expected double crossovers. The greater the coincidence, lesser will be the interference and *vice versa*.

$$c.o.c. = \frac{observed\ double\ crossover\ frequency}{expected\ double\ crossover\ frequency}$$

and $i = 1 - c.o.c.$

When interference is complete (1.0), coincidence becomes zero. When interference decreases, coincidence increases. Coincidence values ordinarily vary between 0 and 1. There is no interference across centromere.

Interference values between 0 and 1 indicate that positive interference has occurred. Positive interference indicates that one crossover interferes with the occurrence of a second crossover nearby.

For closely linked genes, the measure of interference may often be near one, while for widely separated genes on the same chromosome, interference is often near zero. Interference values near zero suggest no interference; that is, different recombination events are independent of each other.

In certain microorganisms, particularly bacteriophages, coefficients of coincidence greater than 1 often occur. Coefficients of coincidence greater than 1 indicate that the occurrence of one crossover increases the likelihood of additional crossover occurring nearby. This second type of interference in called negative interference.

Gene Mapping

Gene mapping, also known as genome mapping, is the sequential allocation of loci to a relative position on a chromosome. Gene maps are species-specific and comprised of genomic markers and/or genes and the genetic distance between each marker and/or gene. Gene mapping offers critical information about the order, position, and distance between genes inside a genome.

Applications of gene mapping

Some of the important applications of gene mapping are as follows:

- ❖ Gene mapping is used to identify and locate genes responsible for specific traits in organisms.
- ❖ Gene mapping plays a crucial role in identifying and studying quantitative trait loci (QTL).

❖ Gene mapping is also used in animal agriculture to identify genes associated with economically important traits. For instance, mapping the genes responsible for milk production in dairy animals can help breeders select animals with higher milk yields, leading to improved productivity in the dairy industry.

❖ One of the significant applications of gene mapping is the identification of disease-causing genes.

Types of Gene Mapping

There are two main types of gene mapping: genetic mapping and physical mapping.

Genetic Mapping

Thomas Hunt Morgan made significant contributions to the modern understanding of chromosome mapping. In the year 1911, Alfred H. Sturtevant, an undergraduate student in Morgan's laboratory, conducted pioneering research on the model organism *Drosophila melanogaster*, which gave rise to the concept of gene mapping and created the first genetic map showing the locations of 6 sex-linked genes on a fruit fly chromosome. He established that genes are arranged in a linear manner on a chromosome, like beads on a necklace and determined that the distance between any two genes remains constant if they are on the same chromosome and the length of the chromosome is the same in all organisms of that species. Sturtevant also discovered that the frequency of crossing over may be used to locate genes on a chromosome.

Sturtevant concluded from his observations that genes that are adjacent to each other on a chromosome are linked and are more likely to inherit together, but genes that are far apart from each other are less likely to be inherited together and thus are not linked. The smaller the distance the more likely two genes will be inherited together. The map constructed by Sturtevant is called the linkage map. Sturtevant's linkage map served as the foundation for subsequent gene mapping studies, including the creation of the first human genome map. The first genetic map of the human genome, known as RFLPs (restriction fragment length polymorphisms), was published in 1987, utilising 400 distinct restriction enzyme markers.

Each gene has definite order and location on a particular chromosome. A genetic map gives the order of genes on a chromosome and the relative distance between them based on recombination frequencies observed in genetic crosses. Genetic maps are like road maps that show the relative locations of towns along a road; that is, they show the arrangements of genes along the chromosomes and the genetic distances between the genes. The method of construction of genetic maps of different chromosomes is called genetic/linkage/chromosome/crossover/recombination/meiotic mapping. Linkage maps can only be constructed for polymorphic genes; monomorphic loci cannot be mapped in this fashion. Genetic maps provide information about which chromosomes contain specific genes and precisely where the genes lie on that chromosome.

The percentage of crossing over is roughly proportional to distance between the genes. Morgan and Alfred Sturtevant suggested that the percentage of recombinants could be used as a quantitative measure of the genetic distance between two loci on a genetic map. The unit of distance in a genetic map is called a **map unit (mu)** or **centimorgan (cM)**, in honor of Thomas Hunt Morgan by his undergraduate student, Sturtevant. One map unit is equal to 1 percent recombination. For example, two genes that recombine with a frequency of 3.5 percent are said to be located 3.5 map units apart.

In fact, later studies have shown that genetic distance measured by this statistical approach is generally similar to cytologically or biochemically measured distances on a chromosome (physical distance). The distance between genes as measured by recombination frequencies is not a precise measure of the physical distance between genes because the frequency of crossing over varies in different parts of the genome. It tends to be lower near centromere and higher near telomeres. It can also differ between males and females. However, gene maps based on linkage data give a reasonably accurate indication of distance between genes and they are most useful in determining the order of genes along the chromosome. Physical maps provide distance between genes in absolute terms.

Genetic map is usually a first step in the identification and isolation of a new gene and the determination of its DNA sequence. Genetic mapping is essential in human genetics for the identification of genes associated with hereditary diseases. It helps breeding programs to know what genes are closely linked. It helps understand evolution on blocks of genes.

The technique of genetic mapping is used to determine the relative positions of genes or DNA sequences on a chromosome or genome. It differs from physical mapping, which directly examines the DNA sequence related to a phenotype or disease. Using techniques such as pedigree analysis or cross-hybridization, the genetic mapping method shows the placements of related genes or DNA sequences associated with a phenotypic trait or disease.

Importance of Chromosome Mapping

1. Chromosome mapping is important for understanding the organization and overall complexity of a particular species.
2. Chromosome mapping is important for identifying the precise positions of genes on chromosomes, which can help to understand gene interactions.
3. Chromosome mapping is also useful to understand the structure and function of genes.
4. Chromosome mapping can be used to find and sequence disease genes, learn what they encode, and develop targeted therapies to treat genetic disease.
5. Chromosome mapping is also useful for understanding evolutionary relationships among different species.
6. Genetic maps allow us to estimate relative distances between linked genes.

Process of Genetic Mapping

The genetic mapping includes following processes/steps:

Determination of Linkage Groups

Before beginning any experiments to construct genetic maps of different chromosomes of a species, one has to know the exact number of chromosomes of that species and then, he has to determine the total number of genes of that species by undergoing hybridization experiments in between wild and mutant strains. Then it is necessary to show that genes under consideration are linked, and thus, the different linkage groups of a species can be worked out. The method used to test for linkage is the testcross, a cross of one individual with an unknown genotype with another individual homozygous recessive for all genes involved. In this case, the distribution of phenotypes is the result of segregation events in only one of the two parents; the other parent contributes only recessive alleles to the progeny and those alleles do not contribute to the phenotype of the progeny. If the two genes are unlinked, then a testcross between $a^+a\ b^+b$ and $aa\ bb$ should result in a $1:1:1:1$ ratio of the four phenotypic classes. A significant deviation from this ratio in the direction of too many parental types and too few recombinant types would suggest that the two genes are linked. Strictly speaking, genes are said to be linked whenever over 50% of the gametes produced contain parental combinations of the genes and less than 50% of the gametes contain recombinant combinations of the genes.

P	$a^+a\ b^+b$	×	$aa\ bb$
Gametes	$a^+b^+,\ a^+b,\ ab^+,\ ab$		ab
Progeny	$a^+a\ b^+b,\ a^+a\ bb,\ aa\ b^+b,\ aa\ bb$ $1:1:1:1$ testcross ratio		

Determination of map distance

The map distance between two linked genes is then calculated based on the frequency of recombination between the two genes. Crossing over frequencies can be converted into map units, *e.g.*, a 5% crossing over frequency equals 5 map units.

Determination of gene order

After determining the relative distances between the genes of a linkage group, it becomes easy to place genes in their proper linear order.

Combining map segments

Finally, the different segments of maps of a complete chromosome are combined to form a complete genetic map.

The easiest way to map genes is to compare them in groups of 3. This allows both the distances between them and their order to be determined. Further genes can be added to the map by using overlapping groups of 3. In an experiment investigating two characteristics A and B it was found that their recombinant frequency was 6.0 percent. In further experiments it was found that the recombinant frequency for characteristics A and C was 18.5 percent and the recombinant frequency between B and C was 12.5 percent. A genetic map of the three genes responsible for these characteristics may be constructed as follows. The values are in centimorgans.

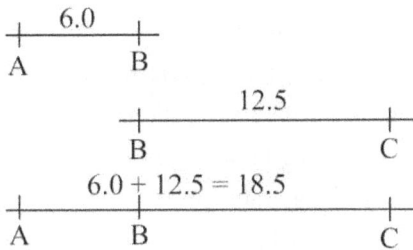

```
   6.0
+--------+
A        B

        12.5
        +-------------+
        B             C
        6.0 + 12.5 = 18.5
+-------+-------------+
A       B             C
```

Gene maps can be created by using the information obtained through a series of test crosses, whereby one of the parents is heterozygous for a different pair of genes and we can calculate the recombination frequencies between pairs of genes. A test cross between two genes is called a **two-point test cross**. First, we will look at **two-point test crosses**, then we will investigate **three-point test crosses**, which are generally more accurate.

Limitations of Genetic Mapping

❖ Genetic mapping does not provide information about the physical distance between genes as it provides only the outline.

❖ Recombination levels can differ across various parts of the genome. Some parts exhibit high rates of recombination, known as recombination hotspots, which can impact the accuracy of genetic mapping.

❖ In organisms where obtaining a large number of progenies is difficult, the resolution of a genetic map is limited because it depends on the number of observed crossovers.

So, it is important to consider mapping information generated using other methods as well. Various physical mapping methods have been developed to address these limitations.

Physical Mapping

Physical mapping is the determination of the actual location of genes on a chromosome in which distances are in physical units of DNA base pairs (bp), kilobase pairs (kb) or megabase pairs (mb). Physical maps show the actual (physical) distance between loci and can be constructed using cytogenetic, radiation hybrid, or sequence mapping. Physical mapping techniques include DNA sequencing, restriction mapping, and hybridization techniques. These methods provide information about the distances between genes and their precise locations on the chromosome. Physical maps are typically represented as linear DNA sequences, where genes are ordered and spaced according to their physical positions. It is one gene mapping approach that has a high degree of accuracy in determining the sequence of DNA base pairs. Physical maps provide a direct representation of the physical structure of a chromosome and the positions of genes along its DNA sequence. Physical mapping is a gene mapping technique that determines the exact location of a gene, disease gene, or DNA sequence of interest. It provides precise information about the physical distances between genetic markers and the actual DNA base pair distances.

Both genetic mapping and physical mapping are important for understanding the organization and structure of genes within a genome. They complement each other, with genetic mapping providing information on the order of genes based on recombination events, and physical mapping providing the precise locations and distances between genes.

Methods of Physical Mapping

There are several methods used in physical mapping, and three commonly employed techniques are:

1. Restriction mapping
2. Fluorescent *in situ* hybridisation (FISH) mapping
3. Sequence tagged site (STS) mapping

Restriction Mapping

Restriction mapping is a physical mapping method used to identify the positions of restriction sites on a DNA fragment. This method utilizes specific restriction endonucleases to cut DNA at its specific restriction site. For example, the restriction enzyme *Eco*RI always cuts at the sequence GAATTC/CTTAAG. If we use *Eco*RI to cut the DNA we know that the DNA sequence either side of the cut will be AATT. The restriction map indicates or locates all the restriction sites from the entire genome or from the piece of DNA. A physical map is generated by aligning the different restriction maps along the chromosomes. The DNA is digested with these enzymes, and the resulting fragments are separated by size using gel electrophoresis, either agarose gel electrophoresis or polyacrylamide gel electrophoresis (PAGE). The sizes of the fragments provide information about the distance between restriction enzyme sites on the DNA.

This technique produces a restriction map that indicates or locates all of the restriction sites present in the entire genome or a specific fragment of DNA. It contains useful information about the placement and distribution of restriction sites.

Restriction mapping is frequently used to map genes on plasmids or shorter segments of DNA. However, mapping the complete genome by restriction method is a time-consuming, labor-intensive, and inaccurate process.

It is important to note, however, that the current mapping method is only applicable to polymorphic restriction sites. Only a tiny number of locations in the genome are highly polymorphic, making it difficult to create a thorough map of the complete genome using RFLP analysis alone.

The restriction mapping method is further divided into two methods:
1. Fingerprint gene mapping
2. Optical gene mapping

Fingerprint mapping. In fingerprint mapping, the genome is broken into fragments, which are then copied in bacteria cells. The DNA copies are then cut by restriction enzymes and the lengths of the resulting fragments are estimated using electrophoresis. Electrophoresis separates the fragments of DNA according to size resulting in a distinct banding pattern. The fingerprint map is constructed by comparing the patterns from all the fragments of DNA to find areas of similarity. Those with similar patterns are then grouped together to form a map.

Optical mapping. Optical mapping is a method to find the locations of a specific base-pair sequence and its reoccurrences on particular single DNA molecules using fluorescence labeling and light microscopy. Optical mapping uses single molecules of DNA that are stretched and held in place on a slide. Restriction enzymes are added to cut the DNA at specific points leaving gaps behind. The fragments are then stained with dye and the gaps are visualised under a fluorescence microscope. The intensity of the fluorescence is used to construct an optical map of single molecules. These can then be combined and overlapped to give a global overview of the genome and aid with assembling a sequenced genome.

Fluorescence *in situ* Hybridization (FISH)

Fluorescence in situ hybridization (FISH) mapping is a technique that uses fluorescent probes to detect the location of specific DNA sequences or disease genes within cells by using fluorescently labeled DNA probes. FISH artificially synthesizes a short sequence of single-stranded complementary DNA to the target DNA sequence or gene. A fluorescent dye is used to label this synthetic DNA sequence, known as a probe. The probe is precisely engineered to hybridize with the complementary sequence on the chromosome. If a labeled probe is applied to cells or tissue samples and allowed to hybridize, it will attach to its corresponding sequence if it exists on a chromosome. This binding produces a fluorescent signal, which can be seen with a fluorescent microscope. The presence and location of the DNA sequence or gene of interest on the chromosome are indicated by the fluorescent signal.

FISH has played an important role in genetic research and diagnostics, allowing the discovery of chromosomal aberrations, gene rearrangements, and disease-related gene localization. In clinical settings, commercially available probes specific to various disease-related loci are increasingly frequently used.

Sequence Tagged Site (STS) Mapping

STS mapping is a powerful gene mapping technique for creating detailed physical maps of large genomes that analyses DNA segments containing highly polymorphic STS regions using PCR amplification and agarose gel electrophoresis. An STS is a short DNA sequences of 100 to 500bp repeatedly present within a genome. One of the most distinguishing features of STS is their high polymorphism, which implies that they vary between individuals or populations. Because STS sections are polymorphic, they can be easily recognised and used as genetic markers. By determining the presence or absence of specific STS markers in a genome, researchers can establish their positions and create a physical map.

The DNA sample of interest is first fragmented using restriction enzymes (RE) and then introduced into a bacterial plasmid. Within the bacterial cells, enough copies of the DNA fragments are produced.

Once enough DNA copies have been recovered, the fragments containing the STS sections are gathered and PCR amplification is performed. The DNA sample was amplified using the known primers specific to the STS regions. This produces PCR amplicons, which are copies of the STS-containing DNA fragments. Using agarose gel electrophoresis, the PCR amplicons are separated and analysed. The amplicons are placed onto a gel matrix and exposed to an electric field in this phase. The fragments migrate through the gel according to their size, allowing them to be separated.

The results of agarose gel electrophoresis can provide vital information on the genome's organisation and the relationships between distinct DNA fragments. If two separate STS markers are found on the same DNA fragment, it indicates that they are physically adjacent to each other in the genome. Similarly, the presence of the same STS marker in two separate DNA fragments shows that those pieces overlap and represent contiguous areas of the genome.

STS mapping reveals information about the physical layout of DNA sequences in the genome. The presence or absence of certain STS markers in distinct DNA fragments allows researchers to establish their relative positions and create a genome map.

CHAPTER SEVEN   SEX DETERMINATION AND SEX RELATED TRAITS

The word sex was derived from the Latin *sexus* meaning section or separation. Sex is one of the fundamental qualities which we recognize in living things, particularly in the higher forms of life. It is the property or quality by which organisms are classified as female or male especially on the basis of their reproductive organs and structures. For most diploid eukaryotes, sexual reproduction is the only mechanism resulting in new members of a species. Meiosis in the gonads of parents produces haploid gametes, which unite during fertilization to restore the diploid state in the offspring. For most organisms, sexual reproduction requires some form of sexual differentiation. Sex determination is the process by which sexually reproducing organisms initiate the pathway to become either male or female. In mammals, sex determination is classically divided into three hierarchical, irreversible stages. The first step is the establishment of chromosomal sex at fertilization. The second is the choice of the bipotential gonad to become an ovary or a testis. The third step is the development of the sexual phenotype of the fetus. Sex differentiation is defined as the phenotypic development of structures consequent upon the action of hormones produced following gonadal determination. In higher forms of life, this is manifested as phenotypic dimorphism between males and females of a species. Traditionally, the symbol ♂ designates male and the symbol ♀ designates female.

In the past, literally hundreds of unscientific theories have attempted to explain sex determination by relating it to phases of the moon, the time of day of fertilization, and other factors. Actually, there is no universal mode of sex determination. Modern geneticists have reported many different mechanisms of sex determination in living organisms. Some important mechanisms of sex determination are following:

1. Genetically controlled sex determining mechanisms.
2. Metabolically controlled sex determining mechanisms.
3. Hormonally controlled sex determining mechanisms.
4. Environmentally controlled sex determining mechanisms.

Genetically Controlled Sex Determining Mechanisms

Sex is determined at fertilization by the combination of genes that the zygote receives. Most of the mechanisms of sex determination are under genetic control and they may be classified into following categories:

1. Sex chromosomal mechanism or heterogamesis.
2. Genic balance mechanism.
3. Male haploidy or haplodiploidy mechanism.
4. Single gene effect.

Sex Chromosomes

The chromosomes in an organism that determine the sex of an individual are called sex chromosomes. The sex chromosomes are of unequal size, shape, and/or staining quality hence are heteromorphic. They carry sex-determining genes. Sex chromosomes are responsible for the determination of separate sexes of individuals during reproduction in some plants, some insects, and in almost all animals. The sex chromosomes are distinct from autosomes in that they differ between males and females in humans and other mammals. They differ from autosomes in form, size, and behavior.

The sex chromosomes typically designated X and Y (sometimes Z and W) chromosomes (Image 7.1). In humans and many other mammals, the sex chromosomes are known as the X and Y chromosomes. The Y chromosome contains a large block of heterochromatin on its q arm comprising noncoding repetitive DNA. This leaves only a short segment of chromosome capable of carrying functional genes. Human body cells have 46 chromosomes: 22 homologous pairs of autosomes plus 2 sex chromosomes. Females have a pair of identical sex chromosomes called the **X chromosomes**. Males have a nonidentical pair, consisting of one X and one Y. The Y chromosome is one of the smallest chromosomes with the least number of genes of any chromosome. Hence, if we let A represent autosomal chromosomes, we can write: females = 44A + XX, and males = 44A + XY

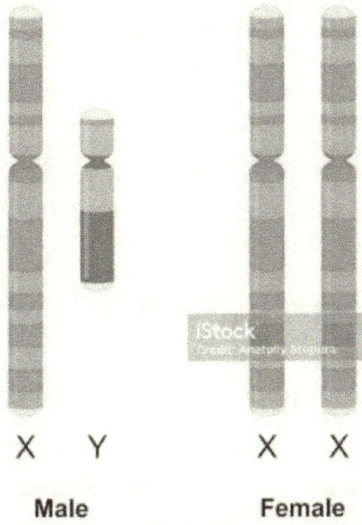

Image 7.1 Sex chromosomes of mammals (source: https://www.istockphoto.com/vector/human-sex-chromosomes-gm1495109323-518230632?searchscope=image%2Cfilm).

At meiosis in females, the two X chromosomes pair and segregate like autosomes, and so each egg receives one X chromosome. Hence, with regard to sex chromosomes, the gametes are of only one type. The female sex, because, produces same types of (identical) gametes with respect to sex chromosome, is called homogametic sex. In case of males there are two types of sperm with respect to sex chromosome, half with an X and the other half with a Y. Such a sex which produces two different types of gametes in terms of sex chromosomes is called heterogametic sex.

In humans, the X chromosome is much larger and more gene-rich than the Y chromosome. The X chromosome is 2.6 times the size of the Y chromosome, containing 155.27 Mb or 5% of the total genome of 3.10 billion base pairs (bp), whereas the Y chromosome contains 59.37 Mb or 1.9% of the total genome. There are 1965 genes in an X chromosome compared with 421 genes in the Y chromosome (Xu Li, 2011). We need to emphasize that the chromosomes themselves are not determining sex, but the genes they carry are. In general, the genotype determines the type of gonad that then determines the phenotype of the organism through male or female hormone production.

Autosomes

Any chromosome other than the sex chromosomes. Humans have 22 pairs of homologous autosomes. Autosomes appear in pairs whose members have the same form but differ from other pairs in a diploid cell.

Sex Chromosomal Mechanism or Heterogamesis

Chromosome number in every species is generally constant. In higher organisms, the number of chromosomes in somatic cell is called somatic number irrespective of the ploidy level and is represented by "2n". In the gametes, the chromosome number is reduced to half, it is known as gametic number or haploid number and is represented by "n". The basic chromosome number is denoted by "x", so that the chromosome number of a diploid cell or individual is expressed as "2x". However, this designation is generally used in polyploid species.

Many eukaryotes (most higher animals and many higher plants) have two copies of each type of chromosome in their nuclei, so their chromosome complement is said to be diploid (2n = 2x). Diploid eukaryotes are produced by the fusion of two gametes, one from the female parent and one from the male parent. Each gamete has only one set of chromosomes and is said to be haploid (n). There is wide variation in the chromosome numbers of different plant and animal species. Chromosome number of some species are given in Table 7.1.

Table 7.1 Chromosome number in different species

Species	Number
Animals	
Aulacantha sp, Protozoan, microscopic unicellular eukaryotes	800 pairs (The organism with the highest number of chromosome pairs)
Hermit crabs	127 pairs (The highest number of chromosome pairs observed in animals)
South America red vizcacha rat, *Tympanoctomys barrerae*	51 pairs (The highest number of chromosome pairs observed in mammals)
Ring-necked pheasant	41 pairs
Domestic duck, Muscovy duck, Goose, Pigeon, Emu, Ostrich, Turkey	40 pairs
Dog, Dingo, Chicken, Japanese quail, Dove, Gray wolf, Golden jackal	39 pairs
Camel, Llamas, Guinea fowl, Sun bear, Sloth bear, Polar bear, Brown bear, Asian black bear, Bush dog	37 pairs
Reindeer, White-tailed deer	35 pairs
Red Deer	34 pairs
Gray fox	33 pairs
Horse, Guinea pig, Fennec fox	32 pairs
Donkey	31 pairs
Cattle, Bison, Yak, Goat, Bengal fox	30 pairs
Elephant, Gaur, Mulberry silkworm	28 pairs
Sheep, Wild silkworm	27 pairs
African/Cape buffalo	26 pairs
River Buffalo	25 pairs
Swamp Buffalo, Chimpanzee, Gorilla, Orangutan, Deer mouse	24 pairs
Human, Nilgai	23 pairs

Rabbit, Golden hamster, Syrian hamster, Dolphin	22 pairs
Rhesus monkey, Rat, Giant panda	21 pairs
Mouse, Hyena	20 pairs
Domestic pig, Tiger, Lion, Foxes, Domestic cat, Cobra	19 pairs
Red panda, Starfish, Yellow mongoose	18 pairs
Red fox	17 pairs
Alligator	16 pairs
Giraffe	15 pairs
Edible frog	13 pairs
Snail, Grasshopper	12 pairs
Kangaroo	8 pairs
House fly	6 pairs
Drosophila	4 pairs
Yellowfever mosquito	3 pairs
Horse round worm, *Ascaris megalocephala univalens*	1 pair (Organism with the lowest number of chromosome pair)
Australian ant, Jack jumper (*Myrmecia pilosula*) 2 for females, males are haploid and thus have 1;	2/1 (Organism with the least number of chromosomes. Other ant species have more chromosomes)
Honeybee (*Apis spp*), European fire ants (*Myrmica rubra*)	32/16
Fishes	
Grass carp and Silver carp	52 pairs
Common carp	50 pairs
Catla catla (Catla), *Labeo rohita* (Rohu), *Cirrhinus mrigala* (Mrigal), *Labeo calbasu*	25 pairs
Haplopappus gracilis (family Compositae)	2 pairs

Plants	
Ophioglossum reticulatum (a fern)	630 pairs
Sugar cane	40 pairs
Cotton	26 pairs
Pineapple	25 pairs
Tobacco, Potato	24 pairs
Oats, Wheat	21 pairs
Sorghum, Soybean, Mango, Peanut	20 pairs
Grape	19 pairs
Cassava	18 pairs
Sunflower	17 pairs
Cultivated Alfalfa (2n = 4x)	16 pairs
Rice, Tomato, Melon, Sweet chestnut	12 pairs
Guava, Bean, Eucalyptus	11 pairs
Maize	10 pairs
Papaya, Sugar beet, Cabbage, Carrot, Radish, Citrus, Passion fruit	9 pairs
Onion, Garlic	8 pairs
Pea, Barley, Rye, Aloe vera, Cucumber	7 pairs
Spinach, Broad bean	6 pairs
Hieracium	4 pairs
Haplopappus gracilis, a flowering plant	2 pairs
Haploid Organisms	
Baker's yeast (*Saccharomyces cerevisiae*)	Diploid cells have 32 chromosomes, haploid 16
Green/black bread mold (*Aspergillus nidulans*)	8 haploid
Pink bread mold (*Neurospora crassa*)	7 haploid

The organism with the lowest number of chromosomes is the horse roundworm, *Ascaris megalocephala univalens*, which has only two chromosomes in its somatic cells (2n = 2). In contrast, the organism with the highest number of chromosomes is the protozoan *Aulocantha* sp., which possesses 1600 chromosomes in its somatic cells (2n = 1600). Within the plant kingdom, the fern *Ophioglossum reticulatum* has the highest number of chromosomes (2n = 1260), while the flowering plant *Haplopappus gracilis* has the lowest (2n = 4). Among animals, the harmit crab has the highest chromosome number, with 254 chromosomes (2n = 254).

Types of Sex Chromosomal Mechanism of Sex Determination

Sex chromosomal mechanism of sex determination varies with the organism. In dioecious diploidic organisms following two systems of sex chromosomal determination of sex have been recognized:

1. Heterogametic males.
2. Heterogametic females.

Heterogametic Males

In this system, the female sex has two X chromosomes and are homogametic, while the male sex has only one X chromosome. Because, male lacks an X chromosome, therefore during gametogenesis produces two types of gametes, 50% gametes carry the X chromosomes, while the rest 50% gametes lack in X chromosome. Therefore, the male is **heterogametic**. The sex of the offspring depends upon the kind of sperm that fertilizes the egg. The heterogametic males may be of following two types:

XX-XO mode of sex determination. This is found in many insect species specially those of the orders Hemiptera (true bugs) and Orthoptera (grasshoppers and roaches). Females contain two X chromosomes. Males have only one X chromosome. The male produces two types of sperms: half with X chromosome and half without X chromosome. The presence of only one X chromosome in the zygote determines the male offspring, and presence of two X chromosomes in the zygote determines female offspring.

XX-XY mode of sex determination. This is found in mammals, certain insects including Drosophila and in some dioecious plants. Here the females have two copies of the X chromosome and males have an X and a Y chromosome. Female gametes all have an X chromosome. The males produce two kinds of sperms: half with an X chromosome and half with a Y chromosome. An egg fertilized by an X-bearing sperm produces a female, but, if fertilized by a Y-bearing sperm, a male is produced.

Heterogametic Females (ZW-ZZ system)

This is essentially the reverse of the XX-XY system, where the female is ZW and the male ZZ. It is found in birds, most reptiles, amphibians, some fishes, some plants and insects (moths, butterflies and silkworm). Here the female is heterogametic and produces two types of ova, 50% ova carry the Z chromosomes, while rest 50% ova carry W chromosomes. The male sex is homogametic and produces single type of sperms, each carries a Z chromosome. The sex of the offspring depends on the kind of egg, the Z-bearing eggs produces males but the W-bearing eggs produces females.

Genic Balance Mechanism

Both Drosophila and mammals have the XX-XY method of sex determination. In 1916, C.B. Bridges reported that in Drosophila, sex is determined by the ratio of the number of X chromosomes to the number of sets of autosomes (X/A ratio). When the ratio is 1.0 or greater flies are female. When it is 0.5 or less flies are male. Intermediate values give rise to intersex flies. Intersexes are sterile individuals and have phenotype in between male and female sexes. Extreme ratios such as 0.33 and 1.5 give rise to flies that are called metamales and metafemales, respectively. Metamales, or supermales are to the male sex as metafemales, or superfemales are to the female sex. They are weak, sterile, underdeveloped, and die early. Some typical examples are given in Table 7.2. In this system, the Y chromosome has no effect on sex determination, but is required for male fertility. A gene on the Y chromosome is essential for the development of functional sperm. Based on these findings, C.B. Bridges proposed that, overall, the X chromosomes were female-determining and the autosomes were male-determining.

Table 7.2 Ratio of X chromosomes to haploid sets of autosomes, and sex determination in Drosophila

Number of X chromosomes (X)	Number of sets of autosomes (A)	X : A ratio	Sex
3	2	1.50	Metafemale
3	3	1.00	Female
2	2	1.00	Female
2	3	0.67	Intersex
1	2	0.50	Male
1	3	0.33	Metamale

Chromosomal Determination of Sex in Drosophila and Mammals

In both Drosophila and mammals including humans, normal females have an XX sex chromosome composition and normal males have an XY sex chromosome composition. However, the mechanism of sex determination in mammals differs from that in Drosophila.

Table 7.3 Chromosomal determination of sex in Drosophila and humans

Species	Sex chromosomes			
	XX	XY	XXY	XO
Drosophila	♀	♂	♀	♂
Humans	♀	♂	♂	♀

In insects, it is the number of X chromosomes that determines sex, with two X's giving a female and one X giving a male, but in mammals it is the presence of the Y that determines maleness and the absence of Y that determines femaleness. This difference is demonstrated by the sexes of the abnormal chromosome types XXY and XO, as shown in Table 7.3.

Sex Determination in Mammals

In mammals, the determination of sex is strictly chromosomal and is not usually influenced by the environment. In humans, the X and Y chromosomes harbor dramatically different numbers and sets of genes. The vast majority of genes on the sex chromosomes are not directly involved in sex determination. The embryonic gonad has the unique ability to differentiate into two completely different organs: testis or ovary. Gonadal differentiation activates one pathway and represses the other, making both developmental processes mutually exclusive. The development of male phenotype requires first and foremost the presence of an active SRY (Sex-determining Region on the male-limited Y chromosome) gene. It has been found only in placental mammals and marsupials (pouched mammals, kangaroo). It is located in the distal region of the short (p) arm of the Y chromosome close to the pseudoautosomal boundary, specifically at position Yp11.2. It is the master gene for male sex determination. If a developing embryo has a functional SRY gene in its cells, it will develop as a male. If there is no functional SRY, the embryo develops as female. SRY gene encodes testis determining factor (TDF) and a cell surface receptor. TDF is a protein which directs the medulla of the embryonic gonads to develop into testes, the primary male reproductive organs. Male and female human embryos develop identically during the first phase of gestation. About 4-6 weeks after fertilization, an embryo that contains the SRY gene develops testes. In the early embryo, two duct systems form. After the gonad differentiates into a testis or ovary, one set of ducts develops further while the other set degenerates. The newly forming testes start to produce two male hormones testosterone and Mullerian Inhibiting Substance/Factor (MIS) (Anti-Mullerian hormone/factor (AMH/AMF) or Müllerian regression hormone) that are responsible for the formation of the male phenotype. Testosterone plays a crucial role in the development of male reproductive organs and secondary sexual characteristics. Testosterone is produced by Leydig cells which causes the male-specific Wolffian ducts to develop into male ductal structures: epididymis, vas deferens, and seminal vesicles. It also promotes the formation of associated sex glands (*e.g.*, prostate) and masculinization of the external genitalia, as well as inhibits the development of the breast premordia. MIS synthesized in Sertoli cells causes regression of the female-specific Müllerian duct. In the absence of testosterone, the Wolffian ducts regress (except a bit becomes the adrenal glands in both sexes). In the absence of MIS, the Mullerian ducts develop into female ductal structures (fallopian tubes, uterus, cervix and upper/proximal vagina). The lack of TDF allows the cortex of the embryonic gonads to develop into ovaries. If ovaries (or no gonads) are present the final phenotype is female; thus, no gonadal hormones are required for female development during embryogenesis. In mammals, the default pathway of sexual development is female, *i.e.*, an embryo will develop as a female unless chemical signals are present that indicate it should develop as a male. Male development in mammals is directed at every step. If there is a loss of direction, the subsequent development will follow the female pathway.

Although the SRY gene is usually on the Y chromosome, it occasionally gets transferred to the X which leads to 46, XX males. Also, sometimes the SRY gene is inactivated by mutation leading to 46, XY females (Swyer syndrome, these individuals usually have a uterus and fallopian tubes, but their gonads (ovaries or testes) are not functional). It is also possible to have a partially inactive SRY gene, leading to ambiguous genitalia. Translocation of part of the Y chromosome containing this gene to the X chromosome causes XX male syndrome.

The Ratio of males to females in humans is not 1.0. **Primary sex ratio** reflects the proportion of males to females conceived in a population. **Secondary sex ratio** reflects the proportion of each sex that is born.

Sex Determination in Birds

Sex is determined genetically in chickens and others birds by the inheritance of sex chromosomes. These chromosomes differ from those in mammals. In birds, the sex-determining chromosomes are called Z and W. The male has two Z chromosomes while the female has one Z chromosome and one W chromosome. Sex determination involves the expression of genes in embryonic gonads to direct testis or ovary development. The gonads then release hormones that masculinise or feminise embryos. No convincing master female gene (ovary-determining gene) has been identified on the W chromosome, which is unique to females. Several lines of evidence suggest that the Z-linked DMRT1 (doublesex and mab-3-related transcription factor 1) gene regulates avian sex determination via a gene dosage mechanism such that males (ZZ) have two doses, while females (ZW) have one. DMRT1 gene is the master regulator of testis development in the chicken. Birds do not have a chromosome-wide dosage compensation mechanism, as seen in mammals, and DMRT1 is more highly expressed in embryonic male chicken gonads compared to females, prior to and during gonadal sex differentiation.

Male Haploidy or Haplodiploidy Mechanism

This is found in Hymenoptera (bees, ants, wasps and sawflies), Coccidia, and the Thysanoptera. A wasp is any insect which is neither a bee nor an ant. It also happens in reptiles and birds, and fish and amphibians. The system also occurs sporadically in some spider mites, Hemiptera, Coleoptera (bark beetles), and rotifers. *Heteronotia binoei*, a species of lizard of the family Gekkonidae in Australia has parthenogenetic populations with a wide geographic range. In this system, males develop through *parthenogenesis* (from the Greek *parthenos*, "virgin," and *genesis*, "creation") from unfertilized eggs and are haploid. Females originate from fertilized egg and are diploid. Because normal males are haploid and normal females are diploid, this mechanism of sex determination is often referred to as haplodiploidy.

Worker and queen honey bees are diploid females, with 32 chromosomes. Drones, on the other hand, are males with 16 chromosomes. The sex ratio of the offspring is under the control of queen. Those eggs which the queen chooses not to fertilize will develop into fertile haploid males. Most of the eggs laid in the hive will be fertilized and developed into females. During investigation, it has been found that the quality and quantity of food available to the diploid larva determines whether that female will become a sterile worker, or a fertile queen. Thus, environment here determines sterility or fertility but does not alter the genetically determined sex.

A queen bee reaches sexual maturity when she is 5-10 days old, at which point she initiates mating flights. She will take one to three mating flights on subsequent days, and mate with an average of 12 drones throughout the course of these flights (Tarpy *et al.*, 2004). Once the queen completes the mating process, she will initiate egg-laying behavior within a few days, and then will never mate again during her lifespan. The young queen stores up to 6 million sperm from multiple drones in her spermatheca. She will selectively release sperm for the remaining 1-7 years of her life.

Bees, ants, wasps, and sawflies, all members of the Hymenoptera order, exhibit diverse chromosome numbers, with males typically being haploid and females diploid. Sawflies, for example, have haploid chromosome numbers (n) ranging from 5 to 22, with most species falling between n=7 and 10. Ants show even greater variation, with haploid numbers ranging from n=1 to n=60. Wasps also display a range of chromosome numbers, with some species having n=9 and others n=34. To date, around 200 bee species have already been karyotyped, with chromosome numbers varying from n = 3 to n = 28.

Single Gene Effect

Many species, particularly eukaryotic microorganisms, do not have sex chromosomes. Here the sexes are specified by simple allelic differences at a small number of loci. For example, the yeast *Saccharomyces cerevisiae* is a haploid eukaryote that has two 'sexes' – a and α, referred to as mating types. The mating types are morphologically indistinguishable, but crosses can occur only between individuals of opposite type. These mating types are controlled by the MATa and MATα alleles, respectively, of a single gene.

In Drosophila, there is an autosomal recessive allele, tra (transformer), which when homozygous in XX individuals causes them to be males, not females as expected. The heterozygotes (tra^+/tra) or homozygous for the wild type allele (tra^+/tra^+) have their sex determined by the genic balance system.

Environmental Sex Determination Systems

Environmental sex determination (ESD) is the establishment of sex by non-genetic (environmental) factors. In many unicellular and some multicellular species, males and females have identical genomes and their sex is determined by environmental factors. ESD is common among unicellular eukaryotes (Beukeboom and Perrin, 2014). Among multicellular organism ESD is found primarily among non-avian reptiles, amphibians, and some fish (Bachtrog *et al.*, 2014). Individuals exhibiting environmental sex determination lack genetic information encoding separate sexes, such as sex chromosomes, and carry genetic information encoding both sexes on autosomes. Generally, when exposed to specific environmental stimuli, epigenetic changes make the developing individual either male or female. Environmental cues that trigger male or female development include temperature, nutrients, water availability, day length, competitive stress, population size, or sex of others after the process of fertilization. Specific mechanisms and cues vary by species.

Incubation temperature: While the sex of most snakes and most lizards is determined by sex chromosomes at the time of fertilization, the sex of all crocodiles, most turtles, and some lizards is determined exclusively by the incubation temperature of eggs during a critical stage of embryogenesis (*i.e.*, temperature-dependent sex determination, or TSD). Small differences in ambient temperature have amazing differences on sex ratio. In certain turtles, eggs incubated above 32^0C produce females, eggs incubated below 28^0C produce males, and eggs incubated in between these temperatures produce a mixture of males and females.

Association with a female: In *Bonellia viridis,* a marine worm, all larvae are genetically and cytologically similar. If a particular larva settles near proboscis (tubular mouthparts used for feeding and sucking) of an adult female, it becomes a male individual. On the other hand, if it has to develop free in water, it becomes a female. Similarly, if a partly developed male is detached from the proboscis it becomes an intersex. Obviously the proboscis secretes a substance suppressing femaleness.

In green spoonworms, females are 200,000 times the size of males. The extremely small male (1 to 3mm long) lives as a parasite in a brood pouch close to the female's genital sac, fertilising her eggs as she produces them. The green spoonworm has a cosmopolitan distribution across all types of sea floors from the surface to more than 1000m deep. Only larvae that encounter females within a few days of hatching will develop into males; the others will develop into females.

Size of the egg: In **Dinophilus**, a small annelid worm, sex determination is influenced by the **size of the egg**, an example of environmental or maternal effect on sex. Larger eggs develop into **females**, while smaller eggs develop into **males**. This system allows the species to adjust the sex ratio according to maternal resources and environmental conditions, as producing larger eggs requires more energy and investment, typically resulting in female offspring, whereas smaller eggs are less costly and tend to produce males.

Social environment: Many fish are sequential hermaphrodites, where they start their life as one sex, but change sex later in development. In the anemone fish (*Amphiprion akallopisos*), which live in social groups with one dominant breeding pair as well as several subordinate males, sex change occurs when the dominant female dies and the largest male in the group becomes the dominant female (Fricke and Fricke, 1977).

Certain plants exhibit **environmental sex determination**, where the development of male or female flowers is influenced more by external factors than by genetics alone. Environmental cues such as day length, temperature, nutrient availability, and overall growth conditions can determine the sex of a plant. For example, in some species, longer days may promote the formation of male flowers, while shorter days favor female flowers. Similarly, higher temperatures can induce more male flowers in certain cucurbits, and nutrient-rich or rapid growth conditions may bias development toward one sex. Even stress factors, such as drought or crowding, can affect sex expression. In these cases, the plant's genetic potential allows for both sexes, but environmental conditions ultimately guide which sex develops, illustrating a flexible strategy that can optimize reproduction under varying conditions.

Hormones in Sex Determination

In birds, hormones play a crucial role in directing sexual development. It has been observed that if the ovary of a genetically female bird is removed, the normal female development process is altered. This can lead to the development of male traits such as a larger comb and male-type plumage. In extreme cases, even testes may develop. Similar phenomena have been reported in pigeons, certain amphibians, fishes, and other lower vertebrates.

Another classic example of hormonal influence on sex determination is the **free martin** phenomenon. When twins of opposite sex are formed in cattle, the male develops normally, but the female twin often becomes sterile and exhibits many male characteristics. This sterile female is referred to as a **free martin**, demonstrating how early exposure to male hormones can affect sexual development.

Sex-linked Inheritance

The X chromosomes of man have been found to be straight, rod-like, and comparatively larger than Y chromosome (Image 7.2). Unlike autosomal pairs of chromosomes, the heteromorphic X and Y are not completely homologous. Some portion of both types of chromosomes has identical or homologous genes, so called homologous or pairing region. Remaining regions of X and Y chromosomes are called nonhomologous or differential regions.

At the time of meiosis in male, the synapsis and crossing over remain restricted to the homologous regions of X and Y chromosomes. A gene that is located on either sex chromosome is called a **sex-linked gene. Sex linkage** applies to genes that are located on the sex chromosomes. In females, segregation for X-linked genes is the same as that for autosomal genes, while in males, half of the gametes carry the X and half carry the Y. The genes which occur in homologous region of X and Y chromosome are called XY-linked genes and their mode of inheritance is called XY-linked inheritance (inheritance like the autosomal genes). They can undergo crossing over and are called **incompletely sex-linked genes**. Examples include bobbed bristles in Drosophila, Xerodermapigmentosum, Nephritis, Retinitis pigmentosa, *etc.* The differential regions of each chromosome hold genes that have no corresponding allele on the other chromosome. Genes in the differential region of X and Y chromosomes are said to be hemizygous (half-zygous) in males. Since only one allele is present, a single copy of a recessive allele determines the phenotype, a phenomenon called pseudodominance. The genes which reside in differential regions of X and Y chromosomes never undergo crossing over. Such genes are called **completely sex-linked genes**. Completely sex-linked genes may be of following types:

Holandric genes. The genes which remain confined to differential region of Y chromosome only, are called holandric or Y-linked genes. The Differential region of Y chromosome is also called the male-specific region of the Y (MSY), because it is only present in males. The Y-linked genes inherit along with Y chromosome. The human Y chromosome harbors genes that are typically responsible for testis development and also for initiation and maintenance of spermatogenesis in adulthood, and thus any conditions involving Y-linked inheritance will typically result in infertility.

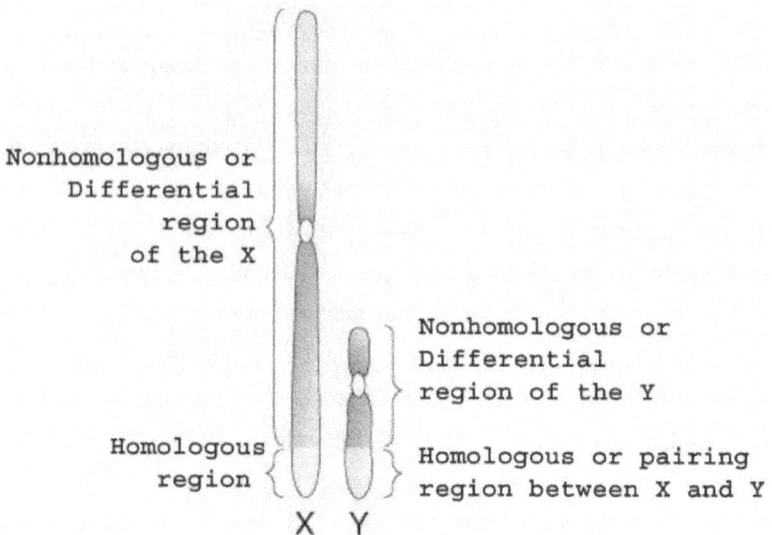

Image 7.2 Differential and pairing regions of sex chromosomes of humans.

X-linked genes. The genes which reside in the differential region of X chromosome only, are called X-linked genes. The X-linked genes are represented twice in female (because, female has two X chromosomes) and once in male (because, male has only one X chromosome). In male, single X-linked dominant or recessive allele expresses it phenotypically.

Genes on the autosome show the kind of inheritance pattern discovered and studied by Mendel. Genes on the differential regions of the sex chromosomes have unique patterns of inheritance that is different from genes located on autosomes. When the results of reciprocal crosses are not the same and different ratios are seen for the two sexes of the offspring, sex-linked characteristics may well be involved. By comparison, the results of reciprocal crosses are always the same when they involve genes located on autosomes.

The sex chromosomes X and Y, besides determining the sex of the offspring, can also carry genes that determine other traits. These genes are considered sex-linked because their expression and inheritance patterns differ between males and females. However, the term sex linked usually refers to genes found only on the X chromosome. Almost all sex-linked traits are found on the X chromosome. Y chromosome contains very few genes and is mainly involved in sex determination. The X chromosome contains genetic information essential for both sexes; at least one copy of an X is required. Human males are then especially vulnerable to defective copies of X-chromosome genes for which there is no 'backup copy' on the other chromosome.

Inheritance of Y-linked Genes

The Y-linked genes inherit directly from father to son. Some characteristics of Y-linked inheritance are:

1. Only males are affected.
2. Trait is always passed from father to son.

Y-linked disorders are extremely rare. Examples of such traits are: hairy ear rims, the sex determining gene SRY, a gene for H-Y antigen, hypertrichosis pinnae auris, azoospermia (no sperm in the ejaculate), retinitis pigmentosa, *etc.*

Hypertrichosis Pinnae Auris (Hairy Ears): Hypertrichosis is defined as an excessive growth of hair on a particular area of the body which is abnormal for the age, sex or race of an individual. The presence of the excessive coarse black hair on the auricle of the human ear is referred to as hypertrichosis pinnae auris or hairy ears. The condition is primarily restricted to older men and occasionally observed in females. According to the available literature, hypertrichosis pinnae auris is a Y-linked character.

Retinitis Pigmentosa (RP): Retinitis pigmentosa (RP) is a group of rare eye diseases that affect the retina. RP makes cells in the retina break down slowly over time, causing vision loss.

H-Y antigen: The most well-known minor histocompatibility antigen in mammals is the H-Y antigen. H-Y antigens are encoded by a gene located on the Y chromosome in both humans and mice. H-Y is a transplantation antigen that can lead to rejection of male organ and bone marrow grafts by female recipients, even if the donor and recipient match at the major histocompatibility locus of humans, the HLA (human leukocyte antigen) locus.

Azoospermia: Azoospermia is the medical term used when there are no sperm in the ejaculate. It can be caused by a variety of genetic factors, including chromosomal abnormalities, gene mutations, and other genetic disorders. These can affect spermatogenesis (sperm production) or cause obstructions in the reproductive tract, leading to azoospermia. Around 10 percent of infertile men and 1 percent of all men have azoospermia. Spermatogenesis is regulated by many Y chromosome specific genes. Most of these genes are located in a specific region known as the azoospermic factor (AZF) region in the long arm of the human Y chromosome (Yq11). A common cause of severe oligozoospermia and azoospermia is microdeletion of the AZF region.

Inheritance of X-linked genes

Diseases caused by mutated genes located on the X chromosome can be inherited in either a dominant or recessive manner.

X-linked Recessive Trait

A trait due to a recessive allele carried on the X chromosome is called an X-linked recessive trait. X-linked recessive traits occur much more frequently among males, who are hemizygous. Affected males often present in each generation. A female would express a recessive X-linked trait only if she were homozygous recessive at that locus. Some characteristics of X-linked recessive inheritance are:

- ❖ Many more males than females exhibit the trait.
- ❖ It is never passed from father to son.
- ❖ All daughters of affected fathers are carriers.
- ❖ All sons of affected mothers (homozygous recessive) are expected to show the trait.
- ❖ Trait or disease is typically passed from an affected grandfather, through his carrier daughters, to half of his grandsons.
- ❖ A carrier female crossed with a normal male will have ½ carrier and ½ normal daughters.

Examples of traits with X-linked recessive inheritance are white-eye phenotype in Drosophila, red-green color blindness, hemophilia A and B, Androgen insensitivity syndrome, Duchenne muscular dystrophy, Lesch-Nyhan syndrome, severe combined immunodeficiency (SCID), Glucose-6-phosphate dehydrogenase deficiency, Hunter's Syndrome, *etc.* In chicken sex-linked (Z-linked) recessive traits are sex-linked dwarfism, sex-linked white skin, sex-linked barring, rapid feathering, sex-linked albinism in turkeys, *etc.*

The Discovery of Sex Linkage (White-Eye Phenotype in Drosophila)

The parallel behavior of genes and chromosomes was strong support that genes actually resided in chromosomes, but it was not direct or definitive evidence that specific genes and specific chromosomes were connected. Thomas Hunt Morgan, an embryologist from Columbia University was the first to associate a specific gene with a specific chromosome in 1910, who described X-linked inheritance in Drosophila.

The normal eye color of Drosophila is bright red. Early in his studies, Morgan discovered a male with completely white eyes. When he crossed this male with red-eyed females from a pure line, all the F_1 progeny had red eyes, showing that the allele for white is recessive. Crossing the red-eyed F_1 males and females, Morgan obtained a 3 : 1 ratio of red-eyed to white-eyed flies, but all the white-eyed flies were males. Among the red-eyed flies, the ratio of females to males was 2 : 1.

First cross

P	X^+X^+ Red-eyed female	×	X^wY White-eyed male
F_1	X^+X^w ½ Red-eyed female	×	X^+Y ½ Red-eyed male

F_2			Male gametes	
			½ X^+	½ Y
	Female gametes	½ X^+	¼ X^+X^+ (red-eyed female)	¼ X^+Y (red-eyed male)
		½ X^w	¼ X^+X^w (red-eyed female)	¼ X^wY (white-eyed male)

Second cross

P	X^wX^w White-eyed female	×	X^+Y Red-eyed male
F_1	X^+X^w ½ Red-eyed female	×	X^wY ½ White-eyed male

F_2			Male gametes	
			½ X^w	½ Y
	Female gametes	½ X^+	¼ X^+X^w (red-eyed female)	¼ X^+Y (red-eyed male)
		½ X^w	¼ X^wX^w (white-eyed female)	¼ X^wY (white-eyed male)

Figure 7.1 Results of Morgan's crosses.

When he crossed white-eyed females with red-eyed males from a pure line, the F_1 generation contains an equal proportion of white and red-eyed individuals, but all males have white eyes and all females have red eyes. Crossing these F_1's again results in 1 : 1 ratio of red and white-eyed individuals, but in the F_2, half the female offspring and half the male offspring have red eyes. Morgan deduced that eye color is linked to sex and the gene for eye color is located only on the X chromosome. Thus, females would have two alleles for this gene, whereas males would have only one. We can represent the two alleles as + (red) and w (white) and diagram the two reciprocal crosses as shown in Figure 7.1.

This was the first solid evidence indicating that a specific gene is associated with a specific chromosome. Morgan received the 1933 Nobel Prize for this work. As we see from the Figure 7.1, the genetic results of the two reciprocal crosses are completely consistent with the known meiotic behavior of the X and Y chromosomes.

Red-Green Color Blindness in Humans

Perhaps the most familiar example of X-linked recessive inheritance is red-green colorblindness in humans. In this case, both red and green cone cells are absent. People with this condition are unable to differentiate between red and green colours. Both these colours appear grey to the colour blind person. It is caused by a recessive mutant allele located on the X chromosome. The normal allele is denoted C and the mutant allele responsible for colorblindness c. The possible genotypes and phenotypes are shown in Table 7.4.

Table 7.4 Possible genotypes and phenotypes at the colorblind locus

Normal male	$X^C Y$
Colorblind male	$X^c Y$
Normal female	$X^C X^C$ or $X^C X^c$
Colorblind female	$X^c X^c$

The c allele is relatively rare in human populations. The progeny from a cross between a woman heterozygous (carrier) for colorblindness, and a man with normal sight are given in Figure 7.2.

The ratio of normal to affected progeny is still 3 : 1, but unlike the F_2 progeny of monohybrid cross for an autosomal gene, all the mutant types are males.

$X^C X^c$ Normal sight but carrier for colorblindness female	×	$X^C Y$ Normal sight male
	Male gametes	
	½ X^C	½ Y
Female gametes ½ X^C	¼ $X^C X^C$ (normal females)	¼ $X^C Y$ (normal males)
½ X^c	¼ $X^C X^c$ (carrier, normal females)	¼ $X^c Y$ (colorblind males)

Figure 7.2 Inheritance of colorblindness.

Hemophilia A and B

Hemophilia (bleeder's disease) is the oldest known hereditary bleeding disorder that slows the blood clotting process. The first person known to carry the disorder was Queen Victoria of England. Thus, all those affected are related to European royalty. People with this condition experience prolonged bleeding or oozing following an injury, surgery, or having a tooth pulled. In severe cases of hemophilia, continuous bleeding occurs after minor trauma or even in the absence of injury (spontaneous bleeding). Serious complications can result from bleeding into the joints, muscles, brain, or other internal organs. Milder forms of hemophilia do not necessarily involve spontaneous bleeding, and the condition may not become apparent until abnormal bleeding occurs following surgery or a serious injury. Even a minor cut or scratch can cause a person with hemophilia to bleed to death. The major types of this condition are hemophilia A (also known as classic hemophilia or factor VIII deficiency, the most common form of hemophilia) and hemophilia B (also known as Christmas disease or factor IX deficiency). Although the two types have very similar signs and symptoms, they are caused by mutations in two different genes. For hemophilia A, the affected gene is F8. The F8 gene encodes blood protein called factor VIII. Hemophilia A is also caused due to inversion of the chromosomal region that involve a portion of the F8 gene (Xq28). This inversion completely splits the F8 gene into two parts and disrupts the F8 transcription, resulting in no factor VIII protein production. For hemophilia B the affected gene is F9. F9 gene codes for a protein called coagulation factor IX (nine). Factor VIII, factor IX, and other coagulation factor proteins circulate in the blood. After an injury, coagulation factors in the area become active. Through a chain reaction, the factors work together to send a signal. In the final step, the signal brings proteins and blood cells together to form a blood clot. The blood clot seals off the injury. Bleeding stops, and healing begins. The severity of hemophilia is related to the amount of the clotting factor in the blood. About 70% of hemophilia patients have less than one percent of the normal amount and, thus, have severe hemophilia. Prevalence of Hemophilia A is estimated at around 1 in 6,000 males. Hemophilia B is much less common than hemophilia A. The incidence of hemophilia B is estimated to be approximately 1 case per 25,000-30,000 male births.

Duchenne Muscular Dystrophy (DMD)

Duchenne muscular dystrophy (DMD) was first described by the French neurologist Guillaume Benjamin Amand Duchenne in the 1860s. DMD is the most common hereditary fatal neuromuscular disorder characterized by progressive muscle weakness and degeneration. It is usually recognized between three and six years of age. The affected child might have difficulty jumping, running, and walking. Generally, by age 10-12 affected individuals become confined to a wheelchair. Affected patients usually die in their twenties due to heart and respiratory muscle weakness. It is inherited as an X-linked recessive trait, boys are more frequently affected than girls. The estimated incidence is 1 in 3600 male live-born infants. The **DMD (dystrophin) gene** is the largest known human gene that contains 2.4 million base pairs and located at the Xp21.2 locus on the X chromosome. It encodes the muscle cell membrane protein **dystrophin**, a crucial structural protein that links the cytoskeleton of muscle fibers to the extracellular matrix, providing stability during muscle contraction. The gene contains **79 exons**, and mutations within it are responsible for **Duchenne muscular dystrophy (DMD)**. These mutations can include deletions, duplications, or point mutations. These mutations result in the severe absence ($< 5\%$) of dystrophin (Rubin, 2024). Muscles without dystrophin are more sensitive to damage, resulting in progressive loss of muscle tissue and function, in addition to cardiomyopathy. Up to 70% of DMD is caused by a single- or multiexon deletion, approximately 20% by a point mutation and 10% by a duplication (Duan *et al.*, 2021). In Becker dystrophy, the mutations result in production of abnormal dystrophin or insufficient dystrophin (Rubin, 2024).

Androgen Insensitivity Syndrome (AIS)

Androgen insensitivity syndrome (AIS), formerly known as testicular feminization syndrome (TFS), is a rare disorder with an incidence of 1 : 20,000-64,000 male births in which a genetic male with internal testes has the external appearance of female. Affected individuals have normal testes with normal production of testosterone and normal conversion to dihydrotestosterone (DHT) but they are unable to respond to either hormone. The syndrome is a form of intersex and is the most common form of pseudohermaphroditism. This syndrome is caused by mutations in the androgen receptor (AR), also called the dihydrotestosterone receptor, gene located on the long arm of the X chromosome in band Xq11-12 that makes XY fetuses insensitive to androgens. The AR gene provides instructions for making a protein called an androgen receptor. The reason for the insensitivity is that there is a malfunction in the androgen receptor, so male hormone testosterone can have no effect on the target organs that are involved in the maleness. Instead, they are born looking externally like normal girls. Internally, there is a short blind-pouch vagina. Patients have testes in the abdomen or inguinal canal. The female internal organs do not develop because the testes produce anti-Müllerian hormone, which prevents males from developing fallopian tubes, uterus, cervix and upper vagina. The low level of estrogen that is produced by the adrenal glands is enough to stimulate female secondary sex characteristics.

Androgen insensitivity syndrome may present as complete form and incomplete/partial form. Individual with complete androgen insensitivity syndrome (CAIS), there is no androgen response, therefore develop typical females external sex characteristics with normal labia, clitoris, and vaginal introitus but lack internal female reproductive organs (ovaries, fallopian tubes, uterus, cervix and upper vagina). The insensitivity to androgens also results in a lack of pubic hair, facial hair and male type muscle development. Affected individuals appear to be girls throughout puberty and is rarely discovered during childhood. Most people with CAIS are not diagnosed until they fail to menstruate or have difficulties becoming pregnant. The partial androgen insensitivity syndrome (PAIS) is associated with wide range of genitalia ambiguity. The degree of sexual ambiguity varies widely in persons with PAIS. They can have genitalia that look typical for females, genitalia that have both male and female characteristics, or genitalia that look typical for males. Treatment of TFS involved surgical removal of the testes, to prevent cancerous change in later life, and therapy with oestrogen drugs. An affected person is not fertile but can live a normal life as a woman.

Severe Combined Immunodeficiency (SCID)

Severe Combined Immunodeficiency (SCID) is a group of very rare and potentially fatal inherited disorders of the immune system that occurs almost exclusively in males. The immune system normally fights off attacks from dangerous bacteria and viruses. People with SCID have a defect in their immune system that leaves them vulnerable to potentially deadly infections. If untreated, infants with X-linked SCID can develop poor growth, chronic diarrhea, a fungal infection called thrush, skin rashes, and life-threatening infections. There are several types of SCID. Mutations in the IL2RG (interleukin-2 receptor) gene cause X-linked SCID. The IL2RG gene provides instructions for making a protein that is used to build a receptor called IL2RG which is critical for normal immune system function. This protein is necessary for the growth and maturation of developing immune system cells called lymphocytes. Lymphocytes defend the body against potentially harmful invaders, make antibodies, and help regulate the entire immune system. Variants in the IL2RG gene prevent these cells from developing and functioning normally. Without functional lymphocytes, the body is unable to fight off infections. Another form of SCID is caused by a mutation on chromosome 20 and is characterized by a deficiency of the enzyme adenosine deaminase.

Lesch-Nyhan Syndrome

Lesch Nyhan syndrome, Lesch-Nyhan disease, Nyhan syndrome, juvenile gout, hypoxanthine-guanine phosphoribosyl transferase (HGPRT or HPRT) deficiency, is a rare severe X-linked recessive lethal disorder that most often affects males. It is caused by mutation of the hypoxanthine-guanine phosphoribosyltransferase (HPRT) gene (Xq26-27) encoding the purine recycling enzyme hypoxanthine-guanine phosphoribosyltransferase (HPRT). Absence or deficiency of the enzyme HPRT causes a build-up of uric acid in all body fluids, which may lead to symptoms such as gouty arthritis, involuntary muscle movements, impaired kidney function, and neurological impairment, the most common of which is self-mutilating behaviors such as lip and finger biting and/or head banging. Lesch-Nyhan syndrome was first described by doctors Michael Lesch and William Nyhan in 1964. The estimated prevalence of Lesch Nyhan syndrome is 1 out of every 100,000 to 380,000 live births.

Glucose-6-phosphate Dehydrogenase Deficiency

Glucose-6-phosphate dehydrogenase (G6PD) deficiency is the most common enzyme deficiency in humans affecting around 400 million individuals worldwide, caused by mutations in the G6PD gene that encodes G6PD enzyme. A lack of this enzyme can cause red blood cells to break down prematurely and develop hemolytic anemia. Symptoms can start as early as a newborn. Most people with this condition don't have symptoms until they are exposed to additional factors to "trigger" the onset of symptoms. Triggers of hemolysis in G6PD-deficient persons include certain infectious diseases, certain drugs, and eating fava beans or inhaling pollen from fava plants. Symptoms can include fatigue, paleness, jaundice, fever, trouble with physical activity, shortness of breath, rapid heartbeat, dark urine and enlarged spleen.

Most important, in the absence of triggering factors, the majority of people with G6PD deficiency are normal.

Hunter's Syndrome

Hunter syndrome, also called Mucopolysarcharidosis type II (MPS II), is a rare progressively debilitating and life-limiting lysosomal storage disorder. It occurs almost exclusively in males. Hunter syndrome is caused by mutation in the iduronate-2-sulfatase (IDS) gene. IDS gene encodes the enzyme iduronate-2-sulfatase (IDS) located in lysosomes, which is responsible for the breakdown of cellular waste called glycosaminoglycans (GAGs, large sugar molecules) – the cell's "garbage." Mutations in the IDS gene result in a deficiency or absence of IDS, which in turn results in an abnormal accumulation of GAGs in various body tissues, which affects many systems. Newborns with Hunter syndrome typically do not have any signs or symptoms at birth. As children grow and develop, signs of disease may become more apparent. Symptoms vary from mild to severe and become apparent between 2 and 4 years of age that may include short stature, large liver and spleen, joint problems, prominent facial features, frequent ear infections, hearing loss, skeletal problems, umbilical hernia, frequent upper respiratory infections, enlarged tonsils and/or adenoids, distinct white skin growths, enlargement of the heart chambers, accumulation of fluid around the brain, intellectual or developmental decline, *etc.* The estimated incidence is 1 case per 100,000 to 170,000 male births.

Sex-linked Dwarfism (SLD) in Chickens

Sex-linked dwarfism (SLD) in chickens is characterized by reduced body weight and longitudinal bone growth, despite normal levels of circulating growth hormone (GH) (Decuypere *et al.*, 1991, Rosenfeld *et al.*, 1994). It is caused by a recessive mutation of the growth hormone receptor (GHR) gene located on the Z chromosome. This mutation results in a number of phenotypic and physiological alterations of the SLD chickens. For example, shorter shanks, lower basal metabolism, lower body weight, higher serum growth hormone (GH), lower serum insulin-like growth factor-1 (IGF-1), higher tolerance to heat and higher feed efficiency of SLD chickens are observed compared to those of normal chickens (Zheng *et al.*, 2007; Guillaume, 1976). Another significant phenotypic change of SLD chickens is that the average body weight of homozygous (dwdw) chickens was about 40% lower than that of normal chickens (Guillaume, 1976; Dodgson and Romanov, 2004). Sex-linked dwarf chickens have been widely used in cross breeding of broilers and laying hens.

Sex-linked White Skin in Chickens

Sex-linked white skin in chickens, often referred to as "white sex-link," refers to a trait where males and females have different skin colors at hatching, making sex determination easier. This is achieved through specific breeding crosses that utilize other sex-linked traits.

Sex-linked Albinism in Turkey

In turkeys, sex-linked albinism is caused by a sex-linked recessive gene, *al*, which is lethal during embryonic development to about 75 per cent of the females carrying it. In that case nearly all melanic pigment is eliminated from the plumage. In the eye, some melanin is found in the retinal portions of the ciliary body and iris, but not in the pigment epithelium of the retina in the posterior part of the eye. Affected poults are blind.

X-linked Dominant Trait (Very Rare)

A trait due to a dominant allele carried on the X chromosome is called an X-linked dominant trait. X-linked dominant conditions are rare, but do exist. Examples of traits with X-linked dominant inheritance are fragile X syndrome, Xg blood group system, Fabry disease, X-linked hypophosphatemia, Rett syndrome, *etc.* In chicken sex-linked (Z-linked) dominant traits are barred plumage, silver color feather, slow-feathering, *etc.* Some characteristics of X-linked dominant inheritance are:

1. The trait is never passed from father to son.
2. All daughters of an affected male and a normal female are affected. All sons of an affected male and a normal female are normal.
3. Mating of affected females and normal males produce ½ the sons affected and ½ the daughters affected.
4. Males are usually more severely affected than females. The trait may be lethal in males.

Fragile X Syndrome (FXS)

Fragile X syndrome (FXS) is a relatively common genetically inherited abnormality of the X chromosome which results various levels of mental impairment – from learning disability to severe retardation. FXS is caused by a mutation in the FMR1 (fragile X messenger ribonucleoprotein 1) gene located on the long arm of the X chromosome at position 27.3. The FMR1 gene provides instructions for making a protein called fragile X messenger ribonucleoprotein (FMRP). FMRP plays a crucial role in brain development and synaptic plasticity. This protein is present in many tissues, including the brain, testes, and ovaries. Most fragile X males have large testes, big ears, narrow faces, and sensory integration dysfunctions that result in learning disabilities. It is likely to occur 1 in 1000 births. Approximately 1 in 700 females are carriers of the gene for this trait.

Xg Blood Group System

The Xg blood group system is best known for its contributions to the fields of genetics and chromosome mapping. Discovered in 1962 by Mann *et al.*, the Xg is the only known blood group system whose gene is located on the X chromosome. It comprises only one antigen specificity. It is a cell-surface antigen present on red blood cells. Inheritance is controlled by two alleles (Xg^+ and Xg^-) with Xg^+ being dominant and Xg^- being recessive. There are two groups in the system, Xg^+ and Xg^-, depending on whether red cells have the antigen Xg or not. The Xg system is not very significant in transfusion medicine because the sole antibody (anti-Xg antibody) in the system is extremely rare. A serum that reacts positively with anti-Xg antibody is said to be Xg-positive, written Xg^+; a serum that does not react is Xg-negative or Xg^-. Corresponding to the two phenotypes, Xg^+ and Xg^-, there are two genotypes for males, Xg^+Y and Xg^-Y, and for females three genotypes, Xg^+Xg^+, Xg^+Xg^- and Xg^-Xg^-. Because the allele Xg^+ is dominant with respect to its allele, it follows that heterozygous Xg^+Xg^- females cannot be identified with certainty since their phenotype is also Xg^+.

X-linked Hypophosphatemia

X-linked hypophosphatemia (XLH), also known as X-linked hypophosphatemic rickets, is an inherited disorder characterized by low levels of phosphate in the blood leading to severe skeletal abnormalities and growth retardation. Features include bowed or bent legs, short stature, bone pain, skeletal deformities, and severe dental pain. XLH is usually diagnosed in childhood. XLH is caused by a mutation in the PHEX (phosphate-regulating gene with homologies to endopeptidases on the X chromosome) gene located on the X chromosome (Xp22.1). This gene is responsible for regulating phosphates in the body. The mutation prevents the kidneys from processing phosphorus correctly which lead to increased production of fibroblast growth factor 23 (FGF 23), which blocks the kidney's ability to hold on to phosphorus. As a result, too much phosphorus leaves the body in the urine, which is also known as "phosphate wasting". It is an X-linked dominant trait with complete penetrance, but variable expressivity.

Fabry Disease

Fabry disease is a rare X-linked dominant lysosomal storage disorder resulting from the absence or deficiency of the lysosomal enzyme, α-galactosidase A (α-Gal A). Fabry disease is caused by mutations in the galactosidase alpha (*GLA*) gene which encode alpha-galactosidase A (alpha-Gal A) enzyme that break down sphingolipids. The mutated gene allows sphingolipids to build up to harmful levels in the blood vessels and tissues that results in progressive organ damage, particularly to heart, kidneys, nervous system and skin. Fabry disease affects an estimated 1 in 1,000 to 9,000 people.

Rett Syndrome

Rett syndrome (RTT) is a progressive neurodevelopmental disorder that occurs almost exclusively in girls, with an incidence of 1 in 9,000-10,000. Only in rare cases males are affected. It is caused by mutations in the methyl CpG binding protein 2 (MECP2) gene (Xq28) that lead to deficiency of the methyl cytosine binding protein 2 (MeCP2), an important protein responsible for normal function in the brain and other parts of the nervous system. Girl appears normal until 6-18 months of age, then gradually lose the ability to crawl, walk, communicate, or use hand, and develop severe intellectual disability, panic-like attacks, microcephaly, feeding and swallowing difficulties, growth retardation, seizures, breathing abnormalities, unusual eye movements, and sleep disturbances. After initial regression, the condition stabilizes and patients usually survive into adulthood.

Nonreciprocity

Nonreciprocity is the condition of being nonreciprocal. A striking characteristic of X-linked inheritance is that fathers cannot pass X-linked traits to their sons (no male-to-male transmission). The X-linked pattern has long been known as the criss-cross pattern of inheritance, also known as skip-generation inheritance, refers to the pattern where traits are passed from one parent to the opposite sex offspring, where a father passes a trait to his daughters, and a mother to her sons. Autosomal traits would show Mendelian inheritance pattern but sex-linked traits would show criss-cross inheritance. Here a character is inherited to the second generation through the carrier of first generation.

X-Chromosome Inactivation (XCI) or Lyonization

Female mammals have two X chromosomes, whereas males have only one. So, females have two copies of all X-linked genes, while males have one. Therefore, females should produce twice as much of the proteins encoded by these genes as males. This would not be a good thing – the overall effect would probably be lethal to female embryos. To compensate for the extra X chromosome (the X chromosome 'dose'), one X chromosome and all of its genes are inactivated/silenced in every cell of early female mammalian embryo. X-chromosome inactivation (XCI) is the process that has evolved in mammals to prevent a gross imbalance in expression of X-linked genes in XX females relative to XY males. Silencing, once established, is stable: the same X chromosome remains inactivated in all subsequent cell generations. It is a classic example of epigenetic regulation. X-chromosome inactivation is the transcriptional silencing of one X chromosome in female mammalian cells that equalizes dosage of gene products from the X chromosome between XX females and XY males. The purpose of dosage compensation is to prevent excessive expression of X-linked genes in Humans and other mammals. Dosage compensation has been studied extensively in mammals, Drosophila, and *Caenorhabditis elegans*. Depending on the species, dosage compensation occurs via different mechanisms. Most X-linked genes show dosage compensation, some X-linked genes do not and the reasons for the difference are not understood.

The X chromosome is inactivated by converting it to heterochromatin. The highly condensed inactive X chromosome that can be observed in stained interphase cells is called a **Barr body**. Which chromosome is inactivated is randomly determined in every cell, so if a female is heterozygous for an X-linked gene, the dominant allele will be expressed in some cells, while the recessive allele will be expressed in others. Because this inactivation occurs randomly, all normal females have roughly equal populations of two genetically different cell types. In roughly half of their cells, the paternal X chromosome has been inactivated, and in the other half the maternal X chromosome is inactive. Female mammals are therefore mosaics with regard to X-linked genes. This has a number of important biological and medical implications, particularly with regard to X-linked genetic diseases.

Mosaicism and Chimerism

Both terms refer to one organism with two or more populations of **genetically Mdifferent** cells – but they are not interchangeable. **Mosaicism** denotes the presence of two or more populations of genetically different cells in one individual who has originated from a single zygote, whereas **chimerism** refers to an individual who has originated from more than one zygote.

Mosaicism is caused when a mutation arises early in development which is propagated to only a subset of the adult cells. The resulting individual will be a mixture of cells, some with the mutation and some without the mutation. This can happen on a gene level or even whole chromosomes. For example, a person may have some of the cells in his or her body with 46 chromosomes, while other cells in his or her body have 47 chromosomes. An example of mosaicism is mosaic Down syndrome. How early in development the mutation occurs will determine what tissue(s) and what percentage of cells will have the mutation. Mosaicism can be germline (affecting only egg or sperm cells), somatic (affecting cells other than egg or sperm cells), or a combination of both.

Chimeras are formed from at least four parent cells (two fertilized eggs or early embryos fused together). Each population of cells keeps its own character and the resulting organism is a mixture of tissues. Chimeras are typically seen in animals; there are some reports on human chimerism. Chimeras can happen with organ transplantation because the donor cells are different from recipient. One should also be a chimera from absorbing a twin while in utero: a pregnancy may start out with two or more zygotes but not all survive and are then absorbed by remaining fetus. Individual could also have cells from mother – a distinct cell line from another zygote.

The different cell lines in chimeras originally come from different zygotes whereas mosaics arise from the same zygote.

Sex Linkage in Poultry

With most animals it is relatively easy to tell the sex of the newborn. This is not the case with poultry. Chickens have no external sex organs; both carry their reproductive organs within their body and transfer genetic material by means of their vent. It is important to identify the sex of chickens at hatching so that, for example, males will not be raised in an egg laying operation. The word "sexing" simply refers to the process by which chicks are sorted into males and females. There are two possible methods of sexing chicks that can be used at hatcheries. They are vent sexing and feather sexing. Vent sexing was developed in Japan and brought to North American poultry producers in the 1930s. It involves holding the chick upside down in one hand, expelling the fecal material and everting (turning outward) the vent area. You can then look for the presence or absence of a rudimentary male sex organ. There are slight differences in the appearance of the cloacal vent, but only highly trained specialists are able to reliably sex dayold chicks on this basis (Martin, 1994). Vent sexing is notoriously difficult and expensive.

For most breeds, day-old male and female chicks are virtually indistinguishable. Some breeds are auto-sexing, meaning that females and males day-old chicks look different. Example includes Amrock, Kennhuhn, Niederrheiner, Norske Jaerhon, Ancobar, Barnebar, Bielefelder, Brockbar, Brussbar, Buffbar, Cambar, Cobar, Dorbar, Gold Legbar, Hambar, Oklabar, Polbar, Rhodebar, Welbar, Whealbar, and Wybar. In certain breeds feather growth rate and plumage colour are sex-linked traits that can be used in sex-linked crosses to identify sex of the offspring; however, this does not avoid the problem of culling one or other sex at hatch. Sex-linked crosses can only be made with male and female chickens of different breeds in specific combinations. It is also important to note that the offspring of a sex-linked cross cannot themselves be used in a second sex-linked cross. Auto-sexing is done either by the sex-linked characteristics of plumage colour (colour sexing) or rate of feather growth (feather sexing).

In poultry, the male has the genotype ZZ, while the female has the genotype ZW. The general principle for autosexing in fowls is that male should have the recessive character in homozygous condition and female should have the dominant character.

Colour Sexing

Black Sex-Links

Black sex-links are often referred to as Rock Reds. Barring means alternate stripes across the feather in some breeds. It can be determined by an autosomal gene, as in Dominiques, or sex-linked dominant gene, as in Barred Plymouth Rocks. Sex-linked barring can be used for autosexing purposes.

P	Z^BW Barred hen	×	Z^bZ^b Non-barred cock
Progeny	Z^BZ^b Barred male (Black with white spot on the top of head)	×	Z^bW Non-barred female (solid color, typically black)

Figure 7.3 Genetic basis of sex-linked cross of barred female to non-barred male.

The barred (black-and-white striped) plumage in Plymouth Rock, Cochin, Holland, Japanese, d'Anver or Old English Game chicken is due to a dominant allele, B located on the Z-chromosome. The non-barred allele, b is recessive. The sex-linked barring gene creates a pattern of white bars or stripes across each feather of the bird. The barring gene inhibits the deposition of melanine, thus causing white bars to be superimposed on a feather that would otherwise be all black. When barred hens (*e.g.,* Barred Plymouth Rock) are mated to non-barred cocks (*e.g.,* Rhode Island Red or New Hampshire), all of the females are black (black sex-links) and males are barred (Figure 7.3). At hatch both sexes have black down, but the males can be identified by the white spots on the top of their head which assume the barred pattern when they become adults. It is this specific cross that must be used. Crossing a barred male with a non-barred female will not work.

Red-Sex Links

Dominant silver color feather and recessive gold color feather in poultry is used in auto-sexing some commercial strains where the sexes are separated by feather color. The silver gene is located on the upper half of the Z chromosome in chickens, making it a sex-linked trait. The recessive allele, s determines gold feather color (either Z^sZ^s or Z^sW), and the dominant allele, S determines silver feather color (either Z^SZ^S, Z^SZ^s or Z^SW). When silver hens (Rhode Island White, White Plymouth Rock, Delaware, Silver-Laced Wyandotte, Light Sussex, Dark Brahma or White Orpington) are mated to gold cocks (Rhode Island Red, New Hampshire, Red Dorking, Barnevelder, Buff Orpington, Buff Rock, Buff Cochin or Brown Leghorn), all of the silver offspring are males, and the female chicks will be a gold-based color, usually red (Figure 7.4). Silver chick is completely yellow at hatch. The silver/gold genes have been used in both the layer and broiler industries. The most common sex link crosses are the red sex link. Close to half of the global chicken population consists of red sex link chickens. ISA Brown, Shaver Brown, Bovans Brown, Babcock Brown, Hisex Brown, Dekalb Brown and Warren are all red sex link crossbreds.

P	Z^SW Silver hen	×	Z^sZ^s Gold cock
Progeny	Z^SZ^s Silver male	×	Z^sW Gold female

Figure 7.4 Genetic basis of sex-linked cross of silver/gold color feathering.

The silver/gold and barred/non-barred genes have been combined and used in some commercial brown-shell egg layers. A gold, non-barred (Rhode Island Red) cock is crossed with a silver, Barred Plymouth Rock hen. The females resulting from the cross are black-red and non-barred while the males are black and white barred (Figure 7.5).

P	$Z^{SB}W$ Silver, barred hen	×	$Z^{sb}Z^{sb}$ Gold, non-barred cock
Progeny	$Z^{SB}Z^{sb}$ Silver (black and white), barred male	×	$Z^{sb}W$ Gold (black and red), non-barred female

Figure 7.5 Genetic basis of sex-linked cross of silver/gold with barring genes.

Feather Sexing

Another sex-linked characteristic that is used commercially for sexing day-old broiler chickens and commercial white egg layers is feather sexing. Feather sexing is the ability to determine the gender of a newly hatched chick based on the rate of growth of its wing feathers. The difference in the length of the primary and covert wing feathers (wing coverts, which are the small downy feathers covering the base of the primary feather shafts) can be seen between one and three days after hatching. After this age, however, it is not possible to use this sex-link cross for sexing chickens. When a chick is a day old, rapid-feathering chicks have primary wing feathers that are thicker and longer than the covert feathers, and in slow-feathering chicks the primary and the covert feathers are similar in length and thickness (Image 7.3). This is caused by a gene located on the Z chromosome where slow feathering allele (K) is dominant to wild type trait rapid feathering allele (k). When slow-feathering hens are crossed with fast-feathering cocks, the male offspring are slow-feathering like their mother, and the female offspring are fast-feathering like their fathers (Figure 7.6).

P	$Z^K W$ Slow-feathering hen	×	$Z^k Z^k$ Fast-feathering cock
Progeny	$Z^K Z^k$ (Slow-feathering male)	×	$Z^k W$ (Fast-feathering female)

Figure 7.6 Genetic basis of feather sexing.

The method is very easy to learn by the poultryman, but the feather appearances are determined by specially selected genetic traits that must be present in the chick strain. Most breeds of chickens do not have these feather sexing characteristics and feathering of both sexes appear identical. Some of the rapid feathering breeds are Andalusian, Campine, Leghorn, Marans, and Minorca. Some of the slow feathering breeds are Cochin, Plymouth Rock, Rhode Island Red, Sussex, and Wyandotte. Breeders today maintain pedigree lines chosen and selected for either slow feathering genotypes or fast feathering genotypes.

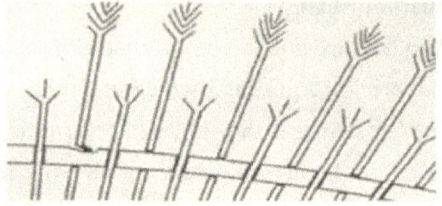

Fast feathered: female
Primary feathers longer than coverts

Slow feathered: male
Primary feathers same size as coverts

Image 7.3 Comparison of the wings of fast- and slow-feathering chicks
(source: Jacquie Jacob, University of Kentucky).

Sex-Influenced and Sex-Limited Traits

Some traits are expressed differently in the two sexes but are not sex-linked. These are sex-limited and sex-influenced traits. The two terms are easily confused, and care must be taken to differentiate between them. Sex-limited and sex-influenced genes are autosomal and the genotypes follow normal Mendelian patterns of inheritance, but the hormonal environment alters the phenotypes. In contrast genes actually situated on the sex chromosomes influence sex-linked traits.

A trait which expression profile is influenced by the sex of the individual is called sex-influenced, or sex-conditioned trait. Sex-influenced traits occur in both sexes but are more common in one sex *i.e.*, express differently based on the sex of the individual but are not limited/restricted to one sex. Both homozygotes express the trait irrelevant of sex. In the heterozygote dominance of the allele depends on the sex of the individual. The dominant phenotype in one sex is recessive in the other. A good example of this is inherited pattern baldness in humans. This is the form of baldness where hairloss spreads out from the crown of the head. One pair of alleles of an autosomal gene is involved in this trait, baldness allele (C^B) and nonbald allele (C^H). The $C^B C^B$ genotype specifies pattern baldness in both males and females, and the $C^H C^H$ genotype gives a nonbald phenotype in both sexes. The difference lies in the heterozygote. In males $C^B C^H$ leads to bald phenotype, and in female it leads to nonbald phenotype. In other words, the baldness allele, C^B acts as dominant in males but recessive in females. This is set out in Table 7.5.

Table 7.5 Expression of pattern baldness genotypes in humans

Genotype	Female Phenotype	Male Phenotype
$C^B C^B$	Bald (Balding crown)	Bald (Balding crown)
$C^B C^H$	Nonbald (Hairy crown)	Bald (Balding crown)
$C^H C^H$	Nonbald (Hairy crown)	Nonbald (Hairy crown)

The expression of the baldness allele is influenced by the hormones testosterone and dihydrotestosterone (DHT), but only when levels of the two hormones are high. In general, males have much higher levels of these hormones than females, so the baldness allele has a stronger effect in males than in females. However, high levels of stress can lead to expression of the gene in women. In stressful situations, women's adrenal glands can produce testosterone and convert it into dihydrotestosterone, which can result in hair loss. The sex-influenced pattern of inheritance and gene expression explains why pattern baldness is far more frequent among men than among women.

Another excellent illustration of sex-influenced trait is horned condition in domestic sheep. In some sheep breeds, such as the Suffolk, neither sex has horns, while in other breeds, such as the Dorset, both sexes have horns. The Dorset sheep are homozygous for the horned gene (*HH*), and the Suffolk sheep are homozygous for the hornless gene (*hh*), so there is no distinction between the sexes in these pure breeds. When these two breeds are crossed, the hybrids (*Hh*) are horned if male but polled if female. In other words, the gene for horn (*H*) acts as dominant in the male and recessive in the female. When two such heterozygous individuals are crossed, we get offspring in the ratio of 3 horned to 1 hornless among the males, but 1 horned to 3 hornless among the females (Figure 7.7). This is the expected ratio of a sex influenced gene.

Parents	*hh* Suffolk (hornless)	×	*HH* Dorset (horned)
Gametes	*h*		*H*
F$_1$	*Hh* (Hornless females and horned males)		
Gametes	½ *H*, ½ *h*		

F$_2$			Male Gametes	
			½ *H*	½ *h*
	Female Gametes	½ *H*	*HH* (horned)	*Hh* (Hornless females and horned males)
		½ *h*	*Hh* (Hornless females and horned males)	*hh* (hornless)

Figure 7.7 Genetic basis of sex-influenced trait.

Sex-limited traits are characteristics expressed in only one sex—either male or female—even though the genes for these traits are present in both sexes. Their expression is therefore restricted to a particular sex, which is why they are called sex-limited traits. A classic example is lactation or milk production in mammals: while both males and females carry the genes responsible for milk production, only lactating females express them. Other examples include breast formation in human females, egg production in chickens, and beard growth in human males, as well as various primary and secondary sexual characteristics. Certain mutant alleles, such as those of the breast cancer susceptibility gene **BRCA1**, are dominant and cause breast cancer in females but not in males, whereas **BRCA2** can cause breast cancer in both sexes and is therefore not sex-limited. When the penetrance of a gene is zero in one sex, the trait is considered sex-limited, as seen in beard growth being restricted to men. Sex-limited traits are commercially the most important among the three types of sex-associated characteristics because economically valuable traits in dairy animals and egg-laying poultry cannot be expressed in males. Nevertheless, males must be thoroughly evaluated genetically for these traits, since they can sire large numbers of offspring, thereby transmitting superior genes widely and contributing significantly to genetic improvement in the population.

CHAPTER EIGHT   MOLECULAR BASIS OF INHERITANCE

Life on earth is very diverse, from single-celled protozoans to complex multicellular plants and animals. But at the molecular level, all life is fundamentally made up of the same building blocks – DNA and RNA. The molecular basis of inheritance refers to the study of how genetic information, encoded in DNA, is stored, organised, copied, and transmitted from one generation to the next, ultimately determining an organism's traits. DNA, RNA and genetic code form the basis of the molecular basis of inheritance.

Nucleic Acids

In 1869, 25-year-old Swiss physiological chemist Johann Friedrich Miescher first identified what he called "nuclein" inside the nuclei of human white blood cells. The term "nuclein" was later changed to "nucleic acid" and eventually to "deoxyribonucleic acid," or "DNA". Somewhat later, he isolated a pure sample known as DNA from the sperm of Salmon. He was the first to identify DNA as a distinct molecule. He also hypothesized that it may serve as the material basis of heredity. In 1899 his student, German chemist Richard Altmann, named it "nucleic acid". Nucleic acids are polymers or chains of nucleotides, linked together by phosphodiester bonds. There are two chemically distinct types of nucleic acids: deoxyribonucleic acid (DNA) and ribonucleic acid (RNA). DNA is a polymer of deoxyribonucleotides and RNA is a polymer of ribonucleotides. Nucleic acids are highly organized macromolecules concerned with the storage, transmission and expression of genetic information. The nucleic acids make up the blue print of life. The function of every protein, and ultimately of every cell constituent, is a product of information programmed into the nucleotide sequence of a cell's nucleic acids. The DNA contained in a fertilized egg encodes the information that directs the development of an organism. The nucleotide sequence of DNA ultimately describe the primary structures of all cellular RNAs and proteins, and through enzymes can indirectly affect the synthesis of all other cellular constituents, determining the size, shape and function of every living thing. DNA molecules are the largest biological molecules for all forms of life.

Nucleosides and Nucleotides

Nucleic acids are polymer of nucleotides. A nucleotide consists of three basic components:

1. A nitrogenous base.
2. A pentose sugar.
3. A phosphate group.

Nitrogenous bases

The nitrogenous bases are derivatives of two parent heterocyclic compounds, purine and pyrimidine (Figure 8.1). Purines have a double-ring structure, whereas pyrimidines have a single-ring structure.

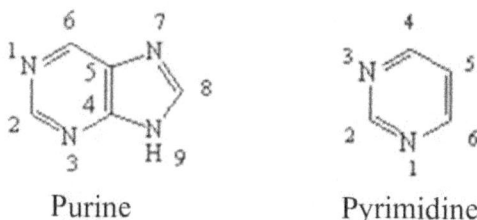

Purine   Pyrimidine

Figure 8.1 Purine and pyrimidine.

DNA and RNA both contain two major purine bases: adenine (A) and guanine (G). DNA and RNA also contain two major pyrimidines; both DNA and RNA contain cytosine (C). The single important difference between the base of DNA and those of RNA is the nature of the second major pyrimidine: DNA contains thymine (T), whereas RNA contains uracil (U). Thymine and uracil differ only in a single methyl group which is present in thymine and absent in uracil. The structures of the two major purine bases are:

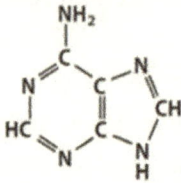

Adenine (A)
(6-aminopurine)

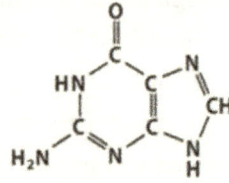

Guanine (G)
(2-amino-6-oxypurine)

The structures of the three major pyrimidine bases are:

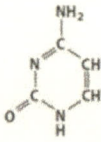

Cytosine (C)
(2-oxy-4-aminopyrimidine)

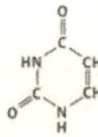

Uracil (U)
(2-4-dioxypyrimidine)

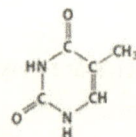

Thymine (T)
(2-4-dioxy-5-methylpyrimidine)

Although nucleotides bearing one of these major bases are most common, both DNA and RNA also contain some minor bases. In DNA the most common of these are methylated forms of the major bases, but in some viral DNAs certain bases may be hydroxymethylated or glucosylated.

Pentose sugar

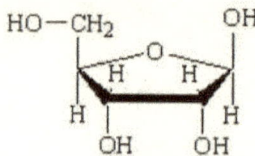

D-ribose

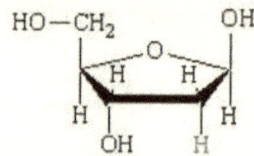

2'-deoxy-D-ribose

Figure 8.2 Ribose and deoxyribose.

Two kinds of pentoses are found in nucleic acids. The recurring deoxyribonucleotide units of DNA contain 2'-deoxy-D-ribose, and the ribonucleotide units of RNA contain D-ribose (Figure 8.2). This diagram misses out the carbon atoms in the ring for clarity. Each of the four corners where there isn't an atom shown has a carbon atom. In deoxyribonucleotides the –OH group on the 2' carbon is replaced with –H i.e., deoxyribose sugar is

230

one oxygen short from ribose. In nucleotides, both types of pentoses are in their β-furanose form (closed 5-member ring form).

The only other thing you need to know about deoxyribose (or ribose) is how the carbon atoms in the ring are numbered. The carbon atom to the right of the oxygen drawn the ring is given the number 1, then CH_2OH side group which is number 5.

Phosphate group

A functional group or radical comprised of phosphorus attached to four oxygen, and with a net negative charge (Figure 8.3). A phosphate group is attached to the sugar molecule in place of the hydroxyl (-OH) group on the 5' carbon.

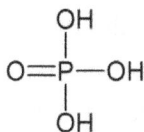

$$O=\overset{\overset{\displaystyle OH}{|}}{\underset{\underset{\displaystyle OH}{|}}{P}}-OH$$

Figure 8.3 Phosphate group.

A nucleoside has a purine- or pyrimidine- derived base attached with a pentose sugar (D-ribose or 2'-deoxy-D-ribose). If the sugar is D-ribose, a ribonucleoside is produced; if the sugar is 2'-deoxy-D-ribose, a deoxyribonucleoside is produced. Numerals 'prime' (*e.g.,* 2' or 5') refer to numbering of the atoms of the sugar. In nucleoside, N-9 of the purine ring, or N-1 of the pyrimidine ring is joined with 1' carbon atom of the sugar through a β-N-glycosidic linkage (Figure 8.4). The β-N-glycosidic bond is formed by removal of the elements of water (a hydroxyl group from the pentose and hydrogen from the base).

A nucleotide is a 5'-phosphate ester of a nucleoside. They are of two types depending upon the type of sugar – ribonucleotides and deoxyribonucleotides. Most commonly, the phosphoryl group is attached to the oxygen of the 5'-hydroxyl. Nucleotides are typically assumed to be 5' – unless otherwise stated. Monophosphates can be further phosphorylated to produce di- and tri-phosphates (Figure 8.4). The phosphate groups are responsible for the negative charges associated with nucleotides, and cause DNA and RNA to be referred to as 'nucleic acids'.

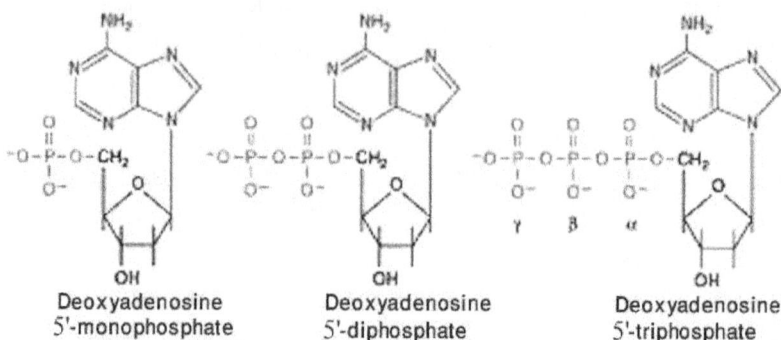

Deoxyadenosine   Deoxyadenosine   Deoxyadenosine
5'-monophosphate   5'-diphosphate   5'-triphosphate

Figure 8.4 Nucleotides and nucleosides.

In addition to serving as precursors of nucleic acids, nucleotides serve unique physiological functions. Nucleoside tri- and di- phosphates, like ATP, GTP, ADP and GDP are energy carriers in metabolic pathways. Nucleotides are also structural components of some essential coenzymes, like coenzyme A, FAD, NAD+, and NADP+. Linked to sugars or lipids, nucleosides constitute key biosynthetic intermediates. The cyclic nucleotides, such as cyclic adenosine monophosphate (cAMP) and cyclic guanosine monophosphate (cGMP), serve as second messengers in signal transduction pathways.

Difference between DNA and RNA

The following are the difference between DNA and RNA:

Parameters	DNA	RNA
Full Form	Deoxyribonucleic acid	Ribonucleic acid
Location	It is found inside the nucleus in the form of chromosomes. A small amount of DNA is found in the cell organelles (mitochondria, chloroplast, and plasmid).	It is found in the cell's cytoplasm but very little is found inside the nucleus.

Structure	It is a double-stranded molecule having longer chains of nucleotides that twists to form a double helix.	RNA is generally single-stranded molecule having shorter chains of nucleotides but intermittently forms a double helix structure.
Function	It stores and transfers genetic information to generate new cells and organisms.	It is used to transfer genetic code from the nucleus to the ribosomes to make proteins.
Type of sugar	A DNA nucleotide contains deoxyribose sugar.	An RNA nucleotide contains ribose sugar.
Base composition	Each DNA nucleotide contains one of the four nitrogenous bases, abbreviated A (adenine), G (guanine), T (thymine), or C (cytosine).	Each RNA nucleotide contains one of the four nitrogenous bases, abbreviated A (adenine), G (guanine), U (uracil), or C (cytosine).
Ratio of bases	In case of DNA: Adenine = Thymine Guanine = Cytosine.	In case of RNA: Adenine $\neq$ Thymine Guanine $\neq$ Cytosine.
Base pairs	DNA base pairs are A::T and G:::C.	RNA base pairs are A::U and G:::C.
Process of replication	DNA is self-replicating	It is synthesised from DNA when needed.
Helix geometry	The helix geometry of DNA is of β-form.	The helix geometry of RNA is of α-form.
Helix produced	DNA produces a regular helix i.e.; it is spirally twisted.	It produces a secondary helix or pseudo helix as its strand may get folded at places.
Occurs as	It occurs in the form of chromosomes or chromatin fibres.	It occurs in ribosomes or forms an association with ribosomes.

Quantity	The quantity of DNA is fixed for the cell.	The quantity of RNA is variable for a cell.
Number	For a particular species, the DNA number remains constant for every cell.	The number of RNA may differ from cell to cell.
Life span	Its life is long.	Its life is short.
Rate of renaturation after melting	Relatively slower.	Fast.
Hydrolyzing enzyme	DNase.	RNase.
Molecular weight	2 to 6 million.	25,000 to 2 million.
Stability	Deoxyribose sugar gives DNA more stability and is stable under alkaline conditions.	Due to ribose sugar RNA is more reactive than DNA and is not stable in alkaline conditions.
Ultraviolet (UV) sensitivity	DNA can be damaged by exposure to UV rays.	RNA is more resistant to damage by UV rays.
Mutation rate	DNA's mutation rate is relatively lower.	RNA's mutation rate is relatively higher.

Phosphodiester Bonds

The chemical bonds by which the sugar components of adjacent nucleotides of both DNA and RNA are covalently linked through the phosphate groups are called phosphodiester bonds. The enzyme that forms the phosphodiester bond between adjacent nucleotides in a nucleic acid chain is called a DNA/RNA polymerase. The 3'-OH (*i.e.*, 3'-OH group of the pentose sugar) of one nucleotide is joined to the 5'-phosphate of the next to form a 3', 5'-phosphodiester bond (Figure 8.5). In the process, a molecule of water is lost (a condensation reaction). This kind of bonding gives rise to the polynucleotide chain with 2 free ends on both sides. That end of the strand which bears a free 5'-phosphate group without phosphodiester bond is called the 5' end. The other end bears a free 3'-hydroxyl group and is called the 3' end. The phosphodiester linkage is not formed by simple dehydration reaction like the other linkages connecting monomers in macromolecules: its formation involves the removal of two phosphate groups.

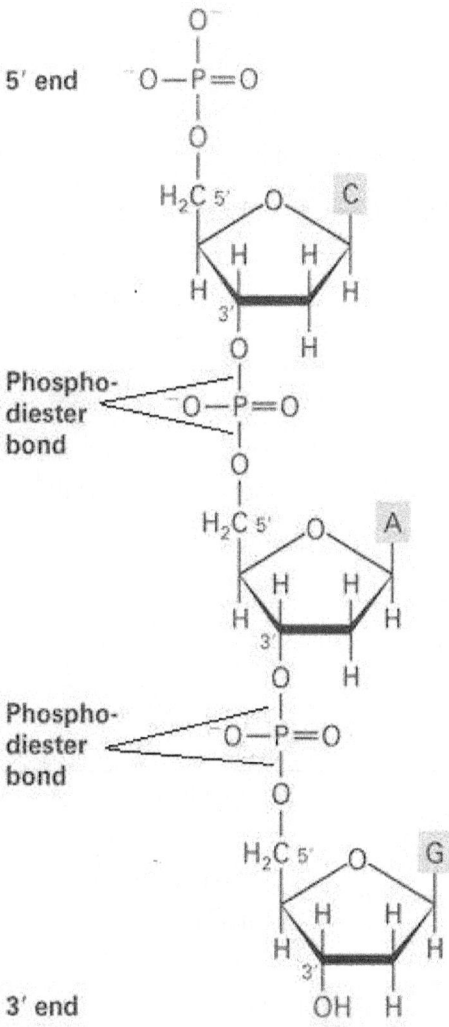

Figure 8.5 Phosphodiester bonds of a polynucleotide.

Nucleoside, Nucleotide and Nucleic Acid Nomenclature

The nomenclature of the nucleoside and nucleotide derivatives of the DNA and RNA bases is somewhat complicated and is summarized in Table 8.1.

Table 8.1 Nucleoside, nucleotide and nucleic acid nomenclature

Base	Nucleoside (Base + Sugar)	Nucleotide (Nucleoside + Phosphate)	Nucleic Acid
Adenine	Adenosine	Adenylate (adenosine 5'-monophosphate), AMP	RNA
	Deoxyadenosine	Deoxyadenylate (deoxyadenosine 5'-monophosphate), dAMP	DNA
Guanine	Guanosine	Guanylate (guanosine 5'-monophosphate), GMP	RNA
	Deoxyguanosine	Deoxyguanylate (deoxyguanosine 5'-monophosphate), dGMP	DNA
Cytosine	Cytidine	Cytidylate (cytidine 5'-monophosphate), CMP	RNA
	Deoxycytidine	Deoxycytidylate (deoxycytidine 5'-monophosphate), dCMP	DNA
Thymine	Deoxythymidine	Deoxythymidylate (deoxythymidine 5'-monophosphate), dTMP	DNA
Uracil	Uridine	Uridylate (uridine 5'-monophosphate), UMP	RNA

Chargaff's Rules

Studying the composition of DNA from many different organisms, Erwin Chargaff, an Austrian-American biochemist established certain empirical rules about the base compositions of DNA in 1950 that helped the discovery of the double helix model of the DNA. These are:

1. The base composition of DNA generally varies from one species to another.
2. DNA specimens isolated from different tissues of the same species have the same base composition.
3. The base composition of DNA in a given species does not change with the organism's age, nutritional state, or changing environment.
4. In all DNAs regardless of the species, the number of adenine bases is equal to the number of thymine bases (that is, $A = T$), and the number of guanine bases is equal to the number of cytosine bases (that is, $G = C$). From these relationships it follows that the sum of the purine bases equals the sum of the pyrimidine bases (1:1 ratio of pyrimidine and purine bases); that is, $A + G = T + C$.

Experimental Proof of DNA as the Genetic Material

In 1914 Robert Feulgen developed a method for staining DNA (now known as the Feulgen stain); however, the connection between DNA and heredity was not made until many years later. A series of experiments beginning in the 1920s finally revealed that DNA was the genetic material.

Frederick Griffith's Transformation Experiment

The discovery of the genetic role of DNA began with research by British bacteriologist Frederick Griffith. In 1928, he designed and performed a series of experiments on mice infected with Streptococcus bacteria. He was working with two strains of the bacterium *Streptococcus pneumoniae* identified as type III-S (smooth) and type II-R (rough) strain. When a bacterial cell is grown on solid medium, it undergoes repeated cell divisions to form a visible clump of cells called a colony. The III-S strain had a polysaccharide coat around each cell, causing the colonies to have a smooth, glossy appearance. The II-R strain lacked the coat, and its colonies had a rough appearance. More importantly, the III-S strain was virulent: when injected into mice, they developed severe pneumonia and died. The II-R strain was avirulent: it did not kill the mice upon injection. If the mice were injected with heat-killed III-S cells, the mice all lived. The surprising result was found when mice were injected with a mixture of live II-R cells and heat-killed III-S cells, each of which by itself did not harm the mice. Unexpectedly, the mice developed an infection and died. The blood of the dead mice was found to contain both live III-S and live II-R bacteria (Image 8.1). Dead changing the living! He concluded that some substance (transforming principle), transferred from the heat-killed, virulent III-S bacteria to the live, coatless II-R bacteria in the mixture made the live II-R bacteria into the virulent III-S bacteria. Griffith called this change of nonvirulent bacteria into virulent pathogens transformation. Today, we know that the 'transforming principle' Griffith observed was the DNA of the III-S strain bacteria. While the bacteria had been killed, the DNA had survived the heating process and was taken up by the II-R strain bacteria.

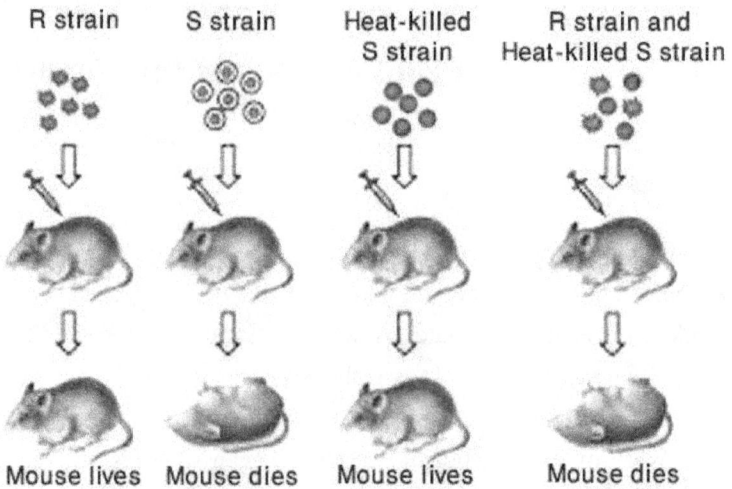

Image 8.1 Diagram illustrating Griffith's transformation experiment.

Avery's Transformation Experiment

Transformation in Streptococcus was originally discovered in 1928, but it was not until 1944 that the chemical substance responsible for transforming the R cells into S cells was identified. In 1944, Oswald Avery, Maclyn McCarty, and Colin MacLeod demonstrated that the component of the cell responsible for the phenomenon of transformation in the bacterium *Streptococcus pneumoniae* was DNA. In doing these experiments, they purified biochemicals (proteins, DNA, RNA, *etc.*) from the heat-killed S strain bacteria

(*Streptococcus pneumoniae*). They mixed various components (such as lipids, polysaccharides, protein, DNA, RNA) of the S cells with live R cells, to determine which component caused the transformation. They found the DNA by itself caused the transformation, and no other component had any effect. They concluded that DNA is the hereditary material, but not all biologists were convinced.

Hershey-Chase Bacteriophage Experiment

In 1952, Alfred D. Hershey and Martha Chase conducted a series of experiments and demonstrated that DNA was the genetic material in the T2 bacteriophage. An important element of the Hershey-Chase experiment is that DNA contains phosphorus but not sulfur, while protein contains sulfur but not phosphorus. Phage contains only DNA and protein, and no other types of molecule. Thus, it is possible to label the two types of molecule independently, with radioactive phosphorus (^{32}P) and sulfur (^{35}S). They used different radioactive isotopes to label DNA and protein coat of the bacteriophage. They grew some bacteriophages on a medium containing radioactive phosphorus (^{32}P) to identify DNA and some on a medium containing radioactive sulphur (^{35}S) to identify protein. Viruses grown in radioactive sulfur (^{35}S) had radiolabeled proteins. Viruses grown in radioactive phosphorus (^{32}P) had radiolabeled DNA. Then, these radioactive labeled phages were allowed to infect *E. coli* bacteria. After infecting, the protein coat of the bacteriophage was separated from the bacterial cell by blending and then subjected to the process of centrifugation. Since the protein coat was lighter, it was found in the supernatant while the infected bacteria got settled at the bottom of the centrifuge tube. They showed that ^{32}P-labeled DNA entered the bacterial cells when the phage infected them, and that the new generation of phage contained a significant amount of ^{32}P-labeled DNA. In contrast, the ^{35}S-labeled protein stayed outside the cells during an infection, and none of it ended up in the new phage. This implies that DNA is necessary for phage replication. They concluded that the injected DNA of the phage provides the genetic information.

A few exceptions were found where the genetic material of some types of viruses is RNA, not DNA.

Tobacco Mosaic Virus (TMV) Experiment

Tobacco mosaic virus (TMV) is a positive-sense single-stranded RNA virus that infects a wide range of plants, especially tobacco and other members of the family Solanaceae. In 1957, Heinz Fraenkel-Conrat and B. Singer conducted experiments on TMV and demonstrated that RNA is the genetic material of TMV. Different strains of TMV can be identified on the basis of differences in the chemical composition of their protein coats and difference in symptoms on the tobacco leaves. By using the appropriate chemical treatments, proteins and RNA of TMV can be separated.

In one experiment protein and RNA components of the TMV were separated and both were used to infect the tobacco leaf separately. It was observed that in case of protein subunits, there was no symptoms on the leaf and no progeny viruses were obtained. RNA part caused the infection and symptoms appeared on the tobacco leaf. Fresh progeny of TMV was also obtained. In the other experiment, two strains of TMV (type A and type B) showing different symptoms (one causing spots in random pattern and the other in a definite ring form) were taken. There protein and RNA parts were separated and chimera (hybrid) viruses were synthesized using RNA of type A and protein of type B and *vice-versa*. These chimera/hybrid viruses were used to infect the tobacco leaves. It was observed that symptoms on the leaf always belonged to the virus strain from which RNA was taken. Fresh progeny also belonged to the same strain. These experiments proved that the genetic information of TMV is stored in the RNA and not in the protein.

Nucleic Acid Structure

Nucleic acid structure refers to the structure of nucleic acids such as DNA and RNA. Nucleic acid structure is often divided into four different levels: primary, secondary, tertiary and quaternary. Primary structure consists of a linear sequence of nucleotides that are linked together by phosphodiester bonds. Secondary structure is the interaction between the bases. The secondary structure of the DNA is predominantly the base pairing of the two polynucleotide strands forming a double helix. Tertiary structure refers to large-scale folding in a linear polymer that is at a higher order than secondary structure. The tertiary structure is the specific three-dimensional shape into which an entire chain is folded. Quaternary structure is the higher-level of organization of the nucleic acids. This structure refers to the interactions of the nucleic acids with the other molecules. The most commonly seen organization is the form of chromatin which shows interaction with small proteins histones.

Forms/Types of DNA

The DNA molecule exists in four different forms under different conditions. These are terms A-form, B-form, C-form, and Z-form DNA that differ in their structural and chemical properties (Image 8.2).

B Form of DNA (B-DNA)

B-DNA is the Watson-Crick's classic, right-handed double helix that most people are familiar with. It is the most commonly found form of DNA under the natural physiological conditions like 92% relative humidity, at neutral p^H, and physiological salt concentrations in the cells. It has 10 base pairs per turn from the helix axis. There is a distance of 3.4Å vertically per base pair with a helical diameter of 20Å. The helix of B-DNA is narrower and longer than A-DNA. B-DNA is the most stable of all the DNA structures in natural physiological conditions.

A Form of DNA (A-DNA)

It is a thicker right-handed duplex with a shorter distance between the base pairs, found at a relative humidity of 75%. In an environment where there is a very high salt concentration or ionic concentrations, such as K^+, Na^+, Cs^+, or in a state of dehydration it endures in a form that contains 11 base pairs per turn with a rise of 2.56Å vertically per base pair. It possesses the broadest helical diameter amongst all DNA forms – 23Å DNA which is a typical helix with a rotation of 32.7^0 per base pair. A major difference between A-form and B-form nucleic acid is the placement of base-pairs within the duplex. In B-form, the base-pairs are almost centered over the helical axis, but in A-form, they are displaced away from the central axis and closer to the major groove. The result is a ribbon-like helix with a more open cylindrical core in A-form. It occurs naturally but is not as common as B-DNA. It is found in hybrid

DNA-RNA structures, including DNA-RNA hybrids formed during DNA replication and transcription. It is also found in some bacterial and viral genomes.

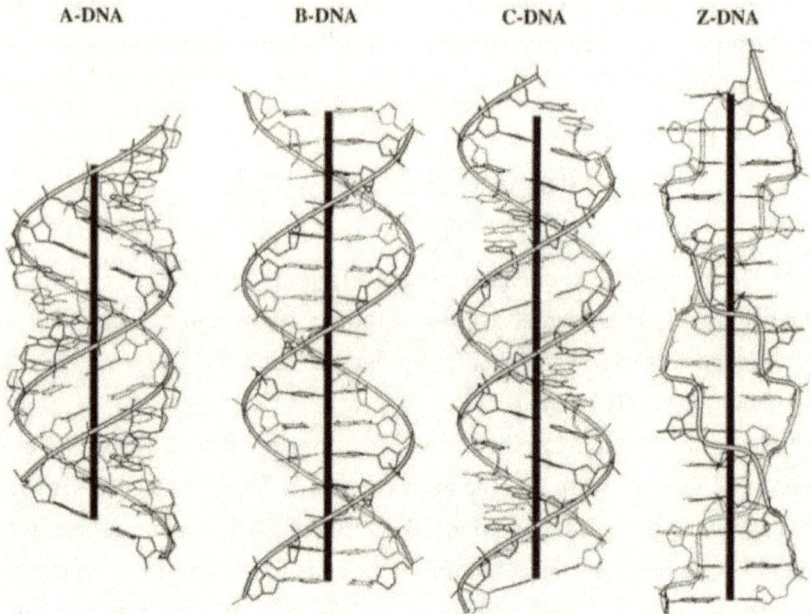

A-DNA   B-DNA   C-DNA   Z-DNA

Image 8.2 Different forms of DNA (source:
https://www.pinterest.com/pin/829577193835856261/).

Z-DNA

A third form of duplex DNA has a strikingly different, left-handed helical structure with a 4.4-nm turn length and 12 base pairs per turn. It is found in an environment with a very high salt concentration by stretches of alternating purines and pyrimidines, *e.g.*, GCGCGC, especially in negatively supercoiled DNA. A small amount of the DNA in a cell exists in the Z-form. The backbone is arranged in a zig-zag (hence the name) pattern formed by the sugar-phosphate linkage wherein the recurrent monomer is the dinucleotide in contrast to the mononucleotide, which is observed in alternate forms. It is long and thin as compared to B-DNA. The Z-form helix's diameter is 18Å, and the base-pair tilt is 7Å, which is lower than the B and A-form. It's major groove's width is 2Å, whereas the width of minor groove is 8.8Å. The Z-DNA has been found in a large number of living organisms including mammals, protozoans and several plant species.

C-DNA

It is rare and occurs only under high salt concentration and low water activity, and it is highly susceptible to mutation or damage. It is found at a relative humidity of 66% and in presence of Li^+ and Mg^{2+} ions. Every turn has 9.33 base pairs. The right-handed helix has a diameter of about 19Å, and the vertical rise for every base pair is 3.32Å . The rotation per base pair is 38.6 degrees, with a base pair tilt of 7.8 degrees. The size and shape of the C-DNA are smaller than the B-DNA and A-DNA.

All four forms of DNA are important in various biological processes and responses, and are characterized by their distinct structural and chemical properties.

The Watson-Crick DNA Double Helix Model

Four scientists are generally recognized as having contributed to the elucidation of DNA structure. Rosalind Elsie Franklin and Maurice Wilkins used X-ray diffraction techniques to identify key properties of the DNA molecule. Wilkins shared this data (without Franklin's permission) with two other scientists, James D. Watson (an American) and Francis H.C. Crick (an Englishman). Watson and Crick in 1953, proposed the three-dimensional model of DNA structure known as the double helix, which explains the gene functions in molecular terms (Image 8.3). Their work was based on X-ray crystallography of Rosalind Franklin and Maurice Wilkins, and the base equivalences in DNA discovered by Chargaff. In 1962, Watson, Crick and Wilkins were awarded the Nobel Prize (Franklin's critical efforts were not recognized) for their determination of the structure of DNA. Watson and Crick DNA model was reconstructed largely from its original pieces in 1973 and donated to the National Science Museum in London. The important features of their model of DNA are as follows:

1.   The DNA molecule consists of two polynucleotide strands twisted around one another to form a right-handed (clockwise) double helix. The two strands are coiled around the same helical axis and are intertwined with themselves. One consequence of this intertwining is

that the two strands cannot be separated without the DNA rotating. Watson and Crick also proposed that DNA is shaped like a long zipper that is twisted into a coil like a spring.

2. The diameter of the helix is about 2.0 nm or 20 Angstroms (1 nm = 10^{-9} meter).

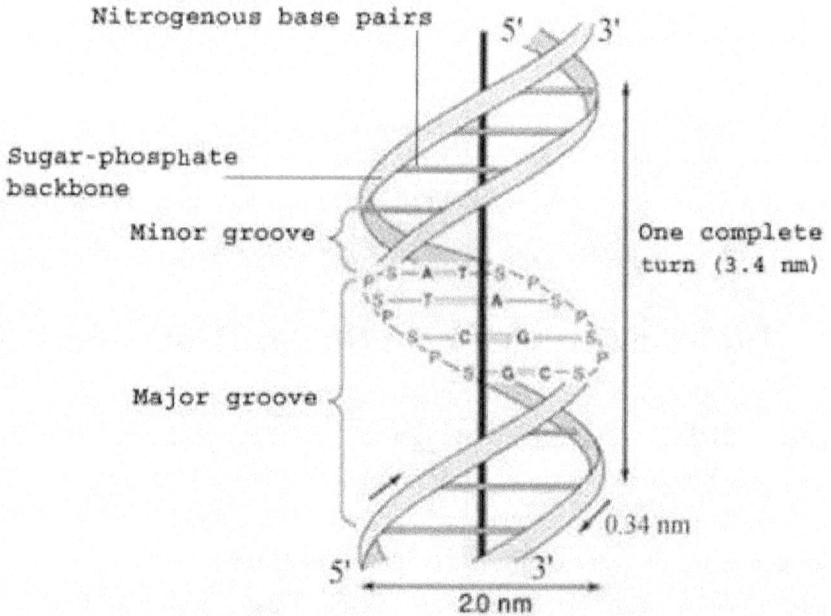

Image 8.3 Features of the DNA double helix.

3. The two strands are antiparallel in nature. They are oriented in opposite directions with one strand runs in the 5' to 3' direction and the other in the 3' to 5' direction so that the 3' end of one strand faces the 5' end of the other strand.

4. The sugar and phosphate part of the molecule forms a sugar-phosphate backbone which is on the outside of the helix. The nitrogenous bases are stacked in the interior in pairs, like the steps of a staircase.

5. Double-stranded DNA is held together by two types of chemical bonds: phosphodiester bonds and hydrogen bonds. Nucleotides are joined by phosphodiester bonds, and the strands are held together by hydrogen bonds. The phosphodiester bond is much stronger of the

two because it is a covalent bond.

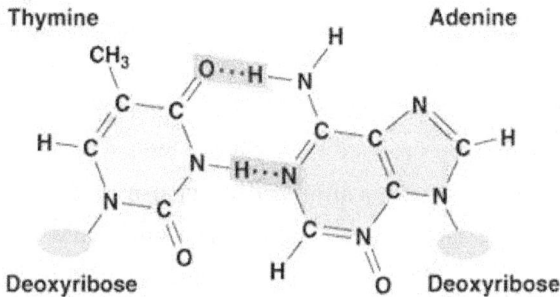

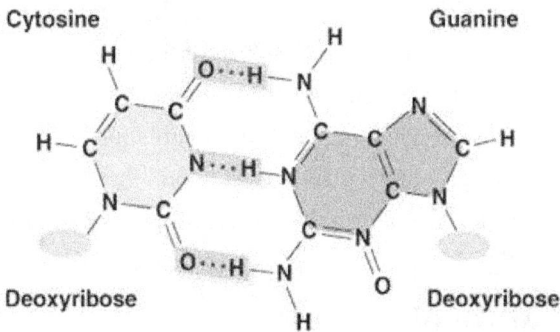

Image 8.4 Complementary base pairing.

6. Each base of one strand is paired in the same plane with a base of the other strand by relatively weak hydrogen bonds. The only specific pairing observed are A with T and G with C. Two hydrogen bonds can form between A and T, and three can form between G and C (Image 8.4). The specific A-T and G-C pairs are called complementary base pairs, so the nucleotide sequence in one strand dictates the nucleotide sequence of the other. For instance, if one chain has the sequence 5'-TATTCCGA-3', the opposite antiparallel chain must bear the sequence 3'-ATAAGGCT-5'. The base pairing scheme immediately suggests a way to replicate and copy the genetic

information.

7. The base pairs are 0.34 nm apart in the double helix. A complete (360°) turn of the helix takes 3.4 nm; therefore, there are 10 base pairs per turn. Each base pair, then, is twisted 36° clockwise with respect to the previous base pair.

8. The beta-glycosidic bonds between sugar and bases which are paired are not directly opposite to each other but are positioned at an angle. This results in unequally spaced sugar-phosphate backbones and gives rise to two grooves along the DNA molecule: the **major (wide) groove** and the **minor (narrow) groove** of different width and depth. The **grooves** are opposite to each other, and each runs continuously along the entire length of the DNA molecule. The major groove is deep and wide, while the minor groove is narrow and shallow. The depths of the minor and major grooves are 0.85 nm and 0.75 nm, respectively. The width of the major groove ranges from ~22 to 26Å, while that of the minor groove ranges from ~12 to15Å. These structural features are essential for DNA-protein interactions, replication, and the regulation of gene expression. Regulatory proteins typically bind within the major groove, whereas the minor groove often serves as an attachment site for histones.

The Nature of Genetic (Hereditary) Material

Genetic material is that substance which not only controls the inheritance of traits from one generation to the next but is also able to express its effect through the formation and functioning of the traits. Long before DNA and RNA were proved to carry genetic information, geneticists recognized that to function as a genetic material, a molecule must have four principal characteristics or abilities:

1. It must contain all the information encoding the organism's structure, function, development and reproduction.

2. It must replicate accurately, so that progeny cells have the same genetic makeup as the parental cell.

3. It must be stable enough to preserve necessary information and yet

unstable enough to allow for changes that are necessary to correct mistakes made during replication and to allow for some mutation to permit evolution.

4. It must be capable of transferring its information to the cell on demand.

A series of experiments proved that the genetic material of organisms consists of one of two types of nucleic acids, DNA or RNA. All cellular life forms (prokaryotes and eukaryotes), and most viruses have DNA as their genetic material. Some viruses have RNA as their genetic material.

Virus Genetic Material

The genetic material of viruses is DNA or RNA (but not both), which may be single-stranded (ss) or double-stranded (ds), or partially double-stranded, linear or circular, and often free of bound proteins. It is circular when packaged inside the virus particle. The form of genetic material varies among organisms and their viruses.

Single-stranded (ss) virus genomes may be positive-sense (+ sense, messenger sense) or negative-sense (- sense, antisense) or ambisense (a combination). The positive sense is directed from 5' to 3' *i.e.*, the same polarity or nucleotide sequence as the **mRNA**, and the negative sense is *vice versa*. Only the positive-sense ssRNA can serve as mRNA and can be directly translated inside the host cell. Therefore, viral proteins are expressed upon entrance of the viral genomic RNA (gRNA) into the cell. Whereas the negative-sense ssRNA serves as templates for the synthesis of mRNA. It should be first transcribed into the positive sense, and then undergoes translation. Those with negative RNA genomes are themselves divided into segmented and non-segmented (*i.e.*, packaged in segments or as one piece).

Genomes of ambisense viruses are partly of positive and partly of negative polarity. The 5' part of an ambisense RNA is of positive polarity containing an open reading frame (ORF) that can theoretically be directly translated. The 3' part of this same RNA is of negative polarity. Indeed, this second part of the ambisense RNA must be transcribed prior to translation. To express this second part, ambisense viruses also encapsidate their RNA dependent RNA

247

polymerase (Ihara *et al.*, 1984).

Although RNA and DNA polymerases make errors at about the same rate, RNA genomes typically accumulate mutations at much higher frequencies than DNA genomes.

Viral genomes exhibit remarkable diversity in both size and gene content. At the smallest end, the **porcine circovirus** has a genome of less than 2 kilobases (kb) and encodes only two genes, while the **satellite tobacco necrosis virus (STNV)** is even more minimal, containing just a single gene. In contrast, giant viruses such as **Pandoravirus salinus** have genomes exceeding 2 megabases (Mb) in length and encode more than 2,500 proteins. Similarly, large DNA viruses like the **poxviruses** possess genomes with around 240 genes, making them vastly more complex than the tiny RNA viruses. This wide range reflects the evolutionary adaptability of viruses, allowing them to exist in highly simplified forms dependent on host machinery or in relatively complex forms with many genes dedicated to replication, host interaction, and immune evasion.

DNA viruses have DNA genomes that are replicated by either host or virally encoded DNA polymerases. The genome of the DNA viruses is usually dsDNA and range in size from 3.2 to 280 kbp; ssDNA is less common ranging from 1.8 to 5 kb, as during replication, ssDNA typically expands to dsDNA. DNA viruses are present everywhere, especially in the marine ecosystem. They infect both eukaryotes and prokaryotes. Examples of DNA viruses are herpes, smallpox, hepatitis B, papilloma, adenoviruses, and warts.

The genomes of the RNA viruses are much smaller than those of most DNA viruses. Nearly all plant viruses (*e.g.*, Tobacco Mosaic Virus (TMV), Turnip Yellow Mosaic Virus (TYMV), Wound Tumor Viruses, Cucumber Mosaic Virus, Tomato Spotted Wilt Virus, *etc.*) utilize RNA as their genetic material. A significant portion of animal viruses have RNA as their genetic material. Examples of RNA animal viruses include those from the families Picornaviridae (like poliovirus), Togaviridae (like Venezuelan equine encephalitis virus), Orthomyxoviridae (like influenza viruses), Rhabdoviridae (like rabies virus), and Coronaviridae (like SARS-CoV-2). Other examples include Reoviridae (like rotavirus), Retroviridae (like HIV), and Bunyaviruses. Many bacteriophages, such as MS2, also employ RNA as their genetic material. In fact, the majority of plant viruses (77%) have single-stranded (ss) RNA genomes, with a significant number possessing positive-sense RNA. The organisms, which have only RNA – employ RNA in their genetic mechanisms, that type of RNA is called genetic RNA. The genetic RNA is self-replicating. The non-genetic RNA (mRNA, tRNA, rRNA) is not self-replicating. It has DNA dependent replication.

Bacterial Chromosome and Prokaryotic Genetic Material

Bacteria represent the vast majority of biological diversity found on Earth. In contrast to the linear chromosomes found in eukaryotic cells, most bacteria have a single, large, covalently closed, circular chromosome. However, many bacteria, such as *Brucella, Rhodobacter sphaeroides, Burkholderia cepacian,* and *Vibrio cholerae*, have multiple chromosomes. Some bacteria, including *Borrelia burgdorferi, Streptomyces coelicolor* and *Agrobacterium tumefaciens*, have linear chromosomes. The chromosomal DNA is located in the nucleoid region of the cytoplasm, where it forms a dense, gel-like mass. Bacterial chromosomes are generally ~1000 times longer than the cells in which they reside. The DNA is then highly compacted through looping and supercoiling, facilitated by DNA-binding proteins to allow the chromosome to fit into the small space inside the cell. Bacterial chromosomes have a single, unique replication origin (named oriC), from which DNA synthesis starts. Unlike eukaryotic chromosomes, bacterial chromosomes lack histones, centromeres, and telomeres, though archaeal DNA shows greater organization and is packaged within nucleosome-like structures. Prokaryotic chromosomes typically contain a single origin of replication, called **oriC**, where DNA synthesis begins, though some archaea may have multiple origins.

Bacteria contain much more DNA than DNA viruses. Nucleus is absent in prokaryotes. The prokaryotic genetic material is usually concentrated in a specific clear region of the cytoplasm called nucleoid. Prokaryotes like the bacteria and archaea typically have a single chromosome. The chromosome consists of a single DNA double helix that is usually circular, lacks telomeres and centromere and has relatively few proteins associated with it. It does not contain any histone protein. The DNA is packaged by DNA-binding proteins. The bacterial DNA is packaged in loops back and forth. Some bacteria in exceptions contain a single linear chromosome. Certain bacteria like the *Borrelia burgdorferi* possess array of linear chromosome like eukaryotes.

The base sequences in prokaryotic chromosomes are less than in eukaryotic cells. The chromosome size of most bacteria is from only 160,000 base pairs to 12,200,00 base pairs. Bacterial chromosomes have a single origin of replication from which the replication starts. In some archaea there are multiple replication origins. The prokaryotic genes are organized into operons and it usually does not contain introns. The DNA of the archaea are more organized, they are packaged within structures similar to eukaryotic nucleosomes. The chromosomes in the prokaryotes and plasmids are generally supercoiled like that of the eukaryotes.

In addition to the main circular chromosome found in the nucleoid, many species of bacteria contain one or more small, circular and closed DNA molecules, called plasmids. They generally remain floated in the cytoplasm and bear different genes based on which they have been studied.

Eukaryotic Genetic Material

In comparison with prokaryotes, the genomes of eukaryotes contain a lot of DNA. The contour length of the entire DNA in a single human cell is about 2m (1m per chromosome set), compared with 1.3 mm for prokaryote *E. coli* DNA. The average human body consists of approximately 75 trillion cells, and therefore contains a total of about 2×10^{10} km of DNA. Compare it with the distance from the earth to the sun, which is 1.5×10^8 km (150 billion meter). The average human has enough DNA to go from the earth to the sun more than 400 times. This peculiar fact makes the points that the DNA of eukaryotes is obviously efficiently packed. In fact, the packing occurs at the level of the nucleus, where the 2m of DNA in a human cell is packed into 46 chromosomes, all in a nucleus 0.006 mm in diameter.

Eukaryotes have, in almost all cases, a number of different chromosomes which are contained within the nucleus. Each eukaryotic chromosome consists of one linear, unbroken, double-stranded DNA molecule running throughout its length, and the DNA is tightly associated with large amount of proteins. Each chromosome has multiple origins of DNA replication, has a telomere at each end and a centromere.

The Components of DNA

There are three basic components of DNA: a phosphate, a 2'-deoxyribose sugar, and a nitrogenous base. From these components cells construct the precursors of DNA, the deoxyribonucleotides, most often referred to simply as nucleotides (deoxyribonucleoside 5'-triphosphate). The phosphate group gives DNA its acidic properties and a negative charge. *In vivo*, unless these charges are neutralized, it could not be possible to pack the huge DNA molecule into the cell's nucleus. Neutralization is brought about by reaction of basic proteins with the acidic DNA in both eukaryotes and prokaryotes. In eukaryotes, histones are the basic proteins involved in packaging DNA, and in prokaryotes, polyamines are the basic proteins.

Some sequences of DNA do not code for amino acids but instead are necessary as signal sequences. Such sequences serve as the origins for DNA replication, as regulatory sequences, and as sites for chromosome folding.

Eukaryotes also show a large range of DNA per diploid nucleus. For example, Drosophila contains 0.2 pg (picogram), human 6.4 pg (1 pg = 10^{-12} gram) DNA per nucleus. Interestingly, only a very small percentage of this DNA appears to actually code for proteins or RNAs. Some estimates place the amount as less as 1.5-2.0% of the total. What function most of the other DNA serves is still a mystery!

DNA functions to store the complete genetic information required to specify the structure of all the proteins and RNAs of each species of organisms, to program in time and space the orderly biosynthesis of cell and tissue components, to determine the activities of an organism throughout its life cycle, and to determine the individuality of a given organism. DNA is the basis for all the processes and structures of organisms. Hereditary information is encoded in DNA and reproduced in all cells of the body. This DNA program directs the development of biochemical, anatomical, physiological, and (to some extent) behavioral traits.

The sequence of bases in DNA molecule carries genetic information whereas their sugar and phosphate groups perform a structural role. Let us now look at what happens to the number of possible nucleotide combinations as the DNA molecule increases in length. The number of possible combinations of nucleotides is given by the term 4^n, where 4 is the number of nucleotides (A, T, G, C) and n is the number of nucleotides in the molecule (single strand); that is, n is the length of the molecule. So, for a molecule made up of 3 nucleotides there are 64 possible combinations ($4^3 = 64$). A molecule of 5 nucleotides could have $4^5 = 1024$ different combinations. A chain just 10 nucleotides long could be sequenced to yield more than a million different molecules (4^{10}). Consider, then, the possible combinations from the billions of nucleotides found in human cells!

Molecular Structure of RNA

RNA (ribonucleic acid) is generally a single-stranded molecule, although in some viruses it occurs in a double-stranded form. Unlike DNA, which forms a stable and uniform double helix, RNA does not adopt a regular helical structure and instead displays greater flexibility. The backbone of RNA is composed of ribose sugars linked to phosphate groups through phosphodiester bonds. A unique feature of ribose is the presence of a hydroxyl group (-OH) at the 2′ carbon, which makes RNA chemically more reactive and less stable than DNA, whose sugar, deoxyribose, lacks this group.

RNA contains four nitrogenous bases: adenine (A), guanine (G), cytosine (C), and uracil (U), with uracil replacing the thymine found in DNA. Although RNA is single-stranded, it is not a simple, linear chain. Through internal base pairing, it can fold back on itself to form secondary structures such as hairpins, loops, and bulges. These structures are stabilized by hydrogen bonds, base stacking, and other interactions, allowing RNA to adopt complex three-dimensional shapes. This structural adaptability is crucial because it enables RNA molecules to perform a wide variety of cellular roles.

RNA-dependent RNA polymerase

RNA-dependent RNA polymerase (RdRp), also known as RNA replicase, is an enzyme that catalyzes the synthesis of RNA from an RNA template. Unlike DNA-dependent RNA polymerases, which transcribe RNA from a DNA template, RdRp uses RNA itself as the template, making it essential for the replication of RNA genomes.

RdRp is a crucial enzyme for most RNA viruses that lack a DNA stage in their life cycle, such as **SARS-CoV-2**, the virus responsible for COVID-19. By enabling the replication of viral RNA, RdRp plays a central role in viral proliferation and infection. Due to its essential function, RdRp is also a major target for antiviral drugs aimed at inhibiting viral replication (Zanotto *et al.*, 1996).

RNA Replication

RNA replication is the process by which new copies of genome-length RNAs are made. RNA replication occurs in the cytoplasm of virus-infected cells during the propagation of RNA viruses with the help of virus-encoded RNA-dependent RNA polymerase (RdRp). The process of RNA replication involves synthesis of full-length genomic complementary RNAs (anti-genomes) that then serve as templates for synthesis of genomic RNAs. These genomic RNAs can subsequently be **translated, further replicated, or packaged** into new virus particles. Despite extensive research, many of the individual steps underlying RNA replication remain **poorly understood**. Notably, viral RNA replication lacks **proofreading activity**, which contributes to the high mutation rates of RNA viruses.

The RNA replication process is a two-step mechanism. First, the initiation step of RNA synthesis begins at or near the 3' end of the RNA template by means of a primer-independent (*de novo*), or a primer-dependent mechanism using a viral protein genome-linked (VPg) primer. In *de novo* initiation, a **nucleoside triphosphate (NTP)** is added to the 3'-OH of the initiating NTP. During the following so-called elongation phase, this nucleotidyl transfer reaction is repeated with subsequent NTPs to generate the complementary RNA strand.

During replication of RNA viruses, at least three types of RNA molecules must be synthesized: **Genomic RNA** (for packaging into new virions), **Complementary copy of the genome** (anti-genome), and **mRNAs** (for translation of viral proteins). Some RNA viruses also synthesize copies of subgenomic mRNAs. RdRp is the central enzyme for all RNA synthesis processes. The RdRp and other proteins required for viral genome synthesis are often called the replicase complex. In addition to the RdRp, the replicase complex may contain RNA-helicases (to unwind highly base-paired regions of the RNA genome) and NTPases (to provide energy for the polymerization process). The composition of the replicase complex **varies among virus families** and may also require **host cell proteins**.

Function of RNA

RNA is formed from DNA in the nucleus. Most of its modifications also occur in the nucleus and then it is transferred to the cytoplasm. Although they are much shorter than DNAs, they are much abundant in most cells. The RNA molecules are essential for cell function in both prokaryotes and eukaryotes. In both prokaryotic and eukaryotic cells, the three major classes of RNA are mRNA, tRNA and rRNA. All three types of cellular RNA are involved in the synthesis of proteins. Each consists of a single strand of ribonucleotides, and each has a characteristic molecular weight, nucleotide sequence, and biological function.

Messenger RNAs (mRNAs)

Messenger RNAs (mRNAs) make up about 2-5% of the total RNA of the cell. They are the only RNA molecules that makes a copy of the genetic information/code in the DNA through the process of transcription. This coded information is then delivered to ribosomes, where it is translated into proteins. It is coded so that every three nucleotides (a codon) correspond to one amino acid. The ribosomes "read" this information and use it to assemble amino acids into proteins. Messenger RNAs get their name because they act as messengers between DNA and ribosomes.

Transfer RNA (tRNA) or soluble RNA (sRNA) – the adapter molecule

Transfer RNA (tRNA) is the second most abundant type of RNA. tRNA constitutes about 15% of the total RNA. It is a small RNA chain of about 80 nucleotides. Each of the 20 amino acids found in proteins has one or more corresponding tRNAs. Each tRNA has an amino acid binding site as well as a specific trinucleotide sequence, called its anticodon on the opposite end that is complementary to the mRNA codon. The anticodon recognizes the codon and thus allows the tRNA to deliver the correct amino acid to the ribosome. Transfer RNA picks up specific amino acids, transfers the amino acids to the ribosomes, and inserts the correct amino acids in the proper place according to the mRNA message. The enzyme aminoacyl tRNA synthetase, also called tRNA-ligase, combines an amino acid to its tRNA. Each tRNA molecule has a well-defined tertiary structure that is recognized by the enzyme aminoacyl-tRNA synthetase which attaches the correct amino acid onto the 3'-end (amino acid acceptor end) of its corresponding tRNA in a process called charging of tRNA. There is a specific enzyme for each amino acid.

Ribosomal RNA (rRNA, ~120-5000 nucleotides) or insoluble RNA

This comprises 80% of the total RNA of the cell. It is one of the structural components of the ribosome. Ribosomal RNA associates with ribosomal proteins to form the complete, functional ribosome, which are the machines used for protein synthesis.

In addition to these major classes, there are wide variety of special function RNAs.

Small nuclear RNA (snRNA, ~60-360 nucleotides)

An abundant class of RNA molecules only found in the nucleus of eukaryotes. Small nuclear RNAs (snRNAs) are critical components of the spliceosome that catalyze the splicing of pre-mRNA. Some are involved in processing heterogenous nuclear RNA.

Small cytoplasmic RNA (scRNA)

Small RNAs found in the cytoplasm usually complexed with proteins in scRNPs (small cytoplasmic ribonucleoproteins), involved in protein trafficking with the cytoplasm.

Heterogeneous nuclear RNA (hnRNA)

Heterogeneous nuclear RNAs (hnRNAs) refer collectively to the variety of

RNAs found in the nucleus, including primary transcripts, partially processed RNAs and snRNA. However, the term hnRNA is often used just for the unprocessed primary transcripts. The complexity of the hnRNA is 4 to 10 times greater than that of the mRNA. hnRNA is processed by spliceosomes, small organelles in the nucleus that are composed of protein and RNA.

MicroRNA (miRNA)

The microRNAs (miRNAs) are short regulatory single-stranded noncoding RNA molecules that are approximately 21-24 nucleotides long. The miRNAs are transcribed in the nucleus as longer pre-miRNAs. These pre-miRNAs are subsequently chopped into mature miRNAs by a protein called dicer. These mature miRNAs recognize a specific sequence of a target RNA through complementary base pairing. miRNAs, however, also associate with a ribonucleoprotein complex called the RNA-induced silencing complex (RISC). RISC binds a target mRNA, along with the miRNA, to degrade the target mRNA. Together, miRNAs and the RISC complex rapidly destroy the RNA molecule. The transcription of pre-miRNAs and their subsequent processing is also tightly regulated. miRNAs are found in many mammalian cell types as well as in extracellular circulating miRNAs. Up to date more than 2500 microRNAs have been described. It is interspersed in whole genome. microRNAs are functional in translational silencing and post-transcriptional regulation of gene expression.

Ribozymes (Ribonucleic Acid Enzymes)

Most biological processes do not happen spontaneously. The central role for many proteins in a cell is to catalyze chemical reactions that are essential for the cell's survival. These proteins are known as enzymes. For many years scientists assumed that proteins alone had the structural complexity needed to serve as specific catalysts in cells. In the early 1980s, however, research groups led by Sidney Altman and Thomas R. Cech independently discovered that some biological catalysts are made of RNA. This class of catalytic RNAs is known as ribozymes. In 1989 they were awarded the Nobel Prize in Chemistry for their discovery of catalytic properties of RNA. Ribozymes, also termed catalytic RNA, are RNA molecules that are capable of catalyzing specific biochemical reactions, similar to the action of protein enzymes. Ribozymes are found in the ribosome of both prokaryotes and eukaryotes where they join amino acids together to form protein chains. Ribozymes are also found in the nucleus, mitochondria, and chloroplasts of eukaryotic organisms. Some viruses, including several bacterial viruses, also have ribozymes. Almost all ribozymes are involved in processing RNA. They act either as 'molecular scissors' to cleave precursor RNA chains (the chains that form the basis of a new RNA chain) or as 'molecular staplers' that ligate two RNA molecules together. Ribozymes also play a role in other vital reactions such as RNA splicing, transfer RNA biosynthesis, and viral replication. Many naturally occurring ribozymes either aid the hydrolysis of their own phosphodiester bonds or cause the hydrolysis of bonds in other RNAs. They also catalyze the aminotransferase activity of the ribosome.

Biological (Genetic) Information

The information carried by the genes of an organism is known as **biological or genetic information**. This information determines the organism's traits and functions by directing the synthesis of proteins and regulating various cellular processes. Essentially, it is the hereditary blueprint that is passed from one generation to the next.

The "Central Dogma" of Molecular Biology (Flow of Genetic Information)

The term **dogma** means a "set of beliefs." Shortly after Watson and Crick discovered the structure of DNA, the mechanism by which genetic information is stored and used to produce proteins became evident. This mechanism is known as the **Central Dogma of Molecular Biology**.

In a landmark presentation in 1957, **Francis Crick** outlined the Central Dogma, describing the relationship between **DNA, RNA, and proteins**. Crick proposed that all biological (genetic) information is encoded in DNA. The flow of genetic information includes the replication of DNA to make more DNA, the transcription of DNA to synthesize complementary RNA, primarily messenger RNA (mRNA), and the translation of mRNA into proteins. This role for DNA is called the 'central dogma' of biology (Image 8.5). Replication copies DNA; transcription uses DNA to make complementary RNAs; translation uses messenger RNAs (mRNAs) to make proteins. In eukaryotic cells, replication and transcription take place within the nucleus while translation takes place in the cytoplasm. In prokaryotic cells, replication, transcription, and translation occur in the cytoplasm.

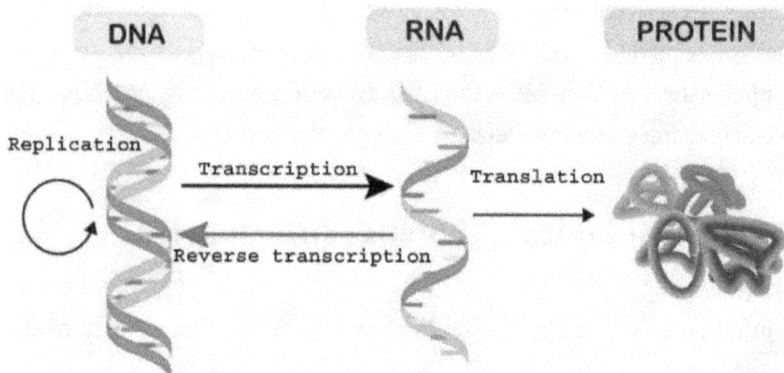

Image 8.5 The central dogma of molecular biology (adapted from: https://stock.adobe.com/search?filters%5Bcontent_type%3Aphoto%5D=1&f ilters).

DNA Replication

Replication is the copying of parental DNA to form daughter DNA molecules having identical nucleotide sequence (Image 8.6). It is a biological process occurring in all living organisms that is the basis for biological inheritance. Replication is a tightly controlled process, occurs in the synthesis (S) phase of Interphase of the cell cycle. During the S phase, the nuclear DNA is completely replicated once and only once. The process of DNA replication is vital for cell growth, repair, and reproduction in organisms.

DNA replication has two requirements that must be met: DNA template and free 3'-OH group. DNA replication occurs in semi-conservative manner, because the double stranded DNA molecule separates and then, each of the separated strands serve as a template for the synthesis of a new anti-parallel complementary strand. As a result, each DNA molecule would have one parental strand (used as the template) and one brand new daughter strand (Image 8.6). Since only one parental strand is conserved in each daughter molecule, it is known as semi-conservative mode of replication.

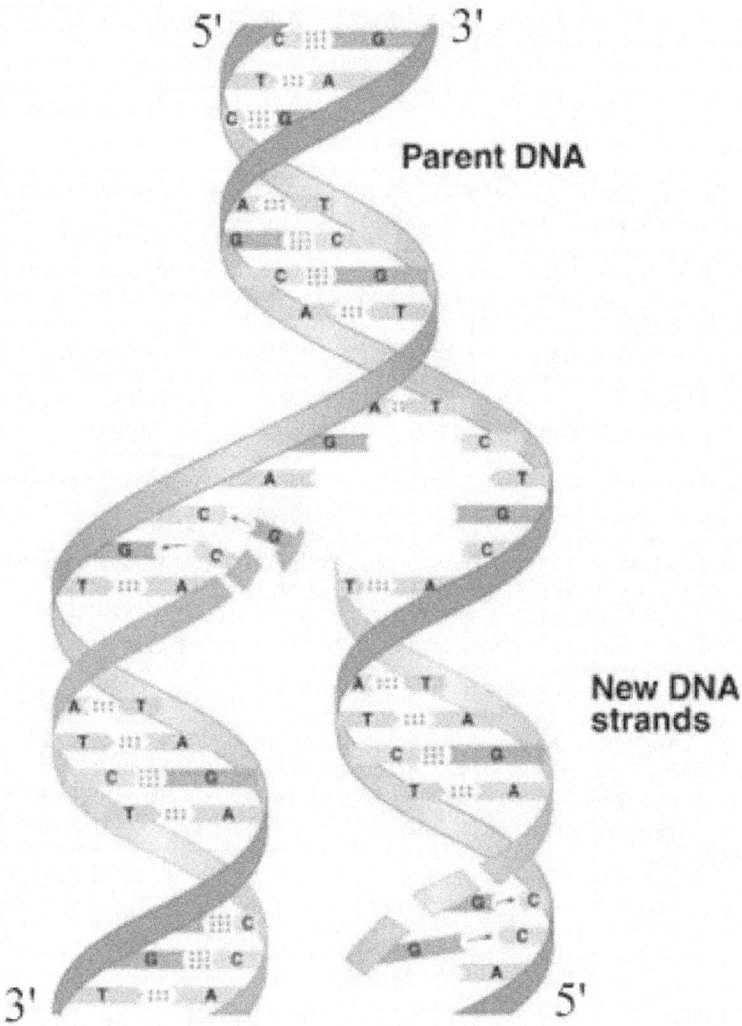

Image 8.6 The model of DNA replication proposed by Watson and Crick.

Firstly, the deoxyribonucleoside monophosphates (dAMP, dGMP, dTMP, and dCMP) occurring in the cell are phosphorylated by the enzyme phosphorylase, converting them into their activated form (dATP, dGTP, dTTP, and dCTP). DNA replication is divided into three stages: initiation, elongation, and termination.

Initiation of DNA replication begins with the assembly of a replication fork, or replication bubble, at a specific region known as the origin of replication (ori) (Image 8.7 and Image 8.8). In *E. coli*, the ori is about 245 base pairs long and is typically rich in A-T base pairs, which are easier to separate due to their weaker hydrogen bonding compared to G-C pairs. Replication cannot proceed in the absence of this ori sequence.

Elongation refers to the phase where nucleotides are sequentially added to synthesize new DNA strands, guided by complementary base pairing.

Termination occurs when replication is complete, resulting in the separation of the newly duplicated chromosomes.

In prokaryotes, plasmids, and many viruses, replication is initiated at a single, unique origin of replication (ori or replicon). By contrast, eukaryotic chromosomes, due to their much greater length, contain multiple origins of replication - often hundreds or even thousands. These eukaryotic origin sites are termed *autonomously replicating sequences* (ARS) or replicators. Multiple origins ensure that replication can be completed efficiently within the time constraints of the cell cycle.

Despite these differences, the underlying process of replication is the same for both prokaryotic and eukaryotic DNA.

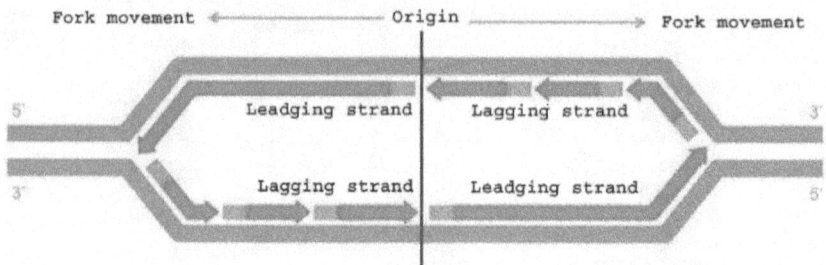

Image 8.7 DNA synthesis at an origin of replication.

Initiator proteins bind to ori sequences, which go on to recruit more proteins including **DNA helicase**, forming a replication complex around the origin and initiate the process of DNA replication. To begin DNA replication, **DNA helicase** within the replication complex travels along the DNA, unwinds/unzips the DNA double helix from its tightly woven form and separates the two strands from one another by breaking the hydrogen bonds between base pairs at the site of ori to form two "Y-shaped" structures called replication forks or spots (Image 8.7). These replication forks are the actual site of DNA copying. Replication fork is the junction between the newly separated strands and unreplicated double-stranded DNA. DNA replication is bidirectional from the origin of replication *i.e.*, replication forks extend bi-directionally as replication continues (Image 8.7 and Image 8.8). The two replication forks move in opposite directions creating a replication bubble. As replication bubbles expand in both directions, they will eventually fuse together as the intervening regions are copied. This functions to greatly limit the time required for DNA replication to occur. As helicase unwinds the DNA at the replication forks, the DNA ahead of it becomes overwound and positive supercoils form. **DNA gyrase (DNA topoisomerases II)** travels just ahead of helicase on the replication fork, produce breaks in the DNA and then rejoin them in order to prevent supercoiling. They release any tension created in the parent DNA due to it's uncoiling at the ori by nicking off and resealing the DNA strand. Now each separated strand serves as a template (or source of instruction) on which a new complementary strand is synthesized. Helicase uses energy from the ATP to break the hydrogen bonds. Helicase is responsible for separating the strands at the replication fork, but is not directly responsible for preventing single-stranded DNA from reannealing. It creates the replication fork, but is incapable of maintaining it. Single-stranded DNA-binding proteins (SSBs) bind to each strand of double-stranded DNA (dsDNA) to prevent the strands from reannealing prematurely. They stabilise the separated parent DNA strands now called the template strands. These proteins are essential for maintaining the replication fork. These proteins also help to prevent the single-stranded DNA from being digested by nucleases. SSB proteins will be dislodged from the strand when a new complementary strand is synthesized by **DNA polymerase III** (Image 8.8). DNA polymerase III is

261

the main enzyme involved in polymerization of deoxyribonucleotides. The number of DNA polymerases in eukaryotes is much more than in prokaryotes: 14 are known, of which five are known to have major roles during replication and have been well studied. They are known as pol α, pol β, pol γ, pol δ, and pol ε.

Because of the anti-parallel nature of DNA, the two parental strands are replicated by different mechanisms during the progression of the replication fork. The parental strand, which is 3' to 5' relative to the direction of unwinding (toward the replication fork), is replicated continuously. This is known as the leading strand. In this case, the nucleotides are added continuously to its growing 3' end. The other strand (referred to as the lagging strand) is synthesized discontinuously in short stretches called Okazaki fragments in the opposite direction that the replication fork moves. They are called Okazaki fragments after their discoverer, Reiji Okazaki. Okazaki fragments are about 1,500 bases in length in prokaryotes, and 150 bases in eukaryotes.

A specialized RNA polymerase called **primase** (DNA-dependent RNA polymerase) synthesises small strand of RNA called RNA primers with free 3'-OH group using single-stranded DNA (ssDNA) as a template in the 5' to 3' direction at the 5' end of the leading strand and at the 5' end of each Okazaki fragments, which acts as a 'kick-starter' for **DNA polymerase**. Primase activity requires the formation of complex of primase and at least six other proteins. This complex is called **Primosome**. RNA primers are necessary because **DNA polymerase III (DNA pol III)** can only attach nucleotides to the 3' end of a pre-existing strand. **DNA pol III** attaches onto the 3' end of each **RNA primer** and adds the complementary nucleotides. **DNA pol III** uses each strand as a template in the 3' to 5' direction to build a complementary strand in the 5' to 3' direction at the same time. Protein from the **DNA pol III holoenzyme complex** holds the DNA polymerase to the DNA through an entire cycle of replication. This enzyme is active only in the presence of Mg^{2+} ions and preexisting DNA. DNA is synthesized from deoxyribonucleoside 5'-triphosphate precursors (dNTPs). DNA polymerase cleaves the outer 2 phosphates that provides energy and uses this energy to form a phosphodiester bond between the 3' OH group of the previous nucleotide of RNA primer and the 5' phosphate group of the new nucleotide. The process of expanding the new DNA strands continues until there is either no more DNA template strand left to replicate (*i.e.*, at the end of the chromosome) or two opposing replication forks meet and the nascent DNA from the two forks is ligated together. The meeting of two replication forks is not regulated and happens randomly along the course of the chromosome.

During termination, primers are removed and replaced with new DNA nucleotides by **DNA polymerase I (DNA pol I)** from the 5' end of the leading strand and from the 5' end of each Okazaki fragment. This enzyme removes the ribonucleotides one at a time with its exonuclease activity (5' → 3' exonuclease) and replaces it with the appropriate complementary deoxyribonucleotide. **DNA ligase** joins the 3' end of the DNA that replaces the primer to the rest of the leading strand. DNA ligase also joins the Okazaki fragments together to form a continuous strand. It does this by covalently joining the sugar-phosphate backbones together with a phosphodiester bond. The termination of DNA replication occurs at specific termination sites in both prokaryotes and eukaryotes. In prokaryotes, a single termination site is present midway between the circular chromosome. The two replication forks meet at this site, thus, halting the replication process. In eukaryotes, the linear DNA molecules have several termination sites along the chromosome, corresponding to each origin of replication. The end result of replication is that a single double-stranded molecule becomes replicated into two copies with identical sequences. The new copies automatically wind up again.

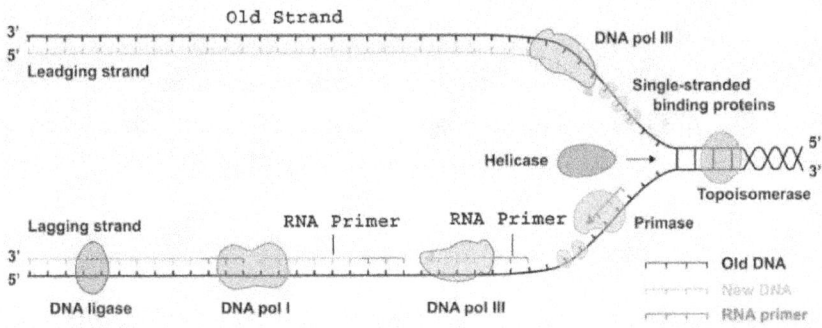

Image 8.8 Replication origin and replication forks (source: https://stock.adobe.com/ Image ID: 500810301).

The fidelity of DNA replication is maintained by (1) the 3' → 5' exonuclease activity (proofreading ability) of the DNA pol I (major repair enzyme), and DNA pol III. Both DNA pol I and III proofread each nucleotide against its template as soon as it is added to the growing strand. Upon finding an incorrectly paired nucleotide, the polymerase removes the nucleotide at the 3' end of the growing strand and replace them with the correct nucleotide, and (2) 5' → 3' exonuclease activity (repair ability) of the DNA pol I to correct errors that escaped proofreading (mismatch repair). It recognizes mis-incorporated nucleotides, removes them from DNA, and replaces them with the correct nucleotides and DNA ligase seals the nick in the sugar phosphate backbone. If DNA polymerase makes a mistake and it is not repaired, a mutation has occurred. On average, for every 5-10 billion nucleotide pairs replicated there is one error. DNA is the only macromolecule for which repair systems exist.

DNA replication is an extraordinarily complex process. It involves many different proteins that open and unwind the DNA double helix, stabilize the single strands, synthesize RNA primers, assemble new complementary strands on each exposed parental strand, remove the RNA primer, and join new discontinuous segments on the lagging strand. At least 30 enzymes and proteins are required to replicate the *E.coli* DNA, each performing a specific task. DNA replication process must be highly precise and accurately timed to prevent any unnecessary loss of energy and to ensure that DNA is faithfully and completely replicated only once per cell cycle (Leonard and Grimwade, 2015).

The purpose of replication is to conserve the entire genome for next generation. Faithful DNA replication is essential for the continuity of life. DNA replication rates in prokaryotes are approximately of the order of 1000 bases per second.

Primer

A primer is a short stretch of RNA with a free 3'-hydroxyl group to which nucleotides can be added. It may be very short, must be present and must be hydrogen-bonded to the template strand. No known DNA polymerase is able to initiate replication without the presence of a primer. In living cells, the primer is a short segment of RNA; in cell-free replication *in vitro*, the primer may be either RNA or DNA. The size of the primer differs according to the initiation event. In *E. coli* the length is typically from 2 to 5 nucleotides; in eukaryotic cells, it is usually from 5 to 10 nucleotides. The RNA polymerase that produces the primer for DNA synthesis is called primase. Primase copies a DNA template strand to make an RNA strand (primer) complementary to it. The 3' end of the primer is called the primer terminus.

Inhibitors of DNA Replication

Inhibitors of DNA replication are molecules that interfere with the process of DNA replication. The first category includes purine and pyrimidine nucleoside analogs that directly inhibit DNA polymerase activity. The second category includes DNA damaging agents including cisplatin and chlorambucil that modify the composition and structure of the nucleic acid substrate to indirectly inhibit DNA synthesis. Other inhibitors of DNA replication include fluoroquinolones, aphidicolin, hydroxyurea, cytarabine, mimosine, gemcitabine, etoposide, *etc.*

Gene

In 1865 Gregor Mendel assumed that each trait is determined by a pair of inherited 'factors' which are now called gene. In 1909 Danish botanist Wilhelm Johannsen coined the term 'gene'. Gene is the unit of inheritance. It is the basic physical and functional unit of genetic information with the potential to be expressed. In molecular terms, a gene is a discrete segment of DNA or RNA that provides instructions for making a functional product, that is, a molecule needed to perform a job in the cell. The functional products of most known genes are proteins, or, more accurately, polypeptides. Genes that specify polypeptides are called protein-coding genes. However, many genes do not code for proteins. They provide instructions for making functional RNA molecules (tRNAs, rRNAs, microRNA, *etc.*). They are called RNA-coding genes. The entire DNA is made up of thousands of genes that directly control all the functions of a cell via protein synthesis.

Gene Structure

Gene structure is the organisation of specialised sequence elements within a gene. Starting from the most upstream (at the 5' end), the DNA sequence of a protein-coding gene comprises in order, the promoter, the 5' untranslated region (5' UTR, also known as the leader sequence), the coding sequence (CDS), the 3' untranslated region (3' UTR, also known as the trailer sequence), and the terminator (Image 8.9).

Promoter Sequences/Regions or Promoter

A promoter is a DNA sequence (100 to1000 base-pairs long) located immediately upstream (at the 5' end) of the transcription unit. The promoter signals the start of a gene. The promoter's function is to serve as an indicator of where and in which direction transcription should proceed, and influence how much RNA is produced and, therefore, the amount of protein. A gene cannot be expressed without a promoter sequence. A promoter acts as 'molecular switches' to turn on transcription. Promoters are specific to genes. The promoter is not actually transcribed; its role is purely regulatory.

Promoters are associated with genes in both prokaryotes and eukaryotes. In prokaryotes, there is often more than 1 gene under control of the same promoter. This is called an operon. All of the genes under control of this same promoter will be transcribed to give a single RNA (a polygenic or polycistronic RNA). In eukaryotes, generally every gene has its own promoter. Transcription factors bind to specific nucleotide sequences in the promoter region and helps RNA polymerase to bind and start transcription. The promoter is composed of a core promoter sequence and a proximal promoter sequence. In both eukaryotes and prokaryotes, within the promoter region, just upstream of the transcriptional start site, the TATA box resides. This box is simply a repeat of thymine and adenine dinucleotides (literally, TATA repeats). Transcription factors bind to the TATA box, assembling an initiation complex. Once this complex is assembled, RNA polymerase binds to its upstream sequence and becomes phosphorylated. This releases part of the protein from the DNA, activates the transcription initiation complex, and places RNA polymerase in the correct orientation to begin transcription. Promoter can be short (only a few nucleotides in length) or quite long (hundreds of nucleotides long). The longer the promoter, the more available space for proteins to bind. This also adds more control to the transcription process. The length of the promoter is gene-specific and can differ dramatically between genes. Consequently, the level of control of gene expression can also differ quite dramatically between genes. In eukaryotes, the promoters have more complexity. This is partly because there are three classes of RNA polymerase in eukaryotes; contrast to one in prokaryotes.

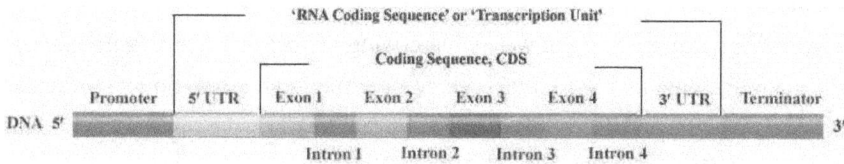

Image 8.9 Structure of a protein-coding gene.

The Untranslated Regions (UTRs)

The untranslated regions (UTRs) are the regulatory regions of genes located at either side (at the 5′ and 3′ end) of the coding sequence (CDS) which are the part of the transcribed mRNA but not translated into protein. UTRs are known to vary substantially across genes, both in size, and in the composition of regulatory elements within them. The UTRs typically encode some regulatory elements critical for regulating transcription or steps of gene expression that occur post-transcriptionally. Eukaryotic transcripts generally contain longer UTRs than do those of prokaryotes.

The region just before (upstream of) the coding sequence of a gene *i.e.*, just before the 5′ end of the first exon is called the **5′ UTR or leader sequence** (Image 8.9). It lies between the **transcription start site (TSS)** and the **translation initiation codon (AUG)**. 5′ UTR is a critical regulatory region as it serves as a binding site for ribosomes to initiate translation. It plays a crucial role in regulating the translation process and impacts the protein expression level. Essential promoter elements can be located within the 5' untranslated region (UTR) of some genes.

The region after (downstream of) the coding sequence of a gene is called the **3′ UTR or trailer sequence** (Image 8.9). It starts immediately after the stop codon (UAA, UAG, or UGA) at the 3′ end of the last exon. The 3′ UTR is a powerful regulatory element that contains the polyadenylation signal and addition site, which allows poly-A polymerase to recognize the 3′ end of the transcript and add poly-A tail to the end of the mRNA. The 3′ end also contains miRNA-binding sites. It can contain sequences that are recognized by specific RNA-binding proteins. Together, these directly or indirectly influence RNA stability, folding, transport, localization, and translation efficiency and consequently RNA and protein levels.

Coding Region

The coding region, also known as the coding sequence (CDS), is the part of the gene that contains the actual code to be turned into a functional molecular product, which can be a protein or RNA. The heart of the gene – the coding sequence (CDS) – lies between the 5′ and 3′ UTR. The coding region of most prokaryotic genes is a continuous stretch of DNA that specifies a protein, with no intervening sequences. It starts with the initiation codon and ends with the termination codon. Though in some Eubacteria and Archaebacteria noncoding sequences have been found. Eukaryotic genes are more complicated than prokaryotic genes and the coding region is often discontinuous: it contain exons interrupted by introns. Exons are nucleotide sequences that code for the amino acids. The term exons refers to the expressed region of the genome. Introns are intervening sequences between two exons that do not code for amino acids. The term exon was coined by Walter Gilbert in 1978. Introns were first discovered in 1977, independently in the laboratories of Phillip Sharp and Richard Roberts, during studies of the replication of adenovirus in cultured human cells. Introns can contribute to numerous biological processes, including gene silencing, gene imprinting, transcription, mRNA metabolism, mRNA nuclear export, mRNA localization, mRNA surveillance, RNA editing, translation, protein stability, ribosome biogenesis, cell growth, embryonic development, apoptosis, molecular evolution, genome expansion, and proteome diversity through various mechanisms (Haddad-Mashadrizeh *et al.*, 2023). Exons and introns are numbered in the 5′ to 3′ direction of the coding strand. Both exons and introns are transcribed to yield a long primary transcript RNA. The size and arrangement of exons and introns are characteristic for every eukaryotic gene. Intron sequences contain some common features. Most introns begin with the sequence GT (GU in RNA) and end with the sequence AG. Otherwise, very little similarity exists among them. The sequences at the 5' end of the intron beginning with GT are called splice donor site and at the 3' end, ending with AG, are called the splice acceptor site. The number of introns varies greatly, from zero to more than 50 in some genes. The number of introns appears to increase as size of the gene increases. Introns also vary in size, location, and nucleotide sequence from one gene to another. Often the total length of the introns of a gene exceeds the total length of the exons of that gene by anywhere from 2 to 10 times and more.

Terminator

A transcription termination site, also known as a terminator, is a specific DNA sequence (~ 50 to over 1000 nucleotides long) located downstream of the gene to be transcribed. It signals the end of transcription, causing the RNA polymerase to detach from the DNA template and initiate the process of releasing the newly synthesized RNA transcript from the transcription machinery. This should not be confused with termination codons that are the stopping signal for translation. Terminator modulates mRNA stability, localization, translation, polyadenylation, and ultimately gene-expression.

Regulatory Sequences/Regions

Regulation of gene expression is an essential feature of all living organisms and viruses. A regulatory sequence is a region in the genome that do not code for proteins but instead control the expression of specific genes within an organism. In order to read the information in a specific gene and transcribe it into RNA, the strands of the DNA double helix must be unwound in the vicinity of that gene. Regulatory regions occur in the vicinity of coding regions. These regulatory regions consist of promoters, enhancers, silencers, insulators, repressors and other regulatory sequences. Promoters and enhancers are the primary regulatory components of gene expression.

Enhancers

In some eukaryotic genes, there are regions that help increase transcription. These regions are called enhancers. Enhancers bind some transcription factors called activators and enhance the rate of transcription. Enhancers are not necessarily close to the genes they enhance. They are most often found far upstream of the promoter (as far as 10,000 base pairs), but they can occur anywhere within genes or downstream of genes, or even on a different chromosome. When an enhancer is far away from the target gene, the DNA folds such that the enhancer is brought into proximity with the promoter, allowing interaction between the activators and the transcription initiation complex.

Enhancers are positive transcriptional control elements which are particularly prevalent in the cells of complex eukaryotes but which are absent or very poorly represented in simple eukaryotes such as yeast. An enhancer can regulate more than one gene. Two different genes may have the same promoter but different enhancers, enabling differential gene expression.

Silencers

A silencer is a DNA sequence that reduces the transcription of a gene. Silencers function as a "turn-off" switch for gene expression, acting in contrast to enhancers which are "turn-on" switches. They provide binding sites for transcriptional suppressor proteins and reduce or shut down transcription, thereby acting as negative regulatory elements. They are typically located in close proximity to the target gene, either upstream (before the gene) or within the gene itself, including in introns or exons. Some silencers are also found within the 3' untranslated region (3' UTR) of messenger RNA (mRNA).

Insulators

Insulators are naturally occurring DNA sequences, 0.2-3.0 kb in length, that establish independent domains of transcriptional activity within eukaryotic genomes. Insulators provide binding sites for proteins that control transcription in a number of ways. An insulator located between the enhancer and the promoter can block enhancer-promoter communication (anti-enhancer insulators). Others prevent the spread of nearby condensed chromatin that might otherwise silence expression (anti-silencer or barrier insulators). Some insulators are able to act both as enhancer blockers and barriers. Others, particularly in yeast, serve primarily as barriers. Insulators have been detected in the genomes of all well-studied higher eukaryotes (Phillips and Corces, 2009; Wallace and Felsenfeld, 2007).

Repressors

Like prokaryotic cells, eukaryotic cells also have mechanisms to prevent transcription. A repressor is a (transcription factor) protein that turns off the expression of one or more genes. Repressors bind to silencers that can bind to promoter or enhancer regions and block transcription (*i.e.*, turning genes off). Like the transcriptional activators, repressors respond to external stimuli to prevent the binding of activating transcription factors. Repression might largely function at the chromatin modification level. Both activators and repressors respond to external stimuli to determine which genes need to be expressed.

Gene Expression

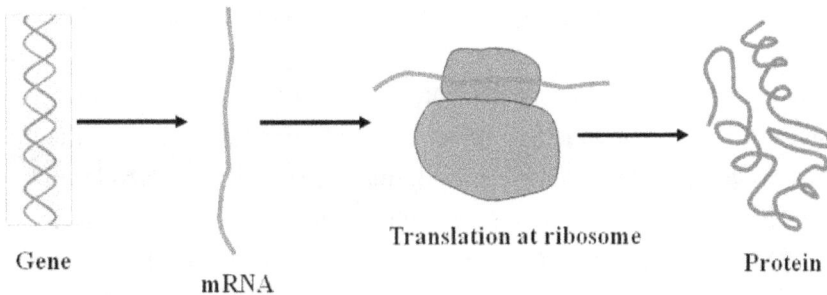

Translation at ribosome

Gene

mRNA

Protein

Image 8.10 Gene expression.

Biological information is encoded in the **base sequence of DNA** as a series of **genes**. **Gene expression** is the process by which a gene is "turned on" to direct the production of specific gene products, which can be **proteins** or **non-coding RNAs** (Image 8.10). This is a tightly coordinated process which allows a cell to respond to its changing environment. It involves two main stages: Transcription and Translation. In prokaryotes where there is no nuclear membrane, both transcription and translation occur in the cytoplasm whereas in eukaryotes, transcription occurs inside the nucleus and the translation occurs in the cytoplasm. Sometimes, the term **gene expression** is used specifically to refer to **transcription** alone.

Transcription

Transcription, in its broadest sense, is the act of converting information from one form to another. In biology, it refers to the process of creating an RNA copy of a gene's DNA sequence. In other words, transcription is the transfer of genetic information from DNA into RNA. This process is called transcription because the RNA is like a transcript, or copy, of the gene's DNA code. It is the first stage of gene expression.

Genes can be located on either strand of the double-stranded DNA – there is no single strand that exclusively contains all genes. For example: Gene A might be on the top strand, Gene B might be on the bottom strand. A gene is indeed located at a specific position on a particular strand of DNA. But, regardless of which strand contains a particular gene, all genes are read in a 5' to 3' direction, and the strand containing a particular gene is referred to as the coding strand (also called the sense strand or plus strand or non-template strand). The opposite strand, complementary to the coding strand, is called the noncoding strand or antisense strand or minus strand or template strand, is used as a template to synthesize RNA during transcription by DNA-dependent RNA polymerase (Image 8.11). The coding strand has the same sequence as the RNA, except with T (thymine) instead of U (uracil). The process of transcription starts at the promoter region of the template DNA and terminates at the terminator region. The segment of DNA between these two regions that is transcribed into RNA is known as transcription unit (Image 8.12).

Transcription is very similar to replication in terms of chemical mechanism, polarity (direction of synthesis) and use of a template. Transcription differs from DNA replication in six aspects. First, only a selected stretch of one DNA strand, not the entire DNA molecule, serves as the template so that only those bases are exposed that constitute the gene to be transcribed. Second, instead of DNA polymerase, a different enzyme, RNA polymerase, synthesizes RNA complementary to a segment of the template strand of duplex DNA (Image 8.11). Third, unlike replication, transcription does not need to build on a primer. Instead, transcription starts at the **promoter region**. Fourth, transcription results in a single, free strand of RNA, not in a double helix. Fifth, the raw materials for the new RNA are the 4 ribonucleoside triphosphates (NTPs) instead of dNTPs. Sixth, there is no highly active, efficient proofreading function during RNA transcription. The transcription also requires certain cofactors such as Mg^{2+}.

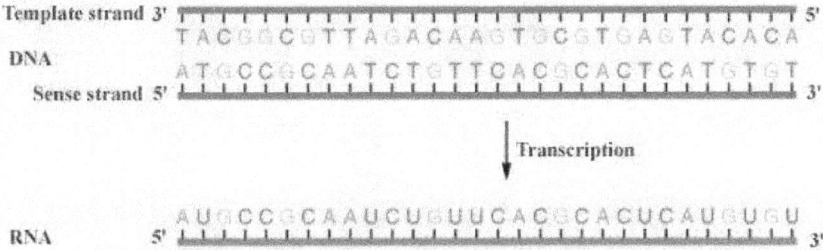

Template strand 3'
T A C G G C G T T A G A C A A G T G C G T G A G T A C A C A
DNA
A T G C C G C A A T C T G T T C A C G C A C T C A T G T G T
Sense strand 5' 3'

Transcription

A U G C C G C A A U C U G U U C A C G C A C U C A U G U G U
RNA 5' 3'

Image 8.11 Transcription of the DNA template strand produces RNA.

Just before transcription, the four types of ribonucleotides (AMP, GMP, CMP and UMP) occurring in the cell are phosphorylated by the enzyme phosphorylase to form activated or phosphorylated ribonucleoside triphosphates (NTPs), *i.e.*, ATP, GTP, CTP and UTP. Transcription occurs in three phases: initiation, elongation and termination. The main enzymes responsible for transcription is DNA-dependent RNA polymerase.

Like prokaryotic cells, the transcription of genes in eukaryotes requires an RNA polymerase to bind to a promoter to initiate transcription. In eukaryotes, RNA polymerase requires **transcription factors (TFs)** to facilitate transcription initiation. TFs are regulatory proteins (~50 different protein) whose function is to activate (or more rarely, to inhibit) transcription by binding to the promoter. They usually contain a DNA-binding domain and a regulatory domain, which will either upregulate or repress transcription. When a transcription factor recognize and bind to the promoter region of the **template strand**, it recruits general transcription factors and mediator proteins that in turn recruit RNA polymerase and forms the transcription initiation complex. This binding only occurs under some conditions: when the gene is 'on'. Other DNA sequences further upstream from the promoter are also involved. RNA polymerase by itself cannot initiate transcription in eukaryotic cells. Transcription factors must bind to the promoter region first and recruit RNA polymerase to the site for transcription to begin. Promotor sequences are typically upstream of a gene and ensure accurate initiation of transcription. Some TFs are generalized to many promoters, while other TFs are site-specific to one promotor and gene. Other transcription factors called activators target enhancer sequences that initiate the bending of the DNA to allow interaction with the proteins bound to the promoter region.

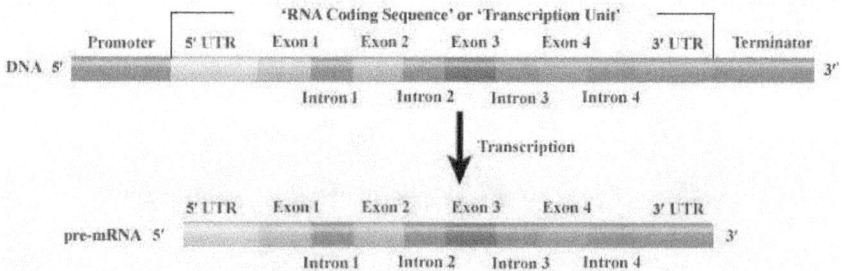

Image 8.12 Transcription process.

Once RNA polymerase is bound to the promoter, it moves down the DNA and leads to the unwinding of DNA duplex progressively into two separate strands. Then, the DNA strand with 3'→5' polarity functions as the template on which the RNA is formed in the 5'→3' direction. Transcription begins at the transcription start/initiation site (TSS). It is the first nucleotide which starts transcribing into RNA is a transcription start site. The enzyme, RNA polymerase, utilizes ribonucleoside triphosphates (NTPs) as raw material and polymerizes them to form RNA copy. Two terminal phosphates are released from each ribonucleotide, which is further hydrolyzed to release energy. The complex of RNA polymerase, DNA template and new RNA is called transcription bubble.

RNA polymerase adds one ribonucleotide at a time as according to the DNA template strand. New ribonucleotides are added to the growing chain at the free 3' end. The order in which the ribonucleotides are added to the growing RNA chain is determined by the order of the bases in the template DNA. Each nucleotide in the newly formed RNA is selected by Watson-Crick base pairing interactions; uridylate (U) residues are inserted in the RNA opposite to adenylate residues in DNA template, adenylate residues are inserted opposite to thymidylate residues. Guanylate and cytidylate residues in DNA specify cytidylate and guanylate, respectively, in the new RNA strand. The 3'-OH of one ribonucleotide reacts with the 5'-phosphate of another to form a phosphodiester bond. As the RNA strand is synthesized it dissociates from the template strand and the template strand reanneals with its original complimentary DNA strand. RNA synthesis usually starts with a GTP or ATP residue.

This process of opening of helix and elongation of polynucleotide chain continues until the enzyme reaches the terminator region which signals the stop of a gene. Once the RNA polymerase reaches the terminator region, the termination-factor helps to release the newly synthesized RNA. This is followed by the release of RNA polymerase to terminate transcription. The DNA molecule re-winds to re-form the double helix. Many copies of an RNA chain are transcribed simultaneously from a single gene. Transcription has some proofreading mechanisms, but they are fewer and less effective than the controls for DNA; therefore, transcription has a lower copying fidelity than DNA replication.

All genes are transcribed into RNA molecules. In prokaryotes three types of genes in DNA are transcribed into complementary RNA molecules – mRNA, tRNA, and rRNA. Eukaryotes encode these three classes of RNA and in addition small nuclear RNA (snRNA). Whereas a single RNA polymerase synthesizes all types of RNA in prokaryotes, there are at least three different RNA polymerases in eukaryotic systems:

1.   The RNA polymerase I (Pol I) transcribes rRNAs.
2.   RNA polymerase II (Pol II) transcribes protein-coding genes into pre-mRNA, and most small nuclear RNA (snRNA) and microRNA.
3.   RNA polymerase III (Pol III) transcribes tRNA and other short RNAs.

The purpose of transcription is to make RNA copies of individual genes that the cell can use in the biochemistry. The brief existence of an mRNA molecule begins with transcription and ultimately ends in degradation.

Gene Fragments

Gene fragments are segments of a gene that include only the **exons**. These fragments **exclude introns**. Gene fragments are typically derived from

complementary DNA (cDNA), which is synthesized from **mRNA**. Since mRNA represents the processed transcript (with introns spliced out), cDNA and the resulting gene fragments contain only the coding sequences. Gene fragments are useful in **cloning, gene expression studies, and protein production**, because they represent the functional coding sequence.

RNA Polymerase versus DNA Polymerase

1. RNA polymerase synthesizes RNA, but does not need a primer to initiate synthesis. DNA polymerase synthesizes DNA, but needs a primer in order to initiate synthesis.
2. DNA polymerase will be continuous till the work is finally done wherein RNA polymerase will continue but eventually may break in the event it will reach a 'stop' cycle.
3. RNA polymerase does not require helicase.
4. RNA polymerase is slower than DNA polymerase.
5. RNA polymerase adds Uracil (U) instead of thymine (T).
6. RNA polymerase proofreading function is less efficient than DNA polymerase.

RNA Processing (Post-transcriptional Modification)

Transcription occurs in a cell's nucleus. The RNA that is synthesized in this process is then transported to the cell's cytoplasm through the nuclear pore complex where it is translated into a protein. The basic features of transcription are the same in prokaryotic and eukaryotic cells, but eukaryotic genes and their mRNA molecules are more complex than those of prokaryotes. In prokaryotes, the RNA that is synthesized during transcription is ready for translation into a protein. In contrast, eukaryotic transcription produces primary transcript RNA or pre-messenger RNA or pre-mRNA or premature RNA. Before eukaryotic products of transcription can be moved into the cytoplasm, they must undergo extensive processing that allow them to become mature messenger RNA. RNA processing comprises of post-transcriptional/ co-transcriptional modification/alterations of the RNA molecule in eukaryotic cells, by which, primary transcript RNA is converted into mature mRNA. If the RNA is not processed, shuttled, or translated, then no protein will be synthesized. Four processes that make up these post-transcriptional modifications are: 5' capping, RNA splicing, tailing, and RNA editing. The additional steps involved in eukaryotic mRNA maturation also create a molecule with a much longer half-life than a prokaryotic mRNA. Eukaryotic mRNAs last for several hours, whereas the typical *E. coli* mRNA lasts no more than five seconds.

5' Capping

The first post-transcriptional modification is 5'-capping. While the pre-mRNA is still being synthesized, a 7-methylguanosine cap (a methylated guanosine triphosphate) is added to the 5' end of the growing transcript by multiple enzymes called the capping enzyme complex (CEC) (Image 8.13). The cap protects the 5' end of the primary transcript RNA from attack by ribonucleases. In addition, initiation factors involved in protein synthesis recognize the cap to help initiate translation by ribosomes. Moreover, the role of the 5'-cap in guiding the transcript for export will directly help to regulate not only the amount of transcript that is made but, perhaps more importantly, the amount of transcript that is exported to the cytoplasm that has the potential to be translated.

Transcript splicing or cleaving

Nascent transcripts must be processed into mature RNAs before protein synthesis by joining exons and removing the introns. The process of removing introns from the primary transcript (premature RNA) and stitching together the exons on either side to produce a continuous protein-coding sequence (mature mRNA) is called RNA splicing (Image 8.13). Splicing must be very precise to avoid an undesirable change of the correct reading frame. The process takes place in the nucleus. Finally, the mature mRNA is exported to the cytoplasm where it acts as a template for protein synthesis. RNA splicing is a complex process catalyzed by a dynamic ribonucleoprotein complex called the spliceosome. This consists of five types of small nuclear RNA molecules (snRNA) and more than 50 proteins (small nuclear riboprotein particles). Spliceosomes bind to the signals that mark the exon/intron border to remove the introns and ligate the exons together. Once the splicing reactions have occurred, the resulting mRNA is freed from the spliceosome machinery.

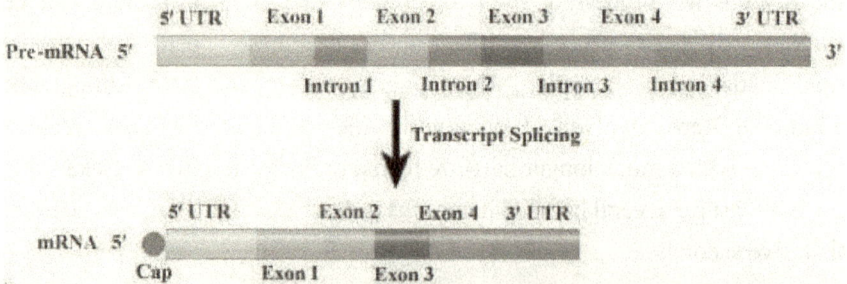

Image 8.13 Pre-mRNA splicing.

Alternative Splicing

Alternative splicing (AS), or alternative RNA splicing, or differential splicing, is a process which produces different combinations of exons that can yield different forms of proteins (isoforms) following translation (Image 8.14). Alternative splicing is a regulated process during gene expression that results in a single gene coding for multiple proteins.

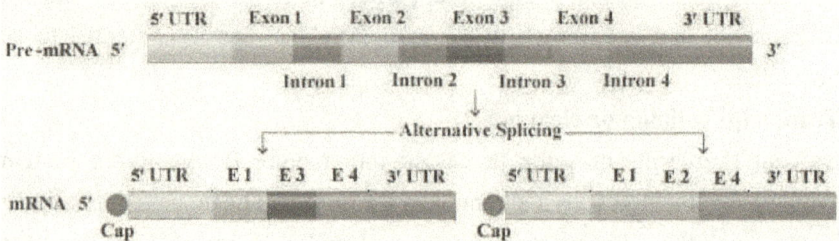

Image 8.14 Pre-mRNA can be alternatively spliced to create different proteins.

In the 1970s, genes were first observed that exhibited alternative splicing. Alternative splicing can act as a mechanism of gene regulation. Differential splicing is used to produce different protein variants that may have different cellular functions or properties in different cells or at different times within the same cell. Alternative splicing is now understood to be a common

278

mechanism of gene regulation in eukaryotes; up to 70 percent of genes in humans are expressed as multiple proteins through alternative splicing. Notably, alternative splicing allows the human genome to direct the synthesis of many more proteins than would be expected from its 20,687 protein-coding genes. Indeed, the cause of many genetic diseases is alternative splicing rather than mutations in a sequence.

Tailing (3' end cleavage and polyadenylation)

One final modification is made to nascent pre-mRNAs before they leave the nucleus is the cleavage of the 3' end and its polyadenylation. Once elongation is complete, the 3' end of the pre-mRNA is modified by the addition of multiple adenine (A) nucleotides (30-500 adenines) by the enzyme poly-A polymerase to produce a structure called **a poly-adenylated (poly-A) tail** (Image 8.15). The length of a poly-A tail is highly variable, but over the lifetime of the mRNA these adenine residues are slowly removed. The purpose and mechanism of polyadenylation vary across cell types, but polyadenylation generally serves to promote transcript longevity in eukaryotes and promote transcript degradation in prokaryotes. Generally, the more adenine residues in the poly-A tail the longer lifetime that transcript has. The poly-A tail protects the mRNA from degradation. Therefore, the 3' poly-A tail plays a role in gene expression by regulating functional transcript abundance and how much is exported from the nucleus for translation. The poly-A tail provides the mRNA with a binding site for a class of regulatory factors called the poly-A tail binding proteins (PABP) that have roles in the regulation of gene expression, including mRNA export, stability and translation (Gorgoni and Gray, 2004; Mangus *et al.*, 2003).

Note that the 5' cap is the 'front' end of the mRNA and the 3' poly-A tail is the 'back' end.

RNA Editing

In the last several years, examples have arisen in which the DNA sequence does not predict the protein sequence even after introns are removed. RNA editing is a process, other than splicing, that changes the mRNA sequence such that it differs from the sequence of the DNA template and as a result alters the protein produced by that mRNA. Editing can occur in two ways. First, by changing one nucleotide to another and second, by inserting or deleting a nucleotide or nucleotides. It results in the substitution, addition, or deletion of amino acids (relative to the DNA template). In rare cases, the mRNA transcript can be edited after it is transcribed.

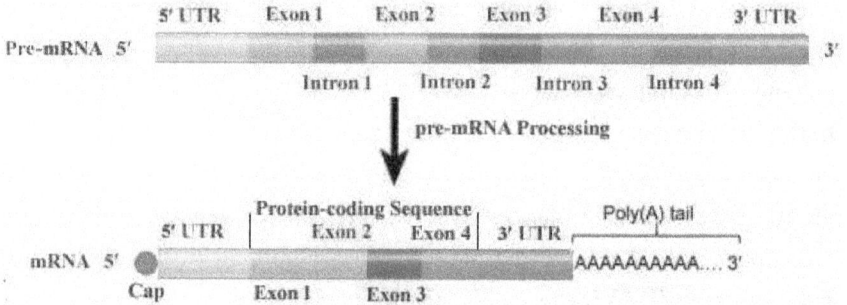

Image 8.15 Pre-mRNA processing.

Once pre-mRNA has been completely processed, it is termed 'mature messenger RNA', 'mature mRNA', or simply 'mRNA', and is exported out of the nucleus. A mature mRNA includes a 5' cap, 5' UTR, protein-coding region, 3' UTR, and poly (A) tail (Image 8.15). The mature mRNA is finally degraded with certain enzymes like nucleases.

Reverse Transcription

Reverse transcription is the process of synthesis of a DNA molecule from an RNA molecule. This process is catalyzed by RNA-dependent DNA polymerases, also known as reverse transcriptases. Reverse transcriptases occur naturally in both prokaryotic and eukaryotic organisms, as well as in retroviruses. This process is utilized by **retroviruses** to use RNA as their genetic material. To replicate the viral genome, reverse transcriptase enzymes copy the (+) ssRNA genome producing (-) ssDNA strands. DNA-dependent DNA polymerase enzymes then copy the (-) ssDNA strands to produce a dsDNA intermediate. This DNA intermediate migrates to the nucleus of the cell where it is integrated into the host cell genome. This process is catalyzed by another viral enzyme called integrase (IN). DNA-dependent RNA polymerase enzymes then copy the (-) DNA strands to produce (+) viral mRNA molecules. The (+) viral mRNA can then be translated into viral proteins by host cell ribosomes. Regarding the enzymes involved, the 'dependent' part of the name tells what type of nucleic acid is being copied. The 'polymerase' part of the name tells what type of nucleic acid is being synthesized, *e.g.*, DNA-dependent RNA polymerase would synthesize a strand of RNA complementary to a strand of DNA.

Complementary DNA (cDNA)

cDNA is a single-stranded DNA molecule with a nucleotide sequence that is complementary to an RNA molecule. cDNA is synthesized from a mature mRNA template in a reaction catalyzed by the enzyme **reverse transcriptase** and the enzyme **DNA polymerase**. cDNA is often used to clone eukaryotic genes in prokaryotes. When scientists want to express a specific protein in a cell that does not normally express that protein, they will transfer the cDNA that codes for the protein to the recipient cell. Some viruses (retroviruses) also use cDNA to turn their viral RNA into mRNA (viral RNA → cDNA → mRNA). The mRNA is used to make viral proteins to take over the host cell.

The Genetic Code (Genetic Dictionary)

The concept of codons was first described by Francis Crick and his colleagues in 1961. A codon is a sequence of three nucleotides in the mRNA that codes for (encodes) a specific amino acid or signals the termination of translation. The four code letters of mRNA (A, U, G, and C) in groups of three can yield $4^3 = 64$ different codons.

Today all 64 codons have been discovered. Sixty-one codons code for the 20 naturally occurring amino acids in proteins; these codons are called the sense codons. The remaining 3 codons (UAA, UAG, and UGA) code for no amino acid. These three codons are the stop codons, also called nonsense codons or termination codons, which normally signal the end of the translation process of a polypeptide chain. Codons dictate where translation starts (AUG), which amino acids to incorporate into the growing polypeptide chain, and when translation stops (UAA, UAG, UGA). The initiation (start) codon AUG (for methionine) signals the beginning of polypeptide chains. All proteins are initially translated with methionine in the first position, although it is often removed after translation. There are also internal methionines in most proteins, coded by the same AUG codon. Codons that specify the same amino acids are called synonyms, *e.g.*, CAU and CAC are synonyms for histidine. In most cases, codons that specify the same amino acid differ only in the third base. Genetic code is a dictionary consists of all codons and the amino acid that each one encodes. It can be expressed in a simple table with 64 entries (Image 8.16). It is the rule for translating from the "language" of nucleic acids to that of proteins. It is the basis of the transmission of hereditary information by nucleic acids in all organisms. The genetic code is traditionally represented as an RNA codon table due to the biochemical nature of the protein translation process. However, with the rise of computational biology and genomics, proteins have become increasingly studied at a genomic level. As a result, the practice of representing the genetic code as a DNA codon table has become more popular.

First letter	Second letter: U	C	A	G	Third letter
U	UUU Phenylalanine	UCU Serine	UAU Tyrosine	UGU Cysteine	U
	UUC	UCC	UAC	UGC	C
	UUA Leucine	UCA	UAA STOP	UGA STOP	A
	UUG	UCG	UAG	UGG Tryptophan	G
C	CUU Leucine	CCU Proline	CAU Histidine	CGU Arginine	U
	CUC	CCC	CAC	CGC	C
	CUA	CCA	CAA Glutamine	CGA	A
	CUG	CCG	CAG	CGG	G
A	AUU Isoleucine	ACU Threonine	AAU Asparagine	AGU Serine	U
	AUC	ACC	AAC	AGC	C
	AUA	ACA	AAA Lysine	AGA Arginine	A
	AUG START Methionine	ACG	AAG	AGG	G
G	GUU Valine	GCU Alanine	GAU Aspartic acid	GGU Glycine	U
	GUC	GCC	GAC	GGC	C
	GUA	GCA	GAA Glutamic acid	GGA	A
	GUG	GCG	GAG	GGG	G

Image 8.16 RNA Codon Table (source:

The genetic code plays a central role in all living organisms. It connects codons in genes and amino acids in proteins, as well as it determines codons responsible for stop signal in synthesis of proteins.

Nature and Characteristics of the Genetic Code

Following properties of the genetic code were proved by definite experimental evidence:

1. The genetic code is a triplet code *i.e.*, a sequence of three nucleotides in the mRNA specifies a single amino acid.
2. The code is continuous. The code is read from a fixed starting point and continues to the end of the coding sequence. There are no gaps between codons.
3. The code is non-overlapping, *i.e.*, a base in an mRNA is not used for two different codons. Each nucleotide is part of only one triplet codon.
4. The code is almost universal (99.9%), *i.e.*, with a very few exceptions present in the mitochondrial codons and in some protozoans, a codon specifies the same amino acid in almost all organisms.
5. The code is degenerate (redundant) *i.e.*, more than one codon can code for a single amino acid. Only 20 amino acids are involved in protein synthesis, while there are 64 codons. With two exceptions (AUG for methionine, and UGG for tryptophan), there is more than one codon specifying a particular amino acid. For example, six different codons, AGA, AGG, CGU, CGC, CGA and CGG, all specify arginine.
6. The code is unambiguous (specific). A particular codon will always code for the same amino acid, wherever it is found.

Open Reading Frame (ORF)

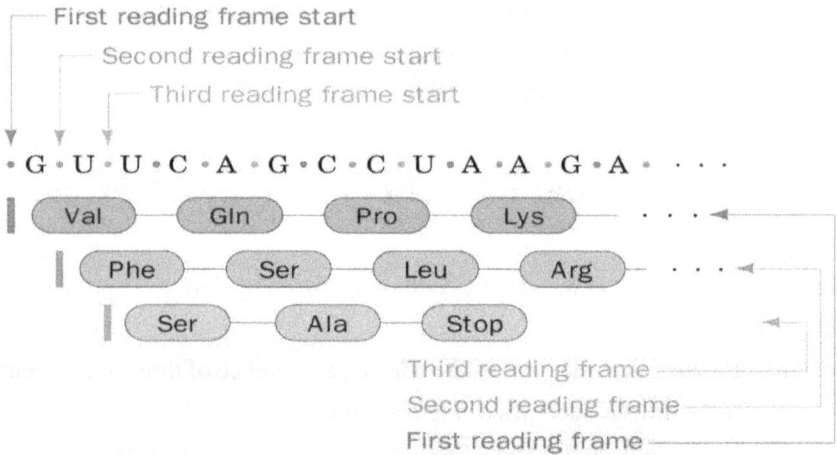

First reading frame start
Second reading frame start
Third reading frame start

· G · U · U · C · A · G · C · C · U · A · A · G · A · · · ·

| Val | Gln | Pro | Lys | · · · ◄
| Phe | Ser | Leu | Arg | · · · ◄
| Ser | Ala | Stop | ◄

Third reading frame
Second reading frame
First reading frame

Image 8.17 Open reading frame.

The genetic code reads DNA sequences in groups of three base pairs, which means that a double-stranded DNA molecule can read in any of six possible ways (three on forward strand and three on complementary strand) (Image 8.17). These are called reading frames. In a triplet, nonoverlapping code, all mRNAs have three potential reading frames. Note that the triplets, and hence the amino acids specified, are very different in each frame.

In molecular genetics, an **open reading frame (ORF)** is a portion of a DNA molecule that has the potential to code for a protein or peptide. An **ORF** is a continuous stretch of DNA codons begins with the 'start' codon, continues with the amino acid codons, and ends at a termination codon. The CDS and ORF of a gene are identical in prokaryotes. Since eukaryotes possess introns, the ORF is the codon sequence that results from RNA splicing.

The Wobble (Non-Specific) Hypothesis

Wobble Hypothesis was proposed by Francis Crick in 1966 to explain how a specific transfer RNA (tRNA) molecule can translate different codons in an mRNA template. The Wobble Hypothesis states that normal base pairing can occur between nitrogen bases in positions 1 and 2 of the codon and the corresponding bases (3 and 2) in the anticodon, while the rules of base pairing are relaxed between the third base of an mRNA codon and the 1st base of a tRNA anticodon. Thus, a single tRNA can pair (bind) with more than one mRNA codon differing in only the third base. In this way, although there are 61 codons specifying amino acids, fewer than those numbers of tRNAs are required to read mRNA. The hypothesis is applicable to most (not all) tRNAs.

Colinearity

The linear sequence of nucleotides in a gene (DNA) determines the linear sequence of amino acids in a protein. This relationship is called colinearity.

Protein Metabolism

Proteins are the end products of most information pathways. A typical cell requires thousands of different proteins at any given moment. These must be synthesized in response to the cell's current needs, transported to the appropriate cellular location, and degraded when the need has passed. Protein synthesis is the most complex of biosynthetic mechanisms. In eukaryotic cells, protein synthesis requires the participation of over 70 different ribosomal proteins; 20 or more enzymes to activate the amino acid precursors; a dozen or more auxiliary enzymes and other specific protein factors for the initiation, elongation, and termination of polypeptides; perhaps 100 additional enzymes for the final processing of different kinds of proteins; and 40 or more kinds of transfer and ribosomal RNAs. Thus almost 300 different macromolecules must cooperate to synthesize polypeptides. Protein synthesis can account for up to 90% of the chemical energy used by a cell for biosynthetic reactions. Both prokaryotic and eukaryotic cells contain thousands of copies of each protein and RNA type per cell. When totaled, the 20,000 ribosomes, 100,000 related protein factors and enzymes, and 200,000 tRNAs present in a typical bacterial cell (with a volume of 100 nm^3) can account for more than 35% of the cell's dry weight. Despite this great complexity, proteins are made at exceedingly high rates. A complete polypeptide chain of 100 residues is synthesized in an *E. coli* cell at 37^0C in about 5s.

The Ribosome

Prokaryotic Ribosome   Eukaryotic Ribosome

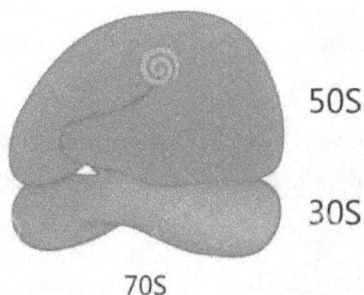

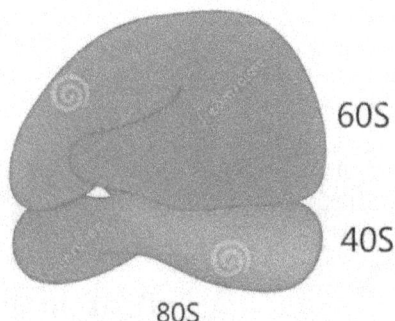

50S

30S

70S

(50S Subunit-5S rRNA,23S rRNA, 34 Proteins)

(30S Subunit-16S rRNA, 21 Proteins)

60S

40S

80S

(60S Subunit-5S rRNA, 5.8S rRNA, 28S rRNA, 50 Proteins)

(40S Subunit-18S rRNA, 33 Proteins)

Image 8.18 Ribosome (source: https://www.dreamstime.com/print-image321055453).

In the cytoplasm, ribosomal RNA (rRNA) and protein combine to form a nucleoprotein particle called a ribosome (Image 8.18). Ribosomes are the cellular organelles that translate the genetic code into proteins. The ribosome carries the enzymes necessary for protein synthesis. Under the electron microscope, a ribosome appears globular. Ribosomes consist of 2 subunits, large and small, which in prokaryotes sediment as 50S and 30S particles and which associate to form a 70S particle. The eukaryotic counterparts are 60S and 40S for the large and small subunits, and 80S for the complete ribosome. The S refers to the sedimentation coefficient of the units. Ribosomal subunits are synthesized in the nucleolus and crosses the nuclear membrane to the cytoplasm through nuclear pores. Both ribosomal subunits join together when the ribosome attaches to mRNA at the time of polypeptide synthesis in the presence of Mg^{2+}. The small subunit is responsible for the formation of the initiation complex, performs the decoding of the genetic information, and controls the fidelity of codon-anticodon interactions. The large subunit catalyzes the peptide bond formation and provides the path for the nascent polypeptide chain. Ribosomes contain specific sites that enable them to bind to the mRNA, the tRNA, and specific protein factors required for protein synthesis. Fully assembled ribosomes have three tRNA binding sites: an A site for incoming aminoacyl-tRNAs, a P site for peptidyl-tRNAs, and an E site where empty tRNAs exit. Each *E. coli* cell contains 15,000 or more ribosomes, which make up almost a quarter of the dry weight of the cell. Bacterial ribosomes contain about 65% rRNA and about 35% protein. The gene that codes for ribosomal RNA in a cell is the 16S rRNA gene. Every cell has a 16S rRNA gene. This gene is only about 1550 bp in length in prokaryotes and every cell has one to more than 25 copies of this gene.

Polysomes

Many ribosomes can translate the same mRNA molecule simultaneously. Multiple ribosomes on the same mRNA molecule form a polyribosome, or polysome.

Translation

Translation refers to the conversion of something from one language to another. In biology, translation is the process in which the genetic information encoded in mature mRNA is decoded by the ribosome to specify the amino acid sequence of a polypeptide that will later fold into an active protein. The name translation reflects that the nucleotide sequence of the mRNA must be translated into the completely different "language" of amino acids. Translation requires ribosomes, amino acids, mRNA, tRNAs and aminoacyl tRNA synthetases. Translation occurs in the cell's cytoplasm, where the large and small subunits of the ribosome are located, and bind to the mRNA. Ribosomes stabilize the mRNA and tRNAs such that the codons of the mRNA can be read 5' to 3' direction and the corresponding tRNAs can deliver the correct amino acids. The mRNA molecule is translated in the 5' to 3' direction, and the polypeptide is made in the N-terminal to C-terminal direction (Image 8.19 and

Image 8.20). The first amino acid on the polypeptide has a free amino group, so it is called the 'N-terminal' amino acid. The last amino acid in a polypeptide has a free carboxyl group, so it is called the 'C-terminal' amino acid. Translation occurs in both prokaryotes and eukaryotes; however, there are several key differences in the overall process between the two groups of organisms. In prokaryotes, translation occurs in the cytoplasm, while in eukaryotes, translation takes place in the endoplasmic reticulum. The process of translation can be divided into four main phases: initiation, elongation, termination, and ribosome recycling. All stages require enzymes and other protein factors; initiation and elongation also require energy provided by GTP (guanosine-5'-triphosphate).

Initiation involves binding ribosome binding site (RBS), also known as the *Shine Dalgarno sequence* in prokaryotes, to the small subunit of ribosome with the aid of special proteins called **initiation factors.** The ribosome binding site is located within the 5' untranslated region (5' UTR) of an mRNA molecule generally ~7 nucleotides upstream from the initiation codon (AUG). It is actually a particular sequence of purine-rich nucleotides. This sequence facilitates the initiation of protein synthesis by aligning the ribosome with the mRNA.

Translation initiation in eukaryotes is a complex, highly regulated process where a pre-initiation complex (PIC) forms on the 5' end of mRNA, scans for a start codon, and initiates protein synthesis. This process involves at least 12 initiation factors that facilitate ribosomal subunit binding, mRNA recognition, and start codon selection. The initiation complex assembles and scans the mRNA until it encounters the start codon, which then triggers a change in the complex and the recruitment of the large ribosomal subunit.

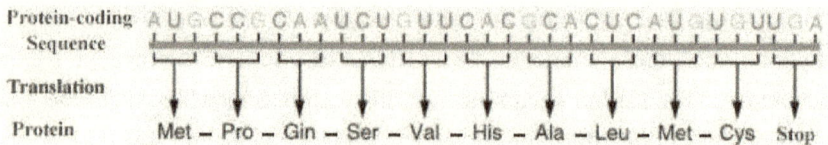

Protein-coding Sequence: A U G C C G C A A U C U G U U C A C G C A C U C A U G U G U U G A

Translation

Protein: Met – Pro – Gin – Ser – Val – His – Ala – Leu – Met – Cys   Stop

Image 8.19 The addition of amino acids to the growing polypeptide chain.

In the presence of ATP and Mg^{2+}, an amino acid combines with its specific aminoacyl tRNA synthetase to form aminoacyl adenylate enzyme complex.

The aminoacyl adenylate enzyme complex reacts with tRNA specific for the amino acid and produces aminoacyl-tRNA-complex (also aa-tRNA or charged tRNA). The enzyme and the AMP are released. The start (initiation) codon (AUG) present on mRNA is recognized only by the charged tRNA. Following this, the methionine-containing charged tRNA ($tRNA^{Met}$) attaches with the start codon on mRNA with its anticodon.

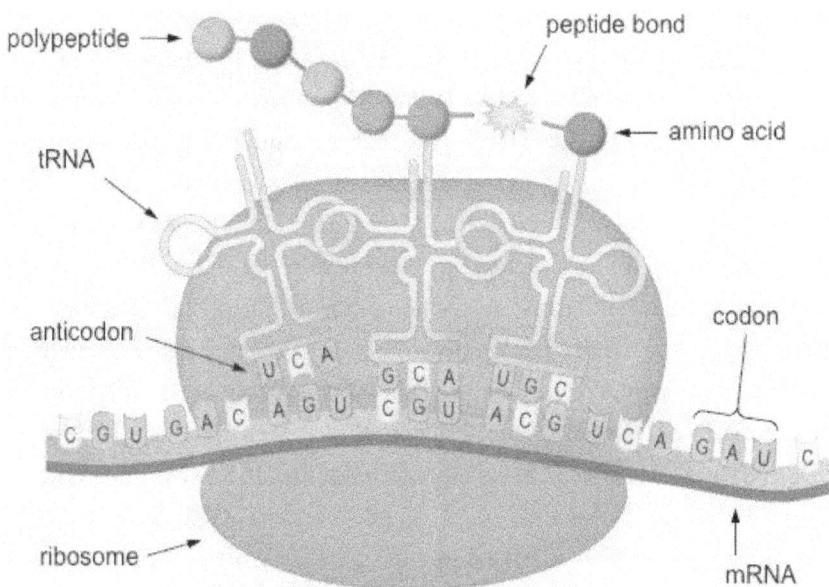

Image 8.20 The addition of amino acids to the growing polypeptide chain (source: https://old-ib.bioninja.com.au/standard-level/topic-2-molecular-biology/27-dna-replication-transcri/translation.html).

In the presence of Mg^{2+}, the large ribosomal subunit, now binds the small ribosomal subunit-mRNA-tRNA^{Met}-complex to produce an **initiation complex** that has all components for beginning protein synthesis. After the large ribosomal subunit joins the complex, the initiation factors are released. Start codon (AUG) signals where the gene begins (at 5' end of mRNA). Translation starts from AUG codon in the mRNA, which encodes methionine. As a result, newly made proteins begin with methionine; in some cases, this methionine is subsequently removed. In eukaryotes, the initiating amino acid is methionine rather than N-formylmethionine. In prokaryotes, a special initiator transfer-RNA, called Met-tRNA, brings the amino acid N-formylmethionine to initiate protein synthesis. The modified amino acid N-formylmethionine is always the first amino acid of the new polypeptide. In all other locations, this codon simply codes for the amino acid methionine. Note that translation does not start at the first base of the mRNA. There is an untranslated region at the beginning of the mRNA after the 5' cap, but before the initiation codon AUG, the **5' untranslated region (5' UTR) or leader sequence.**

Elongation involves repeated addition of amino acids. Elongation consists of three processes: bringing each new aminoacylated tRNA into line, forming the new peptide bond to elongate the polypeptide, and moving the ribosome from codon to codon along with mRNA in 5' to 3' direction so as to leave the space for binding of another charged tRNA. A charged tRNA attaches to the mRNA codon next to the initiation codon with the help of its anticodon. It requires GTP and elongation factors. A peptide bond is established between the carboxyl group (-COOH) and amino group ($-NH_2$). The reaction is catalysed by the enzyme peptidyl transferase. This process continues resulting in the formation of a polypeptide chain. As the mRNA passes through the ribosome, each codon interacts with the anticodon of a specific tRNA molecule by Watson-Crick base pairing. This tRNA molecule carries an amino acid at its 3'-terminus, which is incorporated into the growing protein chain. The tRNA is then expelled from the ribosome. Elongation continues until the polypeptide coded for in the mRNA is completed.

Termination involves the polypeptide chain being released, the ribosomal subunits dissociating from each other, and the mRNA being released. When the ribosome reaches a stop codons (UAA, UAG, or UGA) of the mRNA, proteins called '**release factors**' bind to it, and cause the ribosome, the mRNA, and the new polypeptide to separate. The stop codons are not recognised by any tRNAs. The new polypeptide is completed and released. This polypeptide chain then folds in on itself to form a protein. The two ribosomal subunits dissociate from the mRNA and separate back into a small and a large subunit. The ribosomal subunits can then be recycled for another round of translation. Eukaryotes utilize two releasing factors to carry out the termination step of protein synthesis, while prokaryotes have three different releasing factors. Note that the mRNA continues on past the stop codon. There is also an untranslated region at the end of the mRNA after the termination codon, but before the **3' poly-A tail, the 3' untranslated region (3' UTR) or trailer sequence.** Several roles in gene expression have been attributed to the untranslated regions, including mRNA stability, mRNA localization, and translational efficiency. In prokaryotic cells translation can begin even before transcription is complete.

Post-translational Modification (PTM)

After translation, the addition of other biochemical functional groups to the protein's amino acids extends the range of functions of the protein by modifying the chemical nature of an amino acid. In addition, enzymes may remove amino acids from the amino end of the protein, or even cut the peptide chain in the middle. For instance, most nascent polypeptides start with the amino acid methionine but it is usually taken off during post-translational modification. Post-translational modification (PTM) of proteins refers to the chemical changes that occur after a protein has been produced. These modifications include proteolysis, phosphorylation, glycosylation, ubiquitination, nitrosylation, methylation, acetylation, lipidation, *etc.* It can impact the structure, electrophilicity and interactions of proteins. Post-translational modifications can occur at any step in the 'life cycle' of a protein. For example, many proteins are modified shortly after translation is completed. Other modifications occur after folding and localization are completed to activate or inactivate catalytic activity or to otherwise influence the biological activity of the protein. The post-translational modifications play a crucial role in generating the heterogeneity in proteins and also help in utilizing identical proteins for different cellular functions in different cell types. How a particular protein sequence will act in most of the eukaryotic organisms is regulated by these post-translational modifications. Post-translational modifications are key mechanisms to increase proteomic diversity. While the human genome comprises 20,687 protein-coding genes, the proteome is estimated to encompass over 1 million proteins.

Proteolysis. Peptide bonds are indefinitely stable under physiological conditions, and therefore cells require some mechanism to break these bonds.

Proteases comprise a family of enzymes that cleave the peptide bonds of proteins and are critical in antigen processing, apoptosis, surface protein shedding and cell signaling.

Phosphorylation. Phosphorylation is the addition of a phosphate group, usually to serine, tyrosine, threonine or histidine. Several enzymes or signaling proteins are switched 'on' or 'off' by phosphorylation or dephosphorylation. Phosphorylation is performed by enzymes called 'kinases', while dephosphorylation is performed by 'phosphatases'. Phosphorylation plays critical roles in the regulation of many cellular processes including cell cycle, growth, apoptosis and signal transduction pathways.

Glycosylation. Glycosylation, the attachment of carbohydrate groups by glycosidic bonds to proteins, provides greater proteomic diversity than other PTMs. Protein glycosylation is an enzyme-directed chemical reaction that takes place in the Endoplasmic Reticulum (ER) and in the Golgi Apparatus of the cell. Protein glycosylation is acknowledged as one of the major post-translational modifications, with significant effects on protein folding, conformation, distribution, stability and activity.

Ubiquitination. Ubiquitin is a small but universal regulatory protein. Ubiquitination, also known as ubiquitylation, is an enzymatic process that involves the bonding of a ubiquitin group to a protein. This has sometimes been referred to as the molecular "kiss-of-death" for a protein. Ubiquitin acts like a flag/tag indicating that the protein's lifespan is complete. The tag signals the protein-transport machinery to ferry the protein to the proteasome for degradation.

Nitrosylation. Nitrosylation is the covalent incorporation of a nitrosyl group (NO) into another (usually organic) molecule. Nitrosylation is a physiologically important post-translational modification that affects a wide variety of proteins involved in a number of cellular processes. S-nitrosylation is a post-translational modification of proteins by which thiol groups (SH groups) of cysteine residues are modified by nitric oxide (NO) to generate S-nitrosothiols (protein S-nitrosocysteine), impacting protein function stability, and localization.

Methylation. Methylation refers to addition of a methyl group to lysine or arginine residue of a protein. Arginine can be methylated once or twice, while lysine can be methylated once, twice, or thrice. Methylation is achieved by enzymes called methyltransferases. Methylation has been widely studied in histones wherein histone methylation can lead to gene activation or repression based on the residue that is methylated.

Acetylation. Acetylation refers to addition of an acetyl group, usually at the N-terminus of the protein. It is involved in several biological functions, including protein stability, location, synthesis, apoptosis, cancer, DNA stability. Acetylation and deacetylation of histone form a critical part of gene regulation. Acetylation of histones reduces the positive charge on histone, reducing its interaction with the negatively charged phosphate groups of DNA, making it less tightly wound to DNA and accessible to gene transcription.

Lipidation. Lipidation is the covalent attachment of a lipid group to an amino acid in a protein. Proteins are covalently modified with a variety of lipids, including fatty acids, isoprenoids, and cholesterol. Lipid modifications play important roles in the localization and function of proteins.

Glutamylation. Glutamylation involves the addition of glutamate residues to tubulin and some other proteins. It modulates protein interactions, influences microtubule stability, and can regulate the activity of enzymes involved in microtubule severing.

Glycylation. Glycylation is a post-translational modification where one or more glycine molecules are covalently attached to a protein. Specifically, it involves the addition of glycine to the γ-carboxyl group of glutamic acid residues in the target protein.

Isoprenylation. Isoprenylation, also known as prenylation, is a post-translational modification where proteins are covalently attached to isoprenoid groups, specifically farnesyl (15-carbon) or geranylgeranyl (20-carbon) groups. This modification, typically at the cysteine residue near the C-terminus of the protein, is crucial for protein localization, membrane association, and signaling.

4′-Phosphopantetheinylation. Phosphopantetheine, also known as 4′-phosphopantetheine (4PPTylation), is a prosthetic group of several acyl carrier proteins. The modification of a protein amino acid by the addition of a 4′-phosphopantetheinyl moiety, derived from the hydrolysis of coenzyme A

(CoA), is an essential posttranslational modification of the primary and secondary metabolic pathways in prokaryotes and eukaryotes.

Sulfation. Protein sulfation is a post-translational modification where a sulfate group (SO4) is added to a protein, typically at a tyrosine residue. This process primarily occurs in the Golgi apparatus and affects protein structure, function, and interactions.

Selenation. Protein selenation refers to the incorporation of selenocysteine (Sec), a selenium-containing amino acid, into proteins.

C-terminal amidation. Amidation is the addition of an amide group to the end of the polypeptide chain. C-terminal α-amidation is the final and essential step in the biosynthesis of several peptide hormones.

Hydroxylation: Protein hydroxylation is a post-translational modification (PTM) where a hydroxyl group (-OH) is added to an amino acid residue, typically proline or lysine, within a protein by enzymes hydroxylases. This modification is crucial for regulating protein structure, function, and stability.

Deamidation: Protein deamidation is a chemical process where the amide side chains of asparagine (Asn) or glutamine (Gln) residues are converted to their corresponding acidic forms (aspartic acid or glutamic acid, respectively). This modification can alter protein structure, function, and stability, and is implicated in various biological processes including aging, disease, and protein degradation.

Inhibitors of Protein Synthesis

Protein synthesis inhibitors are substances that stops or slows the growth or proliferation of cells by disrupting the processes that lead directly to the generation of new proteins. It usually refers to antimicrobial drugs and aminoglycosides. Protein synthesis inhibitors usually act at the ribosome level, taking advantage of the major differences between prokaryotic and eukaryotic ribosome structures. Protein synthesis inhibitors work at different stages of translation. On the basis of target organisms inhibitor of protein synthesis is categories as: acting only on prokaryotes such as tetracycline, streptomycin, chloramphenicol, erythromycin, and rifamycin; acting on prokaryotes and eukaryotes such as puromycin, and actinomycin D;

acting only on eukaryotes such as cycloheximide, anisomycin, α-Amaniyin, diphtheria toxin, and ricin.

Control of Gene Expression (Gene Regulation)

Gene expression is the process by which the information encoded in a gene is used to direct the assembly of a protein molecule, in a multistep process involving transcription, messenger RNA (mRNA) maturation and splicing, translation and degradation of mRNAs and proteins.

Some genes are expressed continuously; some genes are expressed as part of the process of cell differentiation; and some genes are expressed as a result of cell differentiation. A gene that is always expressed is called constitutive. Some genes are usually off but can be turned on (induced) while other genes are usually on but can be turned off (repressed). For a cell to function properly, necessary proteins must be synthesized at the proper time. All cells control or regulate the synthesis of proteins from information encoded in their DNA. Whether in a simple unicellular organism or a complex multicellular organism, each cell controls when and how its genes are expressed. For this to occur, there must be a mechanism to control when a gene is expressed to make RNA and protein, how much of the protein is made, and when it is time to stop making that protein because it is no longer needed. The process of turning a gene on and off as per cell's requirement is known as gene regulation. Regulation of gene expression includes a wide range of mechanisms that control which genes, out of the many genes in a cell's DNA are expressed. A complex set of interactions between genes, RNA molecules, proteins (including transcription factors) and other components of the expression system determine when and where specific genes are activated and the amount of protein or RNA product produced. The up-regulation of a gene refers to an increase in expression of a gene whilst down-regulation refers to the decrease in expression of a gene. Regulation of gene expression may occur at various molecular levels due to different metabolic, physiological, or environmental conditions. The control of gene expression is extremely complex. Malfunctions in this process are detrimental to the cell and can lead to the development of many diseases, including cancer.

The regulation of gene expression conserves energy and space. It would require a significant amount of energy for an organism to express every gene at all times, so it is more energy efficient to turn on the genes only when they are required. In addition, only expressing a subset of genes in each cell saves space because DNA must be unwound from its tightly coiled structure to transcribe and translate the DNA. Cells would have to be enormous if every protein were expressed in every cell all the time.

Gene regulation makes cells different

Different cells in a multicellular organism may express very different sets of genes, even though they contain the same DNA. The set of genes expressed in a cell determines the set of proteins and functional RNAs it contains, giving it its unique properties. For example, one of the jobs of the liver is to remove toxic substances like alcohol from the bloodstream. To do this, liver cells express genes encoding an enzyme called alcohol dehydrogenase. This enzyme breaks alcohol down into a non-toxic molecule. The neurons in a person's brain don't remove toxins from the body, so they keep these genes unexpressed, or "turned off." Similarly, the cells of the liver don't send signals using neurotransmitters, so they keep neurotransmitter genes turned off.

Prokaryotic Gene Regulation

Prokaryotic organisms are single-celled organisms that lack a cell nucleus, and their DNA therefore floats freely in the cell cytoplasm. To synthesize a protein, the processes of transcription and translation occur almost simultaneously. When the resulting protein is no longer needed, transcription stops. As a result, the primary method to control what type of protein and how much of each protein is expressed in a prokaryotic cell is the regulation of transcription. All of the subsequent steps occur automatically. When more protein is required, more transcription occurs. Therefore, in prokaryotic cells, the control of gene expression is mostly at the transcriptional level.

The accessibility of promoter regions of prokaryotic DNA is in many cases regulated by the interaction of regulatory proteins with sequences termed operators present adjacent to the promoter elements in most operons and in most cases the sequences of the operator bind a repressor protein. Each operon has its specific operator and specific repressor, *e.g.*, lac operator is present only in the lac operon and it interacts specifically with lac repressor only.

Three kinds of proteins regulate the genes – inducers, repressors, and activators. Repressor proteins bind to the operator region and block the binding capacity of the RNA polymerase. In the case of activator proteins, they bind to the inducer region of the gene and enhance the binding capacity of the RNA polymerase. Inducer molecules bind to the DNA and activate or repress the gene based on the needs of the cell and the availability of substrate. In prokaryotes, the genes required for a particular protein are assembled next to each other, and this arrangement is called an operon. For example – the genes required for the use of lactose are arranged into lac operon. Bacteria require certain amino acids for survival, and these amino acids are produced in the cell with the help of these operons. Tryptophan is one of the essential amino acids required by bacteria. Five genes are required for synthesizing tryptophan, and these genes are placed next to each other. If tryptophan is not available in the environment, bacteria synthesize tryptophan using these genes. When the concentration of tryptophan in the cell is high, two molecules of tryptophan bind to the repressor protein on the operator site and inhibit the transcription of RNA hence genes required for the synthesis of tryptophan are blocked. Trp (tryptophan) operon is a repressible operon. it is negatively regulated by tryptophan molecule. Another example of gene regulation in prokaryotes is the lac operon. This is an inducible operon. When the glucose concentration in a cell is less or glucose is absent, then bacteria can utilize lactose as a source of energy. The lacZ gene in the lac operon encodes an enzyme called β-galactosidase, which breaks down lactose into galactose and glucose. When lactose is present, its isomer allolactose binds to the repressor protein and changes its structure so that it cannot bind to the lac operator site and prevent transcription. The presence of lactose induces the operon to code for proteins and enzymes and so it is an inducible operon. The absence of glucose and the presence of lactose are the important conditions required for a functional lac operon.

Regulation of Gene Expression in Eukaryotes

The control of gene expression is more complex in eukaryotes than in prokaryotes. This is because of the presence of a nuclear membrane in eukaryotes which separates the genetic material from the translation machinery, their genomes are bigger and encode more genes. Gene expression can be regulated by various cellular processes with the aim to control the amount and nature of the expressed genes. Gene expression is controlled with the help of regulatory proteins at numerous levels. These regulatory proteins bind to DNA and send signals that indirectly control the rate of gene expression.

In eukaryotes, gene expression involves many steps, and gene regulation can occur at all stages of the process. Different genes are regulated at different stages, and it's not unusual for a gene to be regulated many times. Gene expression is regulated for proper functioning and differentiation of the cells. Stages of gene regulations are:

Epigenetic Control of Gene Expression

The first level of control of gene expression is **epigenetic** regulation. Epigenetic mechanisms alter the chromosomal structure so that genes can be turned on or off. These mechanisms control how DNA is packed into the nucleus by regulating how tightly the DNA is wound around histone proteins. Regulation may occur when the DNA is uncoiled and loosened from nucleosomes to bind transcription factors. Nucleosomes can slide along DNA. If a gene is to be transcribed, the nucleosomes surrounding that region of DNA can slide down the DNA to open that specific chromosomal region and allow access for RNA polymerase and transcription factors, to bind to the promoter region and initiate transcription. If a gene is to remain turned off, or silenced, the histone proteins and DNA have different modifications that signal a closed chromosomal configuration. In this closed configuration, the RNA polymerase and transcription factors do not have access to the DNA and transcription cannot occur. How the nucleosomes move is dependent on signals found on the histone proteins. These signals are "tags" – in the form of phosphate, methyl, or acetyl groups – that open or close a chromosomal region. These tags are not permanent, but may be added or removed as needed. Since DNA is negatively charged, changes in the charge of the histone will change how tightly wound the DNA molecule will be. When unmodified, the histone proteins have a large positive charge; by adding chemical modifications like acetyl groups, the charge becomes less positive.

Methylation of DNA and histones causes nucleosomes to pack closely together, transcription factors cannot bind the DNA, and gene expression is turned off. Histone acetylation causes nucleosomes to space far apart, transcription factors can bind the DNA, allowing gene expression to occur.

Transcriptional Control of Gene Expression

Transcriptional regulation is control of whether or not an mRNA is transcribed from a gene in a particular cell. Many genes are regulated primarily at the level of transcription. Initiation of transcription is a vital key point of regulation of eukaryotic gene expression. Transcription factors play a role in regulating the transcription of genes. Sets of transcription factors bind to specific regulatory sequences in or near a gene and promote or repress its transcription into an RNA. Here, several factors such as promoters and enhancers alter the ability of RNA polymerase to transcribe the mRNA, thus modulating the expression of the gene. Transcriptional control decides when and how frequently a given gene be transcribed.

Post-transcriptional Control of Gene Expression

Post-transcriptional regulation occurs after the mRNA is transcribed but before translation begins. This regulation can occur at the level of mRNA processing, transport from the nucleus to the cytoplasm, or binding to ribosomes. Regulating the processing of RNA molecules such as polyadenylation, splicing, and capping of the pre-mRNA transcript in eukaryotes can lead to different levels and patterns of gene expression. For example, alternative splicing combinations of the RNA transcript of the same gene can produce different mRNAs from the same pre-mRNA and will generate biologically different proteins following translation.

RNA Transport

After post-transcriptional processing, the mature mRNA must be transported from the nucleus to the cytosol so that it can be translated into a protein. This step is a key point of regulation of gene expression in eukaryotes.

Control of RNA Degradation or Stability

The lifetime of an mRNA molecule in the cytosol affects how many proteins can be generated from it. The longer an mRNA exists in the cytoplasm, the more time it has to be translated, and the more protein is made. Once the RNA is in the cytoplasm, the length of time it resides there before being degraded is called RNA stability. Each RNA molecule has a defined lifespan and decays at a specific rate. Eukaryotic mRNAs are generally more stable than bacterial mRNAs, which typically last only a few minutes before being degraded. Nonetheless, there is great variation in the stability of eukaryotic mRNAs: some persist for only a few minutes, whereas others last for hours, days, or even months. These variations can produce large differences in the amount of protein that is synthesized.

The RNA stability can be increased, leading to longer residency time in the cytoplasm, or decreased, leading to shortened time and less protein synthesis. RNA stability is controlled by RNA-binding proteins (RPBs) and micro RNAs. RNA-binding proteins (RBPs) can bind to the 5′ or 3′ UTR regions. The binding of RBPs to these regions can increase or decrease the stability of an RNA molecule, depending on the specific RBP that binds. **Micro RNAs**, or **miRNAs**, can also bind to the RNA molecule. Typically, miRNAs bind to a complementary sequence in the 3′-untranslated region (UTR) of mRNA, which they target, causing mRNA silencing or degradation. Mature miRNAs recognize a specific sequence of a target RNA through complementary base pairing and cause them to be split up. miRNAs bind to mRNA along with a ribonucleoprotein complex called the **RNA-induced silencing complex (RISC)**. The RISC-miRNA complex rapidly degrades the target mRNA molecule. Many factors contribute to mRNA stability, including the length of its poly-A tail. It is predicted to control the activity of approximately 50% of all the protein-coding genes.

Translational Control of Gene Expression

Translational control decides which mRNAs in the cytoplasm are to be translated by ribosomes and their rate of translation. At this stage, the ability of ribosomes in recognizing the start codon can be modulated, thus affecting the expression of the gene. Translation can also be regulated at the level of binding of the mRNA to the ribosome. Once the mRNA bound to the ribosome, the speed and level of translation can still be controlled. An example of translational control occurs in proteins that are destined to end up in an organelle called the endoplasmic reticulum (ER). The first few amino acids of these proteins are a tag called a signal sequence. As soon as these amino acids are translated, a signal recognition particle (SRP) binds to the signal sequence and stops translation while the mRNA-ribosome complex is shuttled to the ER. Once they arrive, the SRP is removed and translation resumes. Regulators can boost or stifle the translation of an mRNA. For example, miRNAs can occasionally prevent their target mRNAs from being translated, rather than causing them to be chopped up.

Post-translational Control of Gene Expression
The final level of control of gene expression in eukaryotes is post-translational regulation. This type of control involves modifying the protein after it is made, in such a way as to affect its activity. One example of post-translational regulation is enzyme inhibition. When an enzyme is no longer needed, it is inhibited by a competitive or allosteric inhibitor, which prevents it from binding to its substrate. The inhibition is reversible, so that the enzyme can be reactivated later. This is more efficient than degrading the enzyme when it is not needed and then making more when it is needed again.

As proteins are involved in every step of gene expression and regulation, any modification in them affects other processes such as transcription, post-transcription events, translation, stability of the RNA, and also post-translational events. Common modifications in polypeptide chains include glycosylation, fatty acylation, and acetylation – these can help in regulating expression of the gene and offering vast functional diversity.

The activity and/or stability of proteins can also be regulated by adding functional groups, such as methyl, phosphate, or acetyl groups. Sometimes these modifications can regulate where a protein is found in the cell – for example, in the nucleus, the cytoplasm, or attached to the plasma membrane. The addition of a ubiquitin group to a protein marks that protein for degradation via proteasome. One way to control gene expression, therefore, is to alter the longevity of the protein.

Protein Transport and Protein Activity Control
Following translation and processing, proteins must be carried to their site of

action in order to be biologically active. Protein activity control selectively activates, inactivates, degrades, or compartmentalizes specific protein molecules after they have been synthesized. Also, by controlling the stability of proteins, the gene expression can be controlled. Stability varies greatly depending on specific amino acid sequences present in the proteins.

Universality of Genetic Information Transfer

The universality of genetic information transfer lies in the fact that the same fundamental principles of information storage, replication, transcription, and translation operate in all living organisms, from bacteria to humans. DNA serves as the universal genetic material, encoding information in the form of nucleotide sequences. Replication ensures faithful transmission of this information to daughter cells, transcription converts DNA sequences into RNA, and translation deciphers RNA codons into proteins. The universality of the genetic code, where the same codons specify the same amino acids across nearly all organisms, further supports this unity. These conserved mechanisms highlight the common evolutionary origin of life and provide molecular genetics with a powerful unifying framework to understand the diversity of biological systems.

Some Biological Questions

How can we account for the differences between species, between individuals of the same species, and between cells of the same individual if all DNA is made of the same material?

Although all DNA is made of the same chemical material, differences between species, between individuals of the same species, and between cells of the same individual can be explained by variations in DNA sequence, amount, and gene expression. The differences between species are due to variations in the base sequences of their DNA and the total amount of genetic material present. In humans, for example, the genome contains about 3.12 billion base pairs, whereas other organisms may have much smaller or larger genomes, leading to different sets of proteins and traits. Differences between individuals of the same species, such as humans, are largely explained by variation in only about 0.1% of the DNA sequence, meaning that roughly one in every 1000 bases is different, which accounts for unique physical and genetic characteristics. Within a single multicellular individual, all cells contain the same DNA, but they are not identical because different sets of genes are turned on or off depending on the cell's function. For instance, a liver cell has different active genes than a heart cell, allowing them to carry out their specialized roles. Thus, species-level, individual-level, and cellular-level differences all arise from how the universal structure of DNA is organized, varied, and regulated.

CHAPTER NINE   GENES AND CHROMOSOMES

Genes of all true organisms are made of DNA; certain viruses and viroids have genes made of RNA. The gene usually occupies a particular location within the chromosome, has a specific effect on the organism's morphology or physiology, can be mutated, and can recombine with other genes. Genes may be assigned to one of the two broad functional categories: structural genes, or regulatory genes.

Structural Genes

Structural genes are those that code for polypeptides or RNAs essential for the normal metabolic activities of a cell. These include enzymes, structural proteins, and receptors. In addition to protein-coding structural genes, all cells also contain RNA structural genes, which are required for the synthesis of rRNAs and tRNAs. The regulation of protein production is primarily achieved by controlling the access of RNA polymerase to the structural genes during transcription.

Regulatory Genes

Regulatory genes are genes that code for repressor molecules, which play a key role in controlling gene expression. The repressor protein binds to the operator region of an operon, thereby preventing RNA polymerase from attaching to the promoter and transcribing the structural genes. In this way, regulatory genes ensure that structural genes are expressed only when their products are needed, conserving the cell's energy and resources.

Transposons

Although most genes occupy a more or less stable position within the DNA molecule, there is a unique group of genes for which mobility is the normal state of affairs. These are known as **transposable genetic elements**, commonly referred to as *jumping genes*, *mobile genetic elements*, *movable genes*, or *transposons*. These are the transposable genetic elements, or jumping genes, or mobile genetic elements, or movable genes, or transposons, found in diverse life forms, both prokaryotes and eukaryotes. They have been found in viruses, bacteria, fungi, higher plants, and insects. A **transposon** is a segment of DNA that inserts itself into new positions within the genome. They can move to new positions within the same chromosome, or to a different chromosome. The movement of the transposon is called **transposition**. Transposons are often identified by the changes they cause: they may induce mutations by inserting into genes, alter gene expression by inserting into regulatory sequences, or produce chromosomal rearrangements. In 1951, **Barbara McClintock** discovered transposons in Maize and awarded the Nobel Prize in 1983. Transposons represent a powerful natural mechanism for gene movement and genome evolution.

Split or Interrupted Gene

A split or interrupted gene is defined as a gene in which the coding sequence is divided into segments known as **exons** and **introns**. Split genes were first discovered in eukaryotes, such as adenoviruses and higher organisms, and represent a fundamental distinction between eukaryotic and prokaryotic gene organization, since most prokaryotic genes lack introns.

Homeotic Genes

Homeotic genes are a special group of regulatory genes found in multicellular organisms that control the anatomical development and spatial organization of body structures. They determine the identity and arrangement of body parts by directing cells to form specific tissues and organs in the correct locations. Mutations in homeotic genes can result in dramatic changes, such as the transformation of one body part into another (for example, legs growing in place of antennae in fruit flies). A well-known class of homeotic genes is the **Hox genes**, which are highly conserved across species and play a crucial role in establishing the body plan during embryonic development.

Overlapping Genes

Overlapping genes are genes in which the nucleotide sequence of one gene is completely or partially contained within the sequence of another gene. This arrangement allows different proteins to be encoded from the same stretch of DNA by using alternative reading frames or transcription start sites. Overlapping genes are commonly found in viruses, where the compact nature of the genome requires efficient use of genetic material, but they also occur in some eukaryotes. This genetic strategy enables organisms, especially viruses, to maximize the coding potential of their limited genome size.

Housekeeping Gene

Housekeeping genes are genes that are expressed in virtually all cells because they are essential for the maintenance of basic cellular functions required for survival. They encode proteins and RNAs involved in fundamental processes such as metabolism, transcription, translation, and cell structure maintenance. Unlike specialized genes that are expressed only in certain cell types or under specific conditions, housekeeping genes are constitutively expressed at relatively constant levels to ensure the normal functioning of the cell.

Gene Size

Genes vary greatly in size, ranging from fewer than 100 base pairs to several million base pairs. An average prokaryotic gene contains somewhere between 900 and 1500 base pairs. Eukaryotic genes range from several hundred nucleotides (*e.g.*, the human insulin gene) to half a million or more nucleotides. Among human genes, the **SRY (Sex-determining Region on Y)** gene, located on the Y chromosome, is one of the smallest, consisting of only about 14 base pairs. In contrast, the **dystrophin (DMD)** gene, located on the X chromosome, is the largest known human gene, which has at least 2,300 kilobases (kb) and containing 79 exons.

Gene Families

Whereas most prokaryotic genes are represented only once in the genome, many eukaryotic genes are present in multiple copies, called **gene families**. In some cases, multiple copies of genes are needed to produce RNAs or proteins required in large quantities, such as ribosomal RNAs or histones. In other cases, distinct members of a gene family may be transcribed in different tissues or at different stages of development. For example, the α (alpha) and ß (beta) subunits of hemoglobin are both encoded by gene families in the human genome, with different members of these families being expressed in embryonic, fetal, and adult tissues. The human α-globin genes are clustered on chromosome 16, while the β-globin genes are clustered on chromosome 11. Gene families may be arranged in **clusters** within a region of DNA or dispersed across different chromosomes. Some clusters of genes are organized as **operons**, which are characteristic of bacteria. Operons contain genes regulated in a coordinated manner, usually encoding proteins with related functions. A classic example is the *lac* operon in *E. coli*, which includes three genes required for lactose metabolism.

In higher organisms, genes may also be organized into **multigene families**, which are collections of identical or extremely similar genes that likely originated from a common ancestral gene. Multigene families can be **simple** or **complex**.

Simple multigene families contain identical genes arranged in tandem clusters. An example is the 5S ribosomal RNA gene, of which humans have about 2,000 clustered copies to meet the high demand for ribosomal RNA.

Complex multigene families contain genes that are similar but not identical. A classic case is the globin gene families (α and β), which are located on different chromosomes and express different forms of globin proteins at various stages of development.

Thus, gene families provide genetic redundancy, flexibility, and specialization, allowing organisms to efficiently regulate gene expression and adapt to developmental and physiological needs.

Pseudogenes

Pseudogenes are defective copies of functional genes that have lost their ability to code for proteins, making them non-functional. They are especially common in eukaryotic genomes and typically arise through the accumulation of mutations – such as insertions, deletions, or premature stop codons—that disrupt the gene's normal coding potential. Pseudogenes can be formed by processes such as gene duplication, where one copy becomes inactive, or retrotransposition, in which an mRNA is copied back into DNA and inserted into the genome without proper regulatory sequences. An example can be seen in the human α- and β-globin gene families, each of which contains two genes that have been inactivated by mutation. Despite being nonfunctional, pseudogenes are widespread across almost all forms of life and provide important clues about genome evolution and the history of gene families.

The Information in Genes

The gene can be understood as a 'sentence' composed of three-letter words (codons). The 'letters' are the nucleotides, represented by their first letters: A, T, G and C. In different combinations, these letters make up different words, and in different combinations, those words give different information to the cell. In the gene, a word in always of three letters, no more, no less.

The sequence of nucleotides in genes provides two kinds of information during protein synthesis. The first is the particular amino acid that the three-letter words in a gene specify. The second is the position each amino acid will occupy in the protein. These pieces of information – which amino acid to use and its position in the protein – determine the primary structure of the polypeptide, its linear arrangement of amino acids. The primary structure of polypeptide ultimately determines the biological activity of the protein that contains the polypeptide. The primary structure determines the secondary, tertiary, and quaternary structures (that is, the physical shape of the molecule), and the polypeptide's shape, in turn, determines how the molecule will associate with other polypeptides. Depending on whether or not the primary, secondary, tertiary, and quaternary structures are correct, the resulting protein may or may not have biological activity.

Genome

Every cell of a multicellular organism generally contains the same genetic material. Most bacteria and viruses have a single chromosome, whereas eukaryotes usually have multiple chromosomes. The biological information is stored in DNA, organized into a large number of genes. These genes are often dispersed and separated by sequences that do not appear to code for functional products, called **intergenic** or **extragenic DNA**. While some intergenic DNA can regulate nearby genes, most of it has no currently known function and is sometimes referred to as **junk DNA**. Intergenic regions can occupy a substantial portion of the genome; in humans, they make up about 75% of the genome, compared with 30% in yeast and 15% in bacteria.

The **genome** is defined as the total amount of genetic material – including all genes and intergenic DNA – present in a basic chromosome set for a species. Prokaryotes are usually **monoploid**, with a single set of genes (one copy of the genome), while most higher animals and many plants are **diploid**, containing two complete sets of genes (two copies of the genome). Some higher plants are **polyploid**, carrying multiple genome copies. Among known organisms, the symbiotic bacterium *Carsonella ruddii* has one of the smallest genomes, containing only 182 protein-coding genes with just 159,662 base pairs of DNA. In contrast, the tiny (standing just a few inches tall), unassuming fern *Tmesipteris oblanceolata* from New Caledonia possesses the largest known genome, with an astonishing 160 billion base pairs – over 50 times larger than the human genome, which contains about three billion base pairs. If fully unraveled, the fern's DNA would stretch roughly 100 meters, compared to about two meters for the human genome.

Genomics

Genomics is the study of an organism's entire genome and its expression, including the complete set of genes, their nucleotide sequences and organization, and their interactions both within the species and with other species. It is a multidisciplinary field, combining genetics, molecular biology, biochemistry, statistics, and computer science. As a branch of biotechnology, genomics applies techniques such as genetic mapping and DNA sequencing to sets of genes or complete genomes, organizes the results in databases, and explores practical applications of this information. The field focuses on understanding genome structure and function from the molecular level upward, including interactions between genes, between genes and their protein products, and between genes and environmental factors. The starting point of genomics is **genome sequencing**, which allows identification of structural genes that code for proteins and RNAs, regulatory sequences that control gene expression, and non-coding regions. For humans, the genome is estimated to contain approximately 20,000-25,000 genes, which provide the blueprints for building all proteins necessary for life (International Human Genome Sequencing Consortium, 2004).

Genomic DNA

Genomic DNA (gDNA) refers to the chromosomal DNA that makes up the main genetic material of an organism, in contrast to extra-chromosomal DNA elements such as plasmids. It is present in all types of organisms, including both prokaryotes and eukaryotes, and contains the complete set of genes required for the structure, function, and regulation of the cell.

Organization (Packaging) of Eukaryotic DNA in Chromosomes

Each eukaryotic chromosome contains one giant molecule of DNA. DNA molecules are highly organised and tightly packed up in eukaryotic cells to fit inside the nucleus of every cell. DNA is packaged into chromosomes through a hierarchical process of compaction.

The first level of organisation of eukaryotic chromosome

The first level of organization, or packing, is the formation of nucleosomes. A nucleosome is the basic repeating structural and functional unit of chromatin. Each nucleosome consists of a nucleosome core particle (NCP), linker DNA, and a molecule of linker histone (H1). The structure of the NCP is relatively invariant from yeast to metazoans (Luger *et al.*, 1997; White *et al.*, 2001). Two copies each of the four core histones H2A, H2B, H3 and H4 form a histone octamer, around which approximately 146 base pairs (bp) of DNA are wrapped in a left-handed superhelical turns and completes a nucleosome core particle (NCP) which is about 5.5 nm in height and 11 nm in diameter (Richmond *et al.*, 1984). The negatively-charged DNA forms an intense bond with the positively-charged amino acids, keeping the DNA and histones tightly associated. NCP is connected by variable lengths of linker DNA arms on either side. The linker histone H1 binds to the linker DNA at the entry and exit sites of the NCP, connecting adjacent nucleosomes in the chromatin fiber. NCP with the linker DNA and the linker histone H1 builds the nucleosome. The linker histone H1 tightens these DNA-protein complexes together. These full nucleosomes (Image 9.1), also referred to as chromatosomes, are necessary for the formation of higher order chromatin structures *in vivo* (Allan *et al.*, 1981). Roger Kornberg discovered nucleosome in 1974 and won the Nobel prize in 2006. Nucleosome formation is similar from humans to yeast, and even to *Archaea*. The length of DNA that is associated with the nucleosome unit varies between species. But regardless of the size, two DNA components are involved: core DNA and linker DNA. The DNA that wraps around the histone octamer is known as the core or wrapping DNA. This value is invariant and is 146 base pairs (bp). The DNA that is in between each histone octamer is called the linker DNA and can vary in length from 8 to 114 bp. This variation is species specific, but variation in linker DNA length has also been associated with the developmental stage of the organism, or specific regions of the genome. The nucleosomes repeat at regular intervals along the entire length of the chromosome. Under the electron microscope, nucleosomes look like small "beads-on-a-string" structure (Image 9.2). These beads can move along the string (DNA) and change the structure of the molecule.

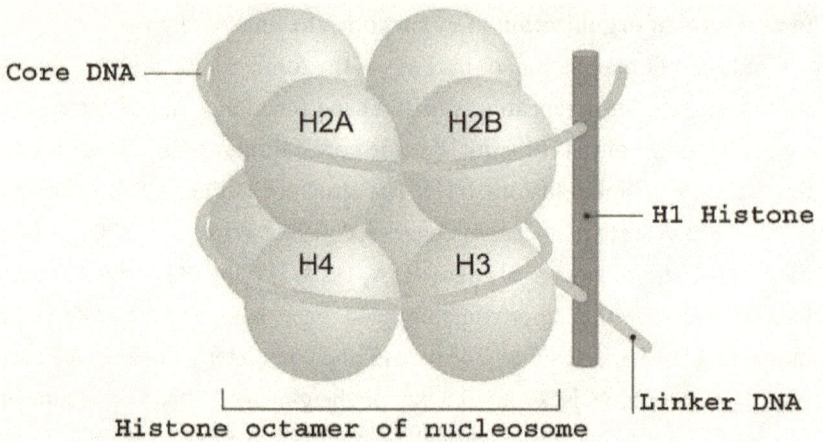

Image 9.1 Nucleosome (adapted from:
https://stock.adobe.com/search?filters%5Bcontent_type%).

Nucleosomes help to supercoil the DNA, resulting in a greatly compacted structure that allows for more efficient storage. A DNA molecule in this form is about seven times shorter than the double helix without the histones, and the beads are about 11 nm in diameter, in contrast with the 2-nm diameter of a DNA double helix.

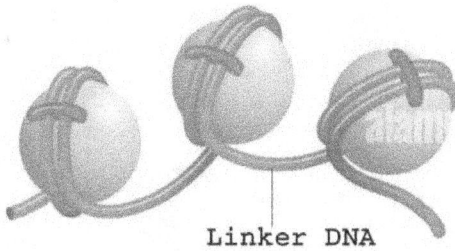

Linker DNA

Image 9.2 Beads-on-a-string form of chromatin (source: www.alamy.com
Image ID: 2J8XYKK).

The second level of organization of chromosomes

At the next level of organization, nucleosomes together with the linker DNA
between them (the classic "beads-on-a-string" structure) are further coiled or
helically wrapped to form a more compact structure known as the **30-nm
chromatin fiber** (Image 9.3), so called because it is about 30 nm in diameter
(solenoid fibre). This solenoid fiber is organized such that each turn contains
six nucleosomes. The coiling greatly shortens the chromatin, making it
approximately 40-50 times more compact than its extended form. Linker
histones play a key role in stabilizing this structure.

The third level of chromosome organisation

The 30-nm fiber develops looped domains, which are compressed and folded
around a nonhistone scaffold proteins to form 300nm chromatin fibers
(chromonema fibers). The 300-nm fibre in turn forms a spiral supercoil (coil
and recoil) with the help of RNAs and non-histone basic proteins to form the
condensed structure of chromosomes with a diameter of 700 nm (super
solenoid) at the metaphase stage of the cell division that are visible (when
stained) under microscope. During mitosis, chromosomes are compacted up
to 10,000 times their original level. The level of compaction of a eukaryotic
chromosome increases as the cell progresses through the cell cycle, with the
most significant compaction occurring during mitotic metaphase. This high
level of compaction allows efficient and precise distribution of chromosomes
into daughter cells. The end result is that DNA are tightly packed into the
familiar structures chromosomes. It is important to realize that chromosomes
are not always present. They form only when cells are dividing. At other times
DNA becomes less highly organized.

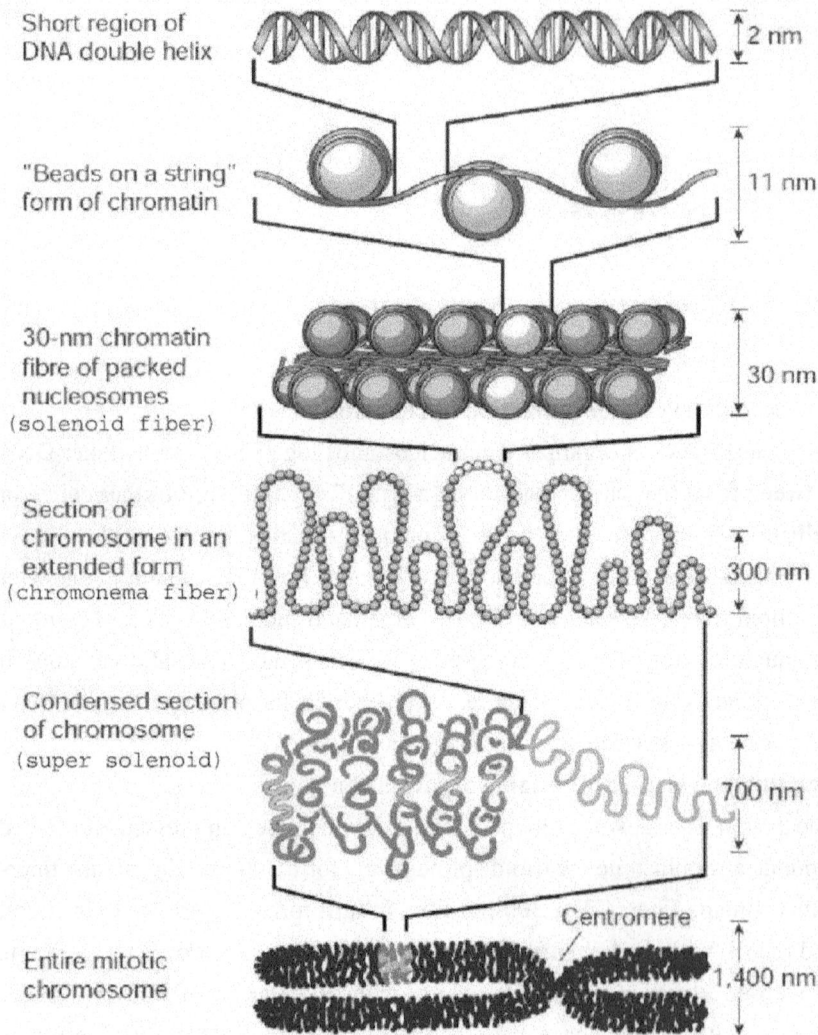

Image 9.3 Chromosome organization (adapted from:
https://www.researchgate.net/figure/ _fig2_325537059).

This remarkable, multiple folding allows 6 feet of DNA to fit into the nucleus of each cell of our body; an object so small that 10,000 nuclei could fit on the tip of a needle.

Supercoiling refers to the additional twisting of a DNA strand. Supercoiling functions to reduce the space required for DNA packaging, allowing for more efficient storage of DNA. Supercoiling helps to protect the DNA from damage and also allows chromosomes to be mobile during mitosis and meiosis. Condensing DNA into chromosomes prevents DNA tangling and damage during cell division.

Packing Ratio

The length of nuclear DNA is far greater than the size of the nucleus. To fit into the nucleus the DNA has to be condensed in some manner. The degree to which DNA is condensed is expressed as its packing/packaging ratio. Packaging ratio is the length of the DNA divided by the length into which it is packaged.

$$\text{Packing Ratio} = \frac{\text{The length of DNA}}{\text{The length into which it is packed}}$$

For example, the smallest of the human chromosome contains 4.6×10^7 bp of DNA (about 10 times the genome size of *E. coli*). This is equivalent to 14,000 μm of extended DNA. In its most condensed state during mitotic metaphase, the chromosome is about 2 μm long. This gives a packing ratio of 7000 (14,000/2).

The packaging of DNA into chromosomes allows for efficient DNA replication, gene expression, and the transmission of genetic information during cell division. This process of packaging DNA is called chromosome compaction.

Chromosomal Cycle

The shape and size of chromosomes change throughout the cell cycle. During interphase, chromosomes exist in the nucleus as a tangled network of fibers known as **chromatin**, a complex of DNA and proteins consisting of a linear, unbroken double-stranded DNA molecule. The fundamental structure of chromatin is essentially the same in all eukaryotes. During cell division, chromatin undergoes condensation. In early prophase, it appears as spirally coiled, thin, continuous filaments called **chromonemata** (singular: chromonema). By metaphase and anaphase, the chromonemata become fully condensed, forming the recognizable **chromatids** of eukaryotic chromosomes.

This cyclical change in chromosome shape and size during the cell cycle is referred to as the **chromosomal cycle**. The degree of chromatin condensation varies throughout the cell's life and plays a critical role in regulating gene expression.

Chromatin Types: Euchromatin and Heterochromatin

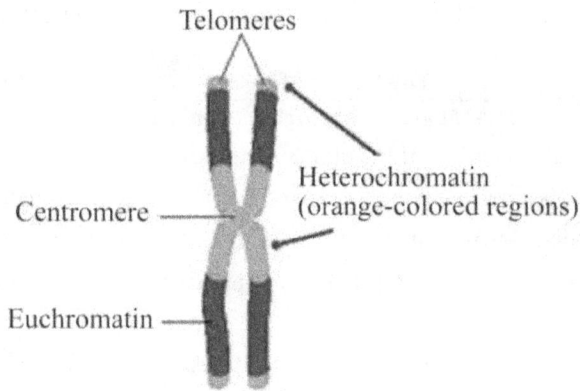

Image 9.4 Euchromatin and heterochromatin.

Chromatin exists in two forms: **euchromatin** and **heterochromatin**. When chromosomes are stained with DNA-binding chemicals, such as the Feulgen reagent, alternating lightly and darkly stained regions can be observed. The lightly stained regions represent euchromatin, which contains single-copy, genetically active DNA, while the darkly stained regions represent heterochromatin, which contains repetitive sequences that are generally genetically inactive (Image 9.4).

Euchromatin is the lightly packed form of chromatin, enriched in genes, and often under active transcription. It constitutes the majority of the genome – about 92% of the human genome – and contains most of the active genes. Euchromatin stains less densely because its DNA is loosely packed, allowing greater accessibility for transcription machinery.

Heterochromatin is the densely packed form of chromatin that is less accessible for transcription, making it largely inactive. About 10% of interphase chromatin is heterochromatin. It is found in all eukaryotes, typically near centromeres, telomeres, and in other species-specific locations. Heterochromatin is classified into two types: **constitutive** and **facultative**.

Constitutive heterochromatin remains condensed throughout the cell cycle. Constitutive heterochromatin is a permanent part of the genome and is always genetically inactive. Examples include centromeric and telomeric heterochromatin.

Facultative heterochromatin can switch between condensed and relaxed states. It may contain genes that are temporarily inactivated when the chromatin becomes condensed. A classic example is the **Barr body**, which is the inactivated X chromosome in female mammals.

Both euchromatin and heterochromatin play crucial roles in regulating gene expression and maintaining genome stability.

Chromosomes

In 1866, **Ernst Haeckel** first described the nucleus as the center responsible for transmitting the elements that determine hereditary characteristics. However, the structures we now know as chromosomes had been observed earlier: in 1842, the Swiss botanist **Karl Wilhelm von Nägeli** identified rod-shaped bodies in pollen, which he called *transitory cytoblasts* – later recognized as chromosomes. Independently, in 1883-1884, the Belgian scientist **Edouard Van Beneden** observed chromosomes in *Ascaris* worms.

In 1870, the German biologist **Walther Flemming**, who discovered mitosis, introduced aniline staining to visualize chromosomes during cell division for the first time. By 1878, Flemming had described the fibrous network within the nucleus, naming it **chromatin** or "stainable material." Finally, in 1888, the German anatomist **Heinrich von Waldeyer** coined the term **chromosome**, derived from Flemming's earlier term chromatin.

A **chromosome** is an organized, thread-like structure of DNA and protein found within the nucleus of cells. Each chromosome consists of a single, long, continuous double-stranded DNA molecule that carries numerous genes, regulatory elements, and other nucleotide sequences. The term *chromosome* is derived from the Greek words *chroma* ("color") and *soma* ("body"), reflecting their ability to take up certain stains. These deeply stained, thread-like structures within the nucleus of plant and animal cells contain the genetic information of an organism, arranged in a linear sequence of genes. They are the physical basis of heredity as they store, replicate, transcribe and transmit the genetic information. Chromosomes are invisible during interphase through light microscope. Under electron microscope (magnification of about 2 lacs), only chromatin fibres of about 300 Å diameter are observed during interphase. Chromosomes assume various shapes and sizes during the different stages of cell division. There occurs a cycle of coiling decoiling (condensation-decondensation) of chromosomes during the cell cycle. In interphase, the chromosomes are in a decoiled (decondensed) state but in metaphase, the maximum coiling is achieved. Therefore, the general morphology of chromosomes can be studied easily at metaphase. For the purposes of identification and distinction, chromosomes can be well studied using specialized staining techniques that yield specific banding patterns. During cell division chromatin organizes/packed itself into chromosomes. Each chromosome contains one DNA molecule and its associated proteins, regardless of whether it is threadlike or condensed form. Although the genomes of prokaryotes are composed of DNA, there DNA is not packed into chromosomes. True chromosomes are found only in eukaryotes. There are major differences between chromosomes in diverse life forms.

Chromosome Size

During the cell cycle, the chromosome length varies due to coiling and de-coiling. The maximum chromosome condensation occurs at metaphase therefore, chromosome size is normally measured at mitotic metaphase. In general, plant chromosomes are larger than animal chromosomes. In general, Orthoptera (Grasshopper) and Amphibia among animals have larger chromosomes. Among the higher plants, monocots generally possess larger chromosomes than dicots. Usually, the mitotic chromosomes range in length from 0.25 μm in fungi and birds to 30 μm in some plant species such as Trillium. On the other extreme, the polytene giant chromosomes of Diptera may be about 300 μm in length and about 10 μm in diameter. Thus, there is a great variation in the size of chromosomes.

Molecular Structure of Chromosomes

Chromosomes are composed of DNA, RNA and proteins. This DNA is efficiently packed to achieve the chromosome shape. A chromosome has about twice the mass of protein as the mass of DNA, and less than 10% of the mass is RNA. Chromosomal proteins are of two types: histones and nonhistones, both of which play important roles in chromatin structure and function. Chromosomal RNA is, for the most part, mRNA, tRNA, and rRNA in various stages of completion.

The histones. Histones are large globular alkaline proteins as they contain large amounts of the basic amino acids, particularly arginine and lysine. In fact, histone molecules consist of approximately 25 percent arginine and lysine. Histones are the main protein component of chromatin that mechanically support the condensation of chromatin into chromosomes. The amount of histone is equal to the total mass of chromosomal DNA. They can be divided into two major groups: core histones and linker histone. There are four core histones: histone 2A (H2A), histone 2B (H2B), histone 3 (H3), and histone 4 (H4); and one linker histone, histone 1 (H1). They are very similar among different species of eukaryotes and have been highly conserved during evolution. They are produced during S Phase, and not during other times of the cell cycle. At physiological pH levels they are cationic, making it easier to interact with negatively-charged DNA in the nucleus with the purpose of packaging the DNA into chromatin. Histones provide a surface around which DNA is wrapped. Chromatin architecture, nucleosomal positioning, and ultimately access to DNA for gene transcription, is largely controlled by histone proteins.

Nonhistone proteins. The nonhistone proteins are all the other proteins associated with the DNA of chromosomes. Most of these proteins are either negatively charged or neutral at physiological p^H. Relative to the mass of DNA in chromosomes, the nonhistone proteins also vary widely, from more than 100 percent of the DNA mass to less than 50 percent. These proteins serve regulatory, enzymatic and stabilizing functions. Non-histone proteins display more diversity and are not as highly conserved as histones across different species. While histones are known for their conserved structure and function in DNA packaging, non-histone proteins encompass a wide range of functions and are more diverse in their sequence and structure. They may also differ between different tissues of same organism.

Specialized Chromosomal Structures

In eukaryotic chromosomes, there are two important special-function repetitive DNA sequence: centromere and telomeres (Image 9.5). They are two essential features of all eukaryotic chromosomes. Each provides a unique function that is absolutely necessary for the stability of the chromosome. In addition, some chromosomes contain nucleolar organizer regions and satellite DNA.

Centromere

The region where two sister chromatids of a eukaryotic chromosome are held together is known as centromere. It generally appears as constriction and is also called the primary constriction. Primary constriction is present in all chromosomes. When chromosomes are stained, they typically show a dark-stained region that is the centromere. Condensation of the chromatin is less at the centromere than in the other parts of the chromosome. The DNA sequence within these regions is called *CEN* **(centromeric) DNA**. Centromeric DNA from mammals is simple-sequence DNA. In human beings, a 170 base pair sequence is repeated over and over again anywhere from 1700 to 2900 times depending on the centromere.

On the surface of the centromere, most species have two small disc-shaped macromolecular protein complexes called kinetochores that govern chromosome movement by binding spindle microtubules during mitosis and meiosis. The kinetochores do not form part of the chromatid but lie one on each side of the chromosome such that each chromatid is having its own kinetochore. It is the actual location where spindle fibers (microtubules) attach during cell division to pull the chromosomes apart. One kinetochore is attached to the spindle fibres towards one pole and the other similarly towards the other pole.

The parts of a chromosome on either side of centromere are called arms. When the arms are unequal, the short arm is called 'p' arm and the long arm is called 'q' arm. These arms are then subdivided by numbers. Each area of chromosome is given number, lowest number closest (proximal) to centromere and highest number at tips (distal) to centromere.

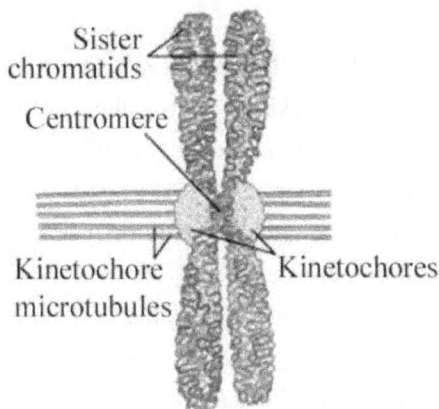

Image 9.5 Chromosome.

The position of centromere is constant for a particular chromosome. Centromeres vary in size within a single species as well as between species. Yeast centromeres are on the order of a few hundred base pairs, and human centromeres range from about 300 to 500 kbp (kilobase pairs).

The structure and function of the centromere is different from that of the rest of chromosome. Usually, each chromosome has only one centromere (monocentric). Sometimes chromosomes may be dicentric (*e.g.*, wheat, maize, *etc.*) or polycentric (*e.g.*, luzula, ascaris, *etc.*). However, in some insects the centromere is diffused along the length of chromosome, *e.g.*, as in *Ascaris megalocephala* and in Hemipteran bug (insects).

The centromere is responsible for the accurate segregation of the replicated chromosomes to the daughter cells during mitosis and meiosis. During mitosis, the centromere that is shared by the sister chromatids must divide so that the chromatids can migrate to opposite poles of the cell. On the other hand, during the first meiotic division the centromere of sister chromatids must remain intact, whereas during meiosis II they must act as they do during mitosis.

Therefore, the centromere is an important component of chromosome structure and segregation. Spindle fibres attach to the centromere and thus it is responsible for chromosome movement during cell division. Centromeres are the first parts of chromosomes to be seen moving towards the opposite poles during anaphase. The chromosome fragment lacking a centromere (acentric fragment) fails to orient itself on the metaphase plate and is unable to move to any of the poles during anaphase and is ultimately eliminated.

Telomeres

The term telomere was coined by Muller in 1938. The tips or the terminal ends of the linear chromosome of eukaryotes are called telomeres. It is a special protective cap and often has DNA sequences different from the rest of the chromosome. The telomere differs in structure and composition from the rest of the chromosome. Telomeres protect the chromosomes from nucleases and other destabilizing influences, and thereby maintain the structural integrity and individuality of linear chromosomes. It has a unique property that it prevents the ends of the chromosomes from sticking/fusing to each or one another. When telomeres are damaged or removed due to chromosome breakage, the damaged chromosome ends can readily fuse or unite with broken ends of another chromosome. Specific proteins bind to the telomeric region, and the resulting nucleoprotein structures are thought to prevent recombination between the ends of different chromosomes. Telomeres are specially modified portion of the chromosomes for attachment to the nuclear membrane. Telomeres consist of hundreds or thousands of repeats of the same short DNA sequence, which varies between organisms but is 5'-TTAGGG-3' in humans and other mammals. During DNA replication, the extreme ends of the telomere (usually 25-200 base pairs) cannot be copied and so the telomere gets marginally shorter. When the telomere becomes too short, the chromosome reaches a "critical length" and can no longer replicate. This means that a cell becomes "old" and dies by a process called apoptosis. Telomere activity is controlled by two mechanisms: erosion and addition. Erosion, as mentioned, occurs each time a cell divide. Addition is determined by the activity of telomerase. Telomerase, also called telomere terminal transferase, is an enzyme made of protein and RNA subunits that elongates chromosomes by adding TTAGGG sequences to the end of existing chromosomes. Telomerase is an RNA-dependent DNA polymerase enzyme that can make DNA using RNA as a template. Telomerase is found in fetal tissues, adult germ cells, and also tumor cells. Telomerase activity is regulated during development and has a very low, almost undetectable activity in somatic cells. Because these somatic cells do not regularly use telomerase, they age. The result of aging cells is an aging body. If telomerase is activated in a cell, the cell will continue to grow and divide. This "immortal cell" theory is important in two areas of research: aging and cancer.

Typically, eukaryotic DNA is a linear double-stranded sequence. However, at the telomere end of each chromosome there is a single-stranded segment. This single-stranded DNA is important for DNA replication because it is where the DNA replication process starts. DNA replication is handled by an enzyme called **DNA polymerase**; however, this enzyme cannot initiate the replication process. Instead, an RNA primer begins the DNA replication process, and it requires single-stranded DNA to work properly.

Secondary Constrictions

The constricted or narrow region other than that of centromere is called secondary constriction. In addition to centromere, a chromosome may have one or more secondary constrictions. Secondary constriction is not present in all chromosomes. The location of secondary constriction is species specific as it plays most important role in karyotyping. Chromosome may possess secondary constriction in one or both arms of it. In 1969, Herbert Lubs discovered secondary constriction on the bottom end of the long arm of the X chromosome. The secondary constriction contains the genes for rRNA synthesis (18S rRNA, 5.8S rRNA, and 28S rRNA). The part of the chromosome separated by secondary constriction is called satellite or trabant, which remains attached to the main part of chromosomes by a thread of chromatin (Image 9.6). The chromosome having satellite is called a sat chromosome or satellite chromosomes.

Satellites show considerable variation in their size. One chromosome of the genome may possess larger satellite, while another (non-homologous) chromosome may possess smaller satellite. Satellites are mostly heterochromatic. But certain genes are located in these regions. Satellites are elongated, round shaped and variable in size. There are at-least two SAT-chromosomes in each diploid nucleus. Many polyploid species, however, have only two SAT-chromosomes.

Secondary constrictions are of two types: **Nucleolar Organizer Region (NOR) and Joint**. They are always constant in their positions and often used as markers. NOR is secondary constriction involved in the formation of nucleolus in the interphase of the cell cycle. The nucleolus is a distinct structure present in the nucleus of eukaryotic cells. However, all the secondary constrictions do not form nucleoli. Nucleoli are formed during telophase, persist throughout the interphase and disappear in middle or late prophase. The nucleolus is the site of ribosomal RNA (rRNA) synthesis. The NOR contains several hundred copies of the gene coding for rRNA. Primarily, it participates in assembling the ribosomes, alteration of tRNA, and sensing cellular stress. NOR represents about 0.3% of the total amount of nuclear DNA. Nucleolar organizers are probably of universal occurrence in several plants and animals. Each nucleolar organizer may be one or two in number and may be attached to one nucleolus.

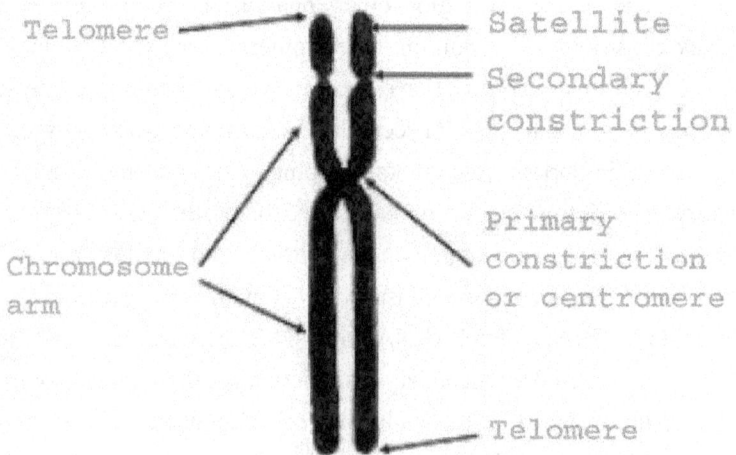

Image 9.6 Secondary constriction (source: https://ask.learncbse.in/t/what-is-secondary-constriction-of-chromosome-explain-with-diagram/68482/2).

In humans NORs are found on the short arms of all acrocentric chromosomes except the Y chromosome. The NOR consists of numerous tandem repeats of a single gene. Each NOR consists of approximately 80-100 repeats.

Joints are areas involved in the breaking and fusion of chromosome segments.

Matrix

The membrane enclosing each chromosome is known as a pellicle. The jelly-like substance found inside the pellicle is called matrix. It is made of non-genetic materials such as RNA, proteins and lipids.

Repetitive DNA

The term **repetitive DNA** (repeats, DNA repeats, repetitive sequences) refers to non-coding sequences that are present in multiple copies within the genome. These sequences occur at various levels throughout the genome and can make up a substantial portion of eukaryotic genomes, ranging from 20% to 90% of the total DNA depending on the species. Repetitive DNA also exhibits considerable variability between individuals in the number of repeats at a given location. In higher eukaryotes, a large fraction of the genome consists of repetitive DNA, with approximately 25% of the mammalian genome made up of these sequences. Repetitive DNA can be broadly classified into two types: **tandem repeats** and **interspersed repeats**.

Tandem repeats

Tandem repeats consist of repetitive units arranged one after another in the genome and can vary in both length and copy number among individuals. Based on the size of the repeating unit, tandem repeats are further categorized into **microsatellites**, **minisatellites**, and **macrosatellites**, with microsatellites being the most widely used in forensic DNA analysis.

Microsatellites (Short Tandem Repeats – STRs): Microsatellites, also known as **short tandem repeats (STRs)** or **simple sequence repeats (SSRs)**, are short, highly repetitive DNA sequences found throughout the genome. They typically consist of motifs of 1-6 base pairs, repeated 5 to 200 times, and are usually located in non-coding regions. Examples include sequences such as (A)15 = AAAAAAAAAAAAAAA, (AT)6 = ATATATATATAT, (CGA)4 = CGACGACGACGA, and (ATCG)3 = ATCGATCGATCG. Microsatellites are highly polymorphic because the number of repeats varies among individuals, making them extremely useful as **molecular markers** in DNA fingerprinting, forensic analysis, and genetic studies. Mutation rates in microsatellites are much higher than in most other DNA sites – 100 to 10,000 times faster than normal base pair substitutions – contributing to substantial genetic variation. In humans, about 90% of microsatellites are found in noncoding "junk" DNA, though they may influence gene expression and provide a source of genetic variability. In bacteria, variation in microsatellite alleles within coding regions can be adaptive to different environments. Individuals inherit **two copies of each microsatellite**, one from each parent, which can be identical or different. The DNA regions between two microsatellites are termed **inter simple sequence repeats (ISSRs)**.

Minisatellites: Minisatellites, also called **variable number of tandem repeats (VNTRs)**, are tandem DNA repeats typically 10 to 100 base pairs in length, repeated 5 to 100 times. They are highly polymorphic, unique, and GC-rich sequences. Minisatellites are commonly found in the **telomeric and centromeric regions** of chromosomes. Due to their high variability among individuals, minisatellites are widely used in **DNA analysis**, particularly for individual identification and forensic applications.

Macrosatellites: Macrosatellites are the largest type of **tandem DNA repeats**, characterized by repeat blocks exceeding 20 kilobases (kb) in size. They are significant in genetic studies because each repeat unit can include functional elements, such as **exons** or even entire genes. Macrosatellites are often associated with **centromeric regions** of chromosomes and may play roles in chromosome structure, stability, and regulation of gene expression.

Interspersed Repeats

Interspersed repeats are repetitive DNA sequences found scattered throughout a genome, distinct from tandem repeats which are clustered together. They are derived from transposable elements (TEs), also known as "jumping genes". These elements can move and integrate randomly within the genome, creating copies of themselves in different locations

Chromosome Morphology

One common method for classifying chromosomes is based on the position of the centromere and the relative lengths of the two arms of the chromosome, which helps in cytogenetic studies and karyotyping. The classes in this system are (Image 9.7):

Metacentric (mediocentric): The centromere is in the middle of the chromosome. The arms on either side of the centromere are equal in length. At anaphase, the chromosomes take the typical V-shape, the centromere being at the apex of V.

Classification of Chromosomes

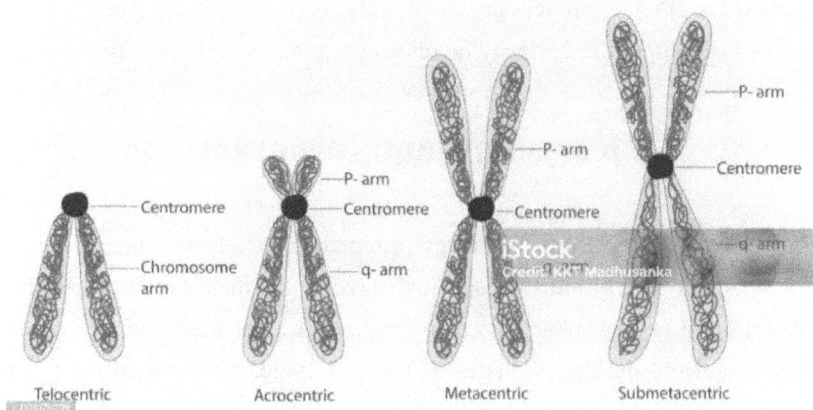

Image 9.7 Chromosome morphology (source: https://www.istockphoto.com/vector/chromosome-classification-gm2125175209-567711310?searchscope=image%2Cfilm).

Submetacentric: The centromere is in the sub median region; *i.e.*, one arm longer than the other. The chromosomes take a J- or L-shape at anaphase.

Acrocentric: The centromere is closer to one end. Each chromatid has two visible arms but the short arm is only noticeable.

Telocentric: The centromere is at the end of the chromosome and there is only one arm. Telocentric chromosomes appear rod-shaped (I-shaped) during anaphase.

C-Value Paradox

The **C-value** refers to the amount of DNA per cell in an organism and is generally constant for all cells of a given species. It is measured in **picograms**, where 1 picogram roughly corresponds to 1 gigabase of DNA. In prokaryotes, the C-value represents the DNA content of a single chromosome, while in eukaryotes, it corresponds to the haploid genome. Although there is often a correlation between DNA content and organismal complexity, a paradox arises because some organisms do not follow this pattern. For example, certain single-celled amoebae possess genomes up to 100 times larger than the human genome. One explanation for this discrepancy is the variation in the amount of **repetitive DNA sequences** present in the genome, which can greatly inflate genome size without necessarily increasing organismal complexity.

Karyotype and Ideogram

Whereas bacteria have only a single chromosome, eukaryotic species have at least one pair of chromosomes. Most have more than one pair. Another relevant point is that eukaryotic chromosomes are detected only during cell division and not during all stages of the cell cycle. They are in their most condensed form during metaphase when the sister chromatids are attached. The chromosomes are most easily seen and identified at the metaphase stage of cell division.

Karyotype is the characterization of metaphase chromosomes of an individual or a species that describes the number, size, and morphology of chromosomes in the normal diploid cell. The process of generating a karyotype begins with the short-term culture of cells derived from a specimen. After a period of cell growth and multiplication, dividing cells are arrested in metaphase stage by addition of colchicine, which poisons the mitotic spindle. The cells are next treated with a hypotonic solution that causes their nuclei to swell and the cells to burst. The nuclei are then treated with a chemical fixative, dropped on a glass slide, and treated with various stains. Karyotypes are prepared using standardized staining procedures that reveal characteristic structural features for each chromosome. The first discriminating parameter when developing a karyotype is the size and number of the chromosomes. In a karyotype, chromosomes are arranged and numbered by size, from the largest to the smallest.

Table 9.1 Karyotypes of some common domestic mammals

| Species | Autosomal pairs | | | Sex chromosomes | |
	Diploid number (2N)	Number of metacentrics	Number of acrocentrics or telocentrics	X	Y
Dog (*Canis familiaris*)	78	0	38	M	A
Horse (*Equus caballus*)	64	13	18	M	A
Donkey (*Equus asinus*)	62	24	6	M	A
Goat (*Capra hircus*)	60	0	29	A	M
Cow (*Bos taurus*)	60	0	29	M	M
Sheep (*Ovis aries*)	54	3	23	A	M
Pig (*Sus scrofa*)	38	12	6	M	M
Cat (*Felis cattus*)	38	16	2	M	M

For most organisms, all cells have the same **karyotype**. Each species is characterized by a karyotype, so a wide range of number, size, and shape of metaphase chromosomes is seen among eukaryotic organisms. Even closely related species may have quite different karyotypes. Table 9.1 describes the karyotypes of some common domestic animals. Karyotypes are used to study chromosomal abnormalities, and may be used to determine other macroscopically visible aspects of an individual's genotype, such as sex. A normal human male chromosome pattern would be described as: 46, XY. Where, 46 = total number of chromosomes, and XY = sex chromosome constitution (XY = male, XX = female). Any further description would refer to any abnormalities or variants found.

Karyotyping is the process of pairing and ordering all the chromosomes of an organism. When viewing a karyotype, it can often become apparent that changes in chromosome number, arrangement, or structure are present. Clinical cytogeneticists analyze human karyotypes to detect gross genetic changes – anomalies involving several megabases or more of DNA. Karyotypes can reveal changes in chromosome number associated with aneuploid conditions, such as trisomy 21 (Down syndrome). Careful analysis of karyotypes can also reveal more subtle structural changes, such as chromosomal deletions, duplications, translocations, or inversions. In fact, as medical genetics becomes increasingly integrated with clinical medicine, karyotypes are becoming a source of diagnostic information for specific birth defects, genetic disorders, and even cancers.

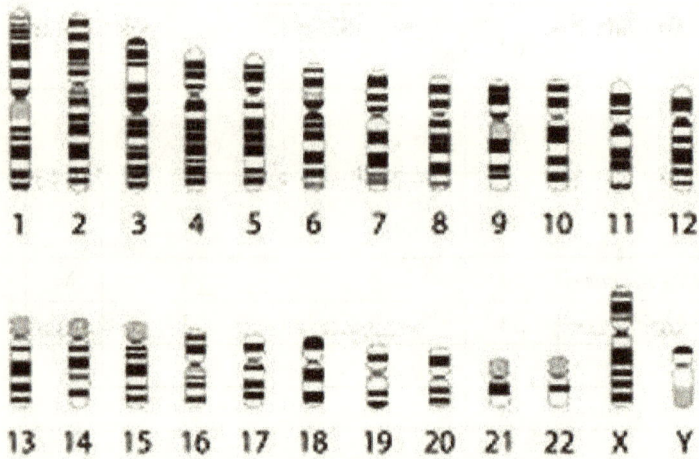

Image 9.8 Ideograms of human chromosomes.

When the karyotype of a species is represented as a **diagram or photograph** showing all the morphological features of its chromosomes in a standard format, it is called an **ideogram** or **ideotype** (Image 9.8). This is the conventional method for displaying karyotypes and provides a clear visual representation of chromosome number, size, shape, and centromere position.

Chromosome Banding Patterns

Chromosomes observed under a light microscope without treatment reveal limited structural detail. To enhance visualization, cytologists developed **staining techniques** that produce characteristic **banding patterns**, allowing identification of individual chromosomes and analysis of their structure.

Chromosome banding refers to the appearance of alternating **light and dark stripes** along the length of a chromosome after staining. A **band** is a region distinguishable from its neighbors by being darker or lighter.

Most banding methods require **mitotic chromosomes**, which are obtained by treating cultured cells with **tubulin inhibitors** (*e.g.*, colchicine or colcemid) to arrest cells in metaphase. The cells are then spread on slides, stained, and observed under a microscope. Dark bands usually correspond to **heterochromatin**, which is more condensed and gene-poor, while light bands correspond to **euchromatin**, which is less condensed, gene-rich, and lightly stained. Different banding techniques highlight different chromosomal features:

- ❖ **G-bands** (Giemsa staining) – dark bands mark heterochromatin.
- ❖ **R-bands** – reverse pattern of G-bands, highlighting euchromatin.
- ❖ **Q-bands** (quinacrine) – AT-rich heterochromatin fluoresces brightly.
- ❖ **C-bands** – centromeric constitutive heterochromatin.
- ❖ **NOR-bands** – nucleolar organizer regions containing rDNA.
- ❖ **T-bands** – telomeric GC-rich regions.

Banding patterns are essential for **detecting chromosomal abnormalities** such as translocations, deletions, and inversions, as each chromosome exhibits a unique pattern. They also serve as a tool for **karyotype analysis, studying chromosome polymorphism**, and exploring **evolutionary relationships**. For instance, comparisons of banding patterns in humans, chimpanzees, gorillas, and orangutans reveal that humans are most closely related to chimpanzees and more distantly related to gorillas and orangutans.

Giemsa banding or G-banding

G-banding is the most widely used chromosome banding technique in cytogenetic analysis, first developed by Seabright in 1971. It involves staining **metaphase chromosomes** with **Giemsa dye**, a mixture of cationic thiazine and anionic eosin dyes. The thiazine molecules intercalate into the negatively charged DNA and stain it blue, while eosin binds to thiazine and produces a purple hue. Prior to staining, chromosomes are typically treated with **trypsin**, which partially digests chromosomal proteins, loosening the chromatin and allowing the dye to access the DNA.

G-banding produces alternating **light and dark stripes** along the length of chromosomes. Lightly stained regions correspond to transcriptionally active **euchromatin**, rich in guanine and cytosine, whereas darkly stained regions correspond to less active **heterochromatin**, rich in adenine and thymine. Each chromosome exhibits a unique banding pattern, providing a distinctive "barcode" that facilitates accurate **karyotyping** and identification of **chromosomal abnormalities**, such as translocations, deletions, inversions, duplications, and changes in chromosome number.

Standard G-banding allows visualization of **400-600 bands** on metaphase chromosomes, while high-resolution techniques can reveal up to **2,000 bands** across the 24 human chromosomes. G-banding offers higher detail than C-banding, is well-suited for animal cells, and can be analyzed using ordinary bright-field microscopy. Along with **R-banding**, G-banding remains one of the most commonly used techniques for chromosome identification and cytogenetic analysis.

Reverse banding or R-banding

R-banding, or reverse banding, is a chromosome staining technique that produces a **reverse pattern** compared to G-banding. Like G-banding, it uses **Giemsa stain**, but metaphase chromosome slides are first incubated in **hot phosphate buffer**. The heat treatment preferentially denatures AT-rich regions, which normally bind Giemsa strongly, leaving the **GC-rich regions** intact and stained. As a result, R-banding highlights **euchromatin regions** that are rich in guanine and cytosine, while AT-rich heterochromatin appears less stained. R-banding can be used for **chromosome identification** and is particularly helpful for analyzing **chromosome ends**, which usually appear lightly stained in G-banding. Although useful, G-banding remains the preferred method for general karyotyping.

Quinacrine banding or Q-banding

Quinacrine banding, or **Q-banding**, was the first chemical chromosome banding method developed, initially using **quinacrine mustard** and later replaced by **quinacrine dihydrochloride**. This technique involves treating chromosomes with quinacrine and visualizing them under **fluorescence microscopy**, producing a characteristic pattern of **bright and dull bands** called **Q bands**. The bright bands correspond to DNA rich in adenine and thymine (AT), while the dull bands are rich in guanine and cytosine (GC). Since AT-rich sequences are more prevalent in **heterochromatin**, Q-banding preferentially highlights these regions, which make up most of the stained DNA. The resulting banding pattern is similar to G-banding, with bright and dull regions corresponding roughly to dark and light G bands, respectively. Q-banding is particularly useful for distinguishing the **human Y chromosome**, as well as identifying chromosome polymorphisms involving **satellites and centromeres**. This technique was first developed by Caspersson *et al.* in 1968 and is widely used for chromosome identification in humans and mice. Like G-, R-, and Q-banding, this method is generally **not effective for plant chromosomes**.

C-banding

C-banding (centromeric or constitutive heterochromatin banding) is a chromosome staining technique that highlights regions of **heterochromatin**, particularly around centromeres, telomeres, and nucleolar organizer regions. The method was developed following the work of Pardue and Gall, who showed that **constitutive heterochromatin** can be specifically stained using Giemsa. Each chromosome has a distinct amount of heterochromatin, allowing for individual chromosome identification. C-banding preferentially stains tightly packed, repetitive, and genetically inactive DNA, such as **AT-rich satellite DNA**.

The technique is complex and involves several steps: treatment with **acid, alkali, or heat** to denature DNA, followed by **renaturation** using sodium-citrate at 60°C. An **alkaline pretreatment** causes depurination, after which the remaining DNA is renatured and stained with Giemsa. These steps allow repetitive heterochromatic DNA to renature and take up the stain, while low-repetitive or unique DNA remains largely unstained. As a result, heterochromatin stains intensely, whereas other chromosome regions stain lightly.

C-banding is particularly **useful for plant chromosomes** and has been applied to characterize chromosomes in both plants and animals, including humans. In mammals, it is especially valuable for identifying the **Y chromosome**, which is largely heterochromatic. Although historically important, C-banding is now less commonly used for diagnostic purposes.

Nucleolar Organizer Region Staining or NOR-banding

NOR-banding, also known as **silver staining**, is a chromosome staining technique that selectively highlights **nucleolar organizer regions (NORs)** using a **silver nitrate solution**. NORs, which contain **active ribosomal DNA (rDNA)**, are located on the **satellite stalks of acrocentric chromosomes**. Treatment with silver nitrate produces **darkly stained regions** corresponding to these NORs. This method is particularly useful for identifying the **human acrocentric chromosomes** 13, 14, 15, 21, and 22, and it allows visualization of rDNA activity in both metaphase chromosomes and interphase nuclei.

N-Banding

N-banding is a chromosome banding technique first described by Matsui and Sasaki in 1973. The method involves treating air-dried chromosome slides with **5% trichloroacetic acid at 95°C for 30 minutes**, followed by **0.1 N HCl at 60°C for 30 minutes**, and then staining with **Giemsa** (diluted 1:10 in 1/15 M phosphate buffer, pH 7.0) for 90 minutes. N-bands are typically located at **secondary constrictions, satellites, centromeres, telomeres, and heterochromatic regions**. These bands are thought to correspond to **structural non-histone proteins** specifically associated with the nucleolar organizer regions (NORs) of eukaryotic chromosomes. N-banding has been used to **locate NORs across a wide range of organisms**, including mammals, birds, amphibians, fishes, insects, and plants. The N-banding patterns vary between species, providing valuable cytogenetic information for comparative chromosome studies.

T-banding

T-banding is a chromosome staining technique used to visualize the **telomeric regions** of chromosomes. It is a modification of **R-banding** and involves **controlled thermal denaturation**. Chromosome slides are incubated in a phosphate or PBS buffer at **87°C**, followed by staining with **Giemsa** or **acridine orange**. The resulting **T-bands** correspond to heat-resistant regions that are particularly rich in **guanine-cytosine (GC) pairs**, allowing specific identification of telomeric DNA. T-banding is primarily used for studying chromosome ends and telomere structure.

CHAPTER TEN   ENVIRONMENTAL EFFECT ON GENE EXPRESSION

——————— ⇒ ⟩⟩⟩⟩ ⟨⟨⟨⟨ ⟨ ——————

No gene by itself totally determines a particular phenotype. The genes give the potential for the development of characteristics; their effects depend not only on their own functions but also on the functions of other genes, as well as on the internal and external environment. The environment influences gene expression in many ways. The environment provides the raw materials for the synthetic processes controlled by genes. For example, animals obtain several of the amino acids for their proteins as part of their diet. Again, most of the chemical synthesis in plant cells uses carbon atoms taken from the air as carbon dioxide. Through genes an organism builds the orderly process that we call life out of disorderly environmental materials. A person's height, for example, is controlled by many genes, the expression of which can be significantly affected by internal (hormones) and external (nutrition) environmental influences.

Genetic Determination

Living organisms mobilize the components of the world around themselves and convert these components into their own living material. An acorn becomes an oak tree, using water, oxygen, carbon dioxide, some inorganic materials from the soil, and light energy. An acorn develops into an oak, while the spore of a moss develops into a moss, although both are growing side by side in the same forest. The two plants that result from these developmental processes resemble their parents and differ from each other, even though they have access to the same narrow range of inorganic materials from the environment. The parents pass onto their offspring the specifications for building living cells from environmental materials. These specifications are in the form of genes in the fertilized egg. Because of the information in the genes, the acorn develops into an oak, and the moss spore becomes a moss. In any environment we can imagine, if the organisms develop, the acorn becomes an oak and the spore becomes a moss. This is an example of purely genetic effect. This implies that the genes are really the dominant elements in the phenotypic determination of organisms; the environment simply supplies the undifferentiated raw materials.

Environmental Determination

The extent of the effect of the environment on a given genotype can be determined by looking at the phenotype of genetically identical individuals raised in different environments. Suppose that two monozygotic twins are born in Bangladesh but are separated at birth and taken to different countries. If one is reared in China by Chinese-speaking foster parents, she will speak Chinese, while her sister reared in England will speak English. Each will absorb the cultural values and customs of her environment. Although the twins begin life with identical genetic properties, the different cultural environments in which they live will produce differences between them and differences from their parents. Obviously, the differences in this case are due to the environment, and genetic effects are of little importance in determining the differences. This is an example of purely environmental effect.

Importance of Heredity and Environment (Nature *versus* Nurture)

Genes do not express themselves outside of an environmental context, and similarly without genes there would be nothing to express. An organism is a highly complex machine in which all functions interact to a greater or lesser degree. As an organism develops and differentiates, its genes interact with its environment at each moment of its life history. The interaction of genes and environment determines what organisms are. Genes set the limit to performance or potential for the phenotype. The extent to which the phenotype of an individual develops within these limits depends on the environment. The best possible inheritance will not result in a superior herd or flock unless the proper environment is also supplied so that the animals can attain the limit set by their inheritance. Nevertheless, the best possible environment will not develop a superior herd or flock unless the proper inheritance is also present in the animals.

It is important to keep in mind that there is a very complex interaction between genes and environment that determines the phenotype of an individual. Many human behavioral traits are the result of interaction between genes and external environment. Numerous studies have shown that alcoholism is influenced by genes. However, there is no gene that forces a person to drink alcohol; that is, one cannot become an alcoholic unless one is exposed to an environment in which alcohol is available and drinking is encouraged. Genes increase or decrease the risk of developing alcoholism. The important point is that genes may influence a behavioral trait such as alcoholism, but the genes alone do not produce the phenotype.

Thus, Genotype + Environment = Phenotype.

Every quantitative character is genetically determined but is subsequently realized in a particular environment. Any attempt to determine exactly the degree to which the phenotypic value of a character in a single individual depends on heredity or environment would amount to biological nonsense, since the two factors cannot operate independently.

Genotype and Phenotype

The complete set of genes of an organism, what is called the genetic constitution of the organism, is its genotype. The physical or chemical manifestation or expression of the genotype is the phenotype. The phenotype describes all aspects of the individual's morphology, physiology, behavior, and ecological relationships. In this sense, no two individuals ever belong to the same phenotype, because there is always some difference between them in morphology or physiology. In practice, we use the terms genotype and phenotype in a more restricted sense. When we use the terms phenotype and genotype, we generally mean 'partial phenotype' and 'partial genotype', and we specify one or a few traits and genes that are the subsets of interest. For example, the genes that control eye color in animals, including human beings, are a part of the animal's genotype. The actual eye color is the phenotype.

If a particular characteristic is a part of an organism's phenotype; that is, if it is expressed – it can be said that the individual has the gene for that characteristic. If, however, a particular characteristic is not expressed, one cannot conclude that the gene is absent. Again, using eye color as the example, we can say that a brown-eyed individual has genes for brown eyes, but we cannot say that the individual does not have genes for blue eyes. Because, gene expression can be repressed, as happens in bacteria and in higher organisms such as animals, one gene may be recessive (not expressed) while another is dominant (expressed).

The genotype is essentially a fixed character of an organism; the genotype remains constant through life and is essentially unchanged by environmental effects. Most phenotypes change continually throughout the life of an organism as its genes interact with a sequence of environments. Fixity of genotype does not imply fixity of phenotype.

The Norm of Reaction

All biological characters are influenced to some degree by the environment as well as by genotype. The environment in which an organism lives will interact with the genotype to produce the phenotype. We can quantify the relationship between the genotype, the environment and the phenotype. If a single genotype is exposed to a range of environments, a range of phenotypes is observed. Such environment-phenotype relationships for a given genotype are called the norm of reaction of the genotype. To obtain a norm of reaction, we must allow different individuals of identical genotype to develop in many different environments. For some genotypes, the norm of reaction is small; the phenotype produced by a genotype is the same in different environments, such as the blood group locus that determines ABO blood type. For other genotypes, the norm of reaction is large and the phenotypes produced by the genotype vary greatly in different environments. For example, an individual's blood cell count varies with environmental factors such as altitude, activity level or infection. A single genotype may produce different phenotypes, depending on the environment in which organisms develop. The same phenotype may be produced by different genotypes, depending on the environment.

Developmental Noise

We have assumed that a phenotype is uniquely determined by the interaction of a specific genotype and a specific environment. However, even when environment and genotype are fixed, different individuals may differ. This uncontrolled variation in phenotype may be due to differences in internal environment during development, and is called developmental noise. Random developmental variation, or developmental noise, is a concept that correspond to the amount of possible phenotypic variance of a given genotype in a given environment. Estimation of random developmental variation requires that both genotype and environment be held constant. A Drosophila of wild type genotype raised at 16^0C has 1000 facets or cells in each eye. In fact, this is only an average value. A typical count may show that a fly has, say, 1018 facets in the left eye and 982 in the right eye. In another fly, the left eye has slightly fewer facets than the right eye. Yet the left and right eyes of the same fly are genetically identical. Thus, we would expect random variation in such phenotypic characters as the number of eye cells, the number of hairs, and the variation of neurons in a very complex central nervous system – even when the genotype and the environment are precisely fixed. It is often difficult to discriminate between developmental noise and variance caused by the environment, but if a character can be scored on each side of the body, both genetic and environmental variance are cancelled and noise is all that remains. Thus, Genotype + Environment + Developmental Noise = Phenotype.

Variation in Gene Expression

All of the genes Mendel examined have a definite genotype-phenotype relationship; for example, a pea plant with genotype *tt* is always dwarf. Although the phenotypes of organisms with a particular genotype are often very similar, this is not always the case. The phenotypic effects of some genes are variable. When the degree of phenotypic expression of a gene varies from individual to individual, the gene is said to have variable expressivity. Two terms, penetrance and expressivity, are frequently used in discussing variability of the effects of specific genes and phenotype.

Penetrance

In some cases, not all individuals who have a particular gene/genotype show the phenotype specified by that gene. Penetrance is defined as the percentage of individuals that exhibit phenotype corresponding to genotype. Penetrance refers to the probability of a gene or trait being expressed. In some cases, despite the presence of a dominant allele, a phenotype may not be present. One example of this is polydactyly in humans (extra fingers and/or toes). A dominant allele produces polydactyly in humans but not all humans with the allele display the extra digits. "Complete" penetrance means the gene or genes for a trait are always expressed in all the population who have the genes. "Incomplete" or 'reduced' penetrance means the genetic trait is expressed in only part of the population. Penetrance is an all-or-none phenomenon: the disease is either present or absent. 100% penetrance means all individuals show phenotype. 50% penetrance means half the individuals show phenotype. For example, in case of retinoblastoma only 75% individuals are affected. It expresses the percentage of observed phenotype over expected phenotype of a given genotype. A genotype that is always expressed has a penetrance of 100 percent. The seven gene pairs in Mendel's experiments and the alleles in the human ABO blood group system are two examples of complete penetrance. One particular example of 100% penetrance is Huntington's disease. There are many examples of incomplete penetrance in humans. For example, brachydactyly, an autosomal dominant trait that causes shortened index fingers and toes, shows 50 to 80 percent penetrance. Other examples are hemochromatosis, familial breast cancer, polydactyly, and retinoblastoma. The penetrance may also change in different age groups of a population.

Reduced penetrance probably results from a combination of genetic, environmental, and lifestyle factors, many of which are unknown. This phenomenon can make it challenging for genetics professionals to interpret a person's family medical history and predict the risk of passing a genetic condition to future generations. Penetrance refers to a population.

Expressivity

Expressivity is the degree or intensity with which a particular gene/genotype is expressed in a given individual *i.e.,* expressivity refers to variation in phenotypic expression when an allele is penetrant. Expressivity may be described either qualitatively or quantitatively, and it may be constant or variable. Even within the same family, an inherited genetic disease can cause massively different degrees of clinical effect from mild to severe symptoms. One patient may be very sick, whereas another with the same disease may be less severely affected (hemophilia). In retinoblastoma, some have both eyes affected, some only one. Polydactyly is also variably expressed, because affected individuals vary in the numbers of extra digits. The digit can be full size or just a stub. Hence, this allele has reduced penetrance as well as variable expressivity. Variable expressivity refers to the range of signs and symptoms (phenotypes) that can occur in different people with the same genetic condition. As with incomplete penetrance, variable expressivity is probably caused by a combination of genetic, environmental, and lifestyle factors, most of which have not been identified. If a genetic condition has highly variable signs and symptoms, it may be challenging to diagnose. Expressivity refers to an individual.

The effects of the environment and developmental noise, as well as genetic background, influence the penetrance and expressivity of simple characters.

Environmental Influences on Gene Expression

The expression of genes in an organism is not determined solely by its genetic makeup but is also influenced by environmental factors. These include both the external world in which the organism lives and the internal environment shaped by its hormones, metabolism, and developmental state. External factors such as drugs, chemicals, temperature, and light can turn genes on or off, thereby affecting how an organism develops and functions.

Effects of the Internal Environment

Complex biochemical responses occur in cells and in the organism as a result of certain stimuli from the internal environment. Two key internal factors are age and sex.

Age: Gene activity changes over time, as not all genes function continuously throughout the life cycle. Certain genes are programmed to activate or deactivate at different stages of development, leading to age-dependent traits. For example, humans exhibit conditions such as pattern baldness and muscular dystrophy that appear at specific ages due to gene regulation linked with aging.

Sex: Gender is another important internal factor influencing gene expression. Sex influences the inheritance and expression of genes in a variety of ways. Sex hormones play a major role in regulating which genes are expressed, resulting in differences between males and females. This leads to the presence of sex-limited traits, expressed only in one sex, and sex-influenced traits, which are expressed differently depending on gender.

Effects of the External Environment

Many factors in the external environment also strongly influence gene expression. Key external factors include nutrition, drugs and chemicals, temperature, infectious agents, and light.

Nutrition: Diet and nutrient availability play a critical role in gene regulation. Nutrient deficiencies or supplements can alter patterns of gene expression, influencing growth, metabolism, and overall health.

Drugs and Chemicals: Exposure to drugs and other chemical agents can affect how genes function. Some chemicals may activate or suppress genes, potentially leading to beneficial effects or harmful consequences.

Temperature and Light: Environmental conditions such as temperature and light also regulate gene activity in some organisms. A classic example is the Himalayan rabbit, which carries the temperature-sensitive C gene responsible for pigment production in the fur, skin, and eyes. At temperatures above 35°C, the gene is inactive, resulting in white fur on the warmer central parts of the body. In cooler regions of the body, such as the ears, nose, and feet (15-25°C), the gene becomes active and produces dark pigmentation, giving the rabbit its distinctive coat pattern.

Phenocopies

Not all the changes in phenotype are the result of genetic causes. When two genotypes produce the same phenotype due to different environments, one is called the phenocopy of the other. A phenocopy is an organism that has a mutant phenotype but a wild type genotype. It got the mutant appearance through an environmental cause. For example, a cat whose tail got run over by a car, looks like a Manx cat (genetically tailless), even though it started out with a normal tail. A phenocopy is a one-time event affecting the phenotype but not causing any transmissible change in the genotype. This is not transmitted to offspring in the way like that of the genetic abnormality.

Genocopy

Genocopy refers to situation when identical phenotype is produced by different mutations or genotypes. In other words, genocopies occur when a genetic mutation or genotype in one locus results in a phenotype similar to one that is known to be caused by a different genotype. Genocopies can be caused by either genetic mutations or environmental changes, and they can be inherited. Normally, polydactyly is caused by mutations in a specific gene, such as *GLI3*. However, a similar polydactyl phenotype can also result from mutations in a different gene, like *ZRS* (a regulatory region affecting limb development). In this case, different genotypes produce the same physical trait, making it a **genocopy**.

CHAPTER ELEVEN   GENE MUTATION AND DNA REPAIR

Genetic change of organisms can result from a number of processes such as mutation, recombination and transposition. The processes themselves are to a large extent under genetic control; *i.e.*, there are genes that affect the efficiency of mutation, recombination and transposition. The term 'mutation' was first coined by Dutch botanist Hugo de Vries, derived from the Latin word *mutātiōnem*, meaning "to change." Mutations are heritable alterations in an organism's genetic material. Both prokaryotic and eukaryotic cells contain enzyme systems capable of correcting replication errors and various forms of damage to DNA. Although these proofreading and repair mechanisms are highly efficient, some errors may remain uncorrected or some damage unrepaired. When this occurs, changes in the DNA sequence can become permanent through replication and are subsequently passed on to daughter cells during cell division. Such permanent alterations in the nucleotide sequence of DNA are known as mutations. Broadly, a mutation is defined as any change in the structure or quantity of an organism's genetic material, whether DNA or RNA.

The process of mutation is called mutagenesis. Discovery of a mutation in a gene can help in identifying the function of that gene. Mutation serves two research purposes: geneticists use mutations to study the process of mutation itself and to increase the yield of mutants for other genetic studies. When a change in genotype produces a change in phenotype, the individual is called a mutant.

There are two general types of mutations: point (gene) mutations and chromosomal mutations. Gene mutations are changes in the nucleotide sequence of a single gene. Because, such a change occurs within a single gene and maps to one chromosomal locus (point), a gene mutation is sometimes called a point mutation. Point mutations affect only one or a few nucleotides within a gene. Gene mutations are never detectable microscopically on the chromosome; a chromosome bearing a gene mutation looks the same under the microscope as one carrying the wild type allele.

A change in the nucleotide sequence of a gene shows up as a change in the amino acid sequence of a polypeptide, or the nucleotide sequence of an rRNA or tRNA molecule. These alterations in proteins can affect the behavior, morphology, and physiology of the organism. The location of the mutation within a gene is important. Only mutations that occur within the coding region are likely to affect the protein. Mutations in noncoding or intergenic regions do not usually have an effect.

Mutation is the ultimate source of genetic variation and can introduce new alleles into populations.

Classification of Mutations

Mutations are changes in the genetic material that can occur at any time and in any cell. They can be classified in several ways, depending on their expression, location, origin, or effect.

Based on Phenotypic Expression
- ❖ **Dominant mutations:** Expressed even if only one allele is mutated.
- ❖ **Recessive mutations:** Expressed only when both alleles are mutated.

Based on Cell Type
- ❖ **Somatic mutations:** Occur in somatic (body) cells after fertilization. They can arise spontaneously, due to errors in DNA repair, or from environmental stressors such as chemicals, radiation, or cigarette smoke. Somatic mutations can cause **mosaicism**, contribute to aging, and sometimes lead to diseases like cancer, but they are **not inherited**.
- ❖ **Germinal (germ-line) mutations:** Occur in cells that give rise to gametes. These mutations **can be inherited** by offspring.

Based on Direction
- ❖ **Forward mutation:** Change from **wild-type** to mutant form.
- ❖ **Backward (reversion) mutation:** Change from **mutant** back to wild-type.

Based on Morphology or Function

- ❖ **Morphological mutants:** Altered shape or structure.
- ❖ **Lethal mutants:** Cause death, usually recessive but can be dominant.
- ❖ **Conditional mutants:** Normal under **permissive conditions** but abnormal under **restrictive conditions**; useful for studying development or DNA replication.
- ❖ **Biochemical mutants:** Defects in biochemical pathways, leading to substance deficiency.

Based on Functional Effects

- ❖ **Loss-of-function mutations:** Cause absence or inactivity of the gene product. Typically **recessive**.
- ❖ **Gain-of-function mutations:** Produce a new function not present in the wild-type, either through an aberrant protein or expression in abnormal conditions. Typically **dominant** and can have severe phenotypic effects.

Other Classifications

Mutations can also be classified as:

- ❖ **Autosomal or sex-linked** (based on chromosome type)
- ❖ **Spontaneous or induced** (based on origin)
- ❖ **Endogenous or exogenous** (based on influencing factors)
- ❖ **Gametic or zygotic** (based on developmental stage)

Mutations provide the raw material for **evolution**, but they can also lead to **diseases or developmental defects** depending on their type and location.

The Molecular Basis of Mutation

Point mutations typically refer to alterations of single base pairs of DNA, or of a small number of adjacent base pairs. Point mutations may occur as a single event in a gene, or as multiple events involving bases at more than one location. At the DNA level, there are two main types of point mutations:

1. Frameshift mutation.
2. Base substitution mutation.

Frameshift Mutation

Frameshift mutations are the result of insertion or deletion of one or more base

pairs in a gene (Image 11.1). There are two types of frameshift mutations:
1. Insertions or duplications.
2. Deletions.

Insertions involve the addition of one or more base pairs in a gene. Deletions involve the loss of one or more base pairs in a gene. Any addition or deletion of base pairs that is not a multiple of three changes the reading frame in DNA segments that code for proteins. The result can be the generation of an entirely different protein from the point within the coding region where it occurs, and frequently chain termination. If deletions or insertions occur in groups of three no frameshift occurs, although an amino acid will be removed or added to the polypeptide. An example of three-nucleotide deletion is seen in **cystic fibrosis.** In this case, three nucleotides deletion in the CFTR gene result in abnormal protein, missing phenylalanine amino acid at the 508th position. This abnormal protein is rapidly destroyed by the action of proteosomes. Because they may involve many bases or just one, frameshifts are sometimes macroalterations and sometimes microalterations. Deletions and insertions cause more dramatic changes in the DNA, and often result in complete loss of gene function. This causes a complete change to the entire amino acid sequence of a protein after the mutation site. An example of a disease that can be caused by a frameshift mutation is Tay-Sach's disease.

An example of a disease caused by a **frameshift mutation** is **Tay-Sachs disease**. This genetic disorder results from insertions or deletions of nucleotides in the **HEXA gene**, which encodes the α-subunit of the enzyme **hexosaminidase A**. Frameshift mutations alter the reading frame of the gene, producing a dysfunctional or truncated enzyme. Without functional hexosaminidase A, the body cannot break down **GM2 gangliosides**, leading to their accumulation in nerve cells. This buildup causes progressive damage to the nervous system, resulting in the severe neurological symptoms characteristic of Tay-Sachs disease.

One form of **β-thalassemia** is caused by an **insertion mutation** in the **HBB gene**, which encodes the β-globin chain of hemoglobin. Insertion of one or more nucleotides shifts the normal reading frame of the gene, disrupting the synthesis of functional β-globin. This results in reduced or completely absent production of β-globin chains, leading to an imbalance between α- and β-globin. The excess unpaired α-globin chains damage red blood cells, causing ineffective erythropoiesis and anemia. Such insertion mutations are a common cause of severe forms of β-thalassemia.

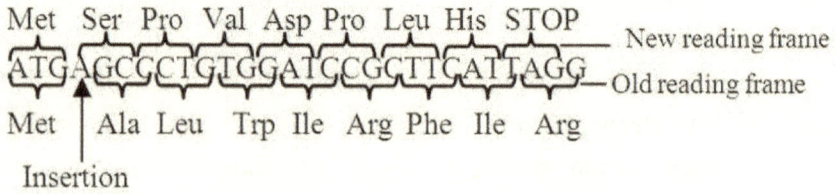

Image 11.1 A frame-shift mutation.

Base Substitution Mutation

Base substitutions are point mutations where one base is replaced by another. There are two types of base substitutions:

1. Transitions.
2. Transversions.

Transitions occur when a purine is replaced by another purine, or a pyrimidine by another pyrimidine. The possible transition mutations are:

$T \rightarrow C$, or $C \rightarrow T$

(pyrimidine $\rightarrow$ pyrimidine)

$A \rightarrow G$, or $G \rightarrow A$

(purine $\rightarrow$ purine)

Transversions occur when a purine is replaced by a pyrimidine or vice versa. The possible transversion mutations are:

$T \rightarrow A$, $T \rightarrow G$, $C \rightarrow A$, or $C \rightarrow G$

(pyrimidine $\rightarrow$ purine)

$A \rightarrow T$, $A \rightarrow C$, $G \rightarrow T$, or $G \rightarrow C$

(purine $\rightarrow$ pyrimidine)

Transitions occur more commonly than transversions even though more theoretical combinations of transversions are possible.

Sickle cell anemia is caused by a **substitution mutation** in the gene that codes for the β-globin chain of hemoglobin. In this mutation, a single nucleotide change occurs in the DNA sequence, where the codon **GAG** (which codes for glutamic acid) is replaced by **GTG** (which codes for valine). As a result, glutamic acid at the sixth position of the β-globin polypeptide chain is substituted by valine. This seemingly small change alters the structure of hemoglobin, causing the red blood cells to become sickle-shaped under low oxygen conditions. These abnormally shaped cells can block blood vessels, leading to anemia, pain, and other complications. Base substitutions can also be classified according to their effects on amino acid sequences in proteins.

Silent, Same-Sense or Synonymous Mutation

A silent mutation is a base substitution in a gene which converts a codon into a synonymous codon leading to same polypeptide sequence. Silent mutations are possible because of the degeneracy of the genetic code. Example: AUU → AUC; both codons specify isoleucine. Silent mutations have no effect on the encoded protein and do not result in a mutant phenotype. They tend to accumulate in the DNA of organisms, where they are known as polymorphisms. They contribute to variability in the DNA sequence of individuals of a species.

A classic example of a **silent mutation** is the change in the DNA codon **GAA → GAG** in the **β-globin gene**. Both codons code for the same amino acid, **glutamic acid**, due to the degeneracy of the genetic code. Since the amino acid sequence of the protein remains unchanged, there is no effect on the structure or function of hemoglobin.

Missense or Nonsynonymous Mutations

A missense mutation is a base substitution which alters the sense of a codon from one amino acid to another nonfunctional amino acid, resulting in a mutant phenotype. Example: AAG (lysine) → GAG (glutamic acid). Such mutations usually occur in one of the first two bases of a codon. An example of a disease caused by a missense mutation is Sickle-cell anaemia. This mutation causes the replacement of glutamic acid by valine in the beta chains of haemoglobin. The haemoglobin thus formed precipitates in the RBCs, changing their shape from biconcave to crescent-like. Such RBCs are rapidly destroyed as they are unable to squeeze through small capillaries.

Nonsense Mutations

A nonsense mutation is a base substitution in DNA which converts a sense codon into a nonsense/stop codon (UAG, UAA, or UGA). Example: GAG (glutamine) → UAG (stop). The mutation causes premature termination of protein synthesis and generates shortened proteins. Nonsense mutations usually have a serious effect on the activity of the encoded protein and often produce mutant phenotype.

An example of a disease caused by a **nonsense mutation** is **Duchenne muscular dystrophy (DMD)**. Nonsense mutations introduce a premature stop codon in the **DMD gene**, which prevents the complete translation of the dystrophin protein. When such mutations occur in any of the fifty-one exons critical for dystrophin production, they result in either a truncated, nonfunctional protein or a complete absence of dystrophin. Since dystrophin is essential for maintaining muscle cell integrity, its loss leads to the progressive muscle weakness and degeneration characteristic of DMD.

Neutral Mutation

A neutral mutation is a base substitution where the new codon codes for a different but functionally equivalent amino acid and hence does not affect the protein's function. For example, the codon **AAA** (which codes for lysine) may mutate to **AGA** (which codes for arginine). Both arginine and lysine are basic amino acids and are sufficiently similar in properties so that the protein's function may not be altered.

Spontaneous and Induced Mutations

Mutations occur in one of two ways: spontaneous or induced. Spontaneous mutations are naturally occurring mutations and arise in all cells without the addition of mutagens. Spontaneous mutations may result from errors in DNA replication and repair mechanisms or from spontaneous lesions in DNA or from transposable genetic elements. They may actually be caused by mutagens present in the environment. Spontaneous mutations are rare events because the DNA replication and repair processes have high fidelity. Most of such mutations are seen in highly proliferating cells such as cells of the intestines, skin cells, *etc*. Mutations occur spontaneously in all living things.

Induced mutations are produced when an organism is exposed to a known mutagen; such mutations typically occur at much higher frequencies than spontaneous mutations do. There are no qualitative differences between spontaneous and induced mutations. The process of inducing mutations is known as mutagenesis.

Errors in DNA Replication

Despite effective proofreading functions of DNA polymerases, occasionally the wrong nucleotide is incorporated. Errors in DNA replication, if not corrected, can lead to mutations. These errors can occur due to various factors, including polymerase mistakes, strand slippage, and external influences like carcinogens or radiation.

1. Tautomeric shifts (Tautomerization).
2. Transitions.
3. Transversions.
4. Deletions and Duplications.

Tautomeric shifts

Each of the DNA nitrogenous bases can occur in one of several forms, called tautomers. For thymine and guanine, one form is the keto (C-O) and the other enol (C-OH). For adenine and cytosine, one is the amino (C-NH2) and the other imino (C=NH) form. The common forms are the keto and amino forms. The other two forms are rare. However, at times, a base will shift from one form to its alternate form. Then it switches back. Usually, there is no significant problem with a tautomeric shift. The base across it is still correct and in the standard tautomeric form. If it happens during replication, it can cause the wrong base to be incorporated. Naturally, A in amino form pairs with T in keto form, whereas A in its imino form pairs with C, and T in enol form match with G.

DNA Repair

DNA in the living cell is subject to many chemical alterations. If the genetic information encoded in the DNA is to remain uncorrupted, any chemical changes must be corrected. DNA repair refers to a collection of processes by which a cell identifies and corrects damage to the DNA molecules. Cells have extensive repair mechanisms that can effectively repair any mutation in the genes. The harmful effects of mutations are only seen when the DNA repair mechanisms fail to correct these mutations. The DNA repair ability of a cell is vital to the integrity of its genome and thus to its normal functioning and that of the organism. Five major DNA repair mechanisms are mismatch repair, base excision repair, nucleotide excision repair, homologous recombination repair and non-homologous end joining.

Mismatch repair (MMR): DNA mismatch repair (MMR) is a highly conserved repair system that maintains the fidelity of the genome by recognizing and correcting erroneous insertions, deletions, and base misincorporations that occur during DNA replication. Although replication errors are the major source of mismatched base pairs, mismatches may also arise from other biological processes such as recombination. Unlike other repair systems that deal with damaged bases, MMR specifically corrects mispaired but otherwise normal nucleotides.

The process involves specialized proteins known as **Mut proteins**. When a mismatch occurs in the newly synthesized daughter strand, Mut proteins recognize the error and excise a segment of DNA containing the mismatch. DNA polymerase then fills the gap with the correct nucleotides, and DNA ligase seals the strand to restore integrity. Through this mechanism, MMR ensures high accuracy of DNA replication and prevents the accumulation of mutations.

Base excision repair (BER): Base excision repair (BER) repairs damage to a single nucleotide caused by oxidation, alkylation, hydrolysis, or deamination. This process removes the abnormal bases present in the gene. The abnormal base is recognized by a specific glycosylase enzyme that hydrolytically cleaves the base from the strand, leaving behind an empty site. The empty site is then filled by the action of DNA polymerase I and DNA ligase.

Nucleotide excision repair (NER): Nucleotide excision repair (NER) is the main pathway responsible for the removal of a wide variety of bulky and helix-distorting lesions induced by UV irradiation, environmental mutagens, and certain chemotherapeutic agents. This mechanism is used to repair the dimers formed by UV rays. The dimers are recognized by a UV-specific endonuclease and exonuclease (a single enzyme with two activities). This enzyme creates a kink on both sides of a small segment containing dimer. This segment is then removed from the strand. The gap is then filled by DNA polymerase I and DNA ligase.

Homologous recombination repair: Homologous recombination repair **(HRR)** is a high-fidelity DNA repair mechanism that corrects **double-strand breaks (DSBs)** – one of the most lethal forms of DNA damage. In this process, the damaged DNA molecule invades an undamaged DNA molecule of identical or very similar sequence, typically the sister chromatid, to serve as a template. Using this intact sequence, the cell accurately resynthesizes the damaged region, thereby restoring the original DNA information. Because HRR relies on homologous sequences, it ensures precise repair without introducing mutations. This mechanism is essential for maintaining genomic stability and preventing chromosomal abnormalities (Stewart *et al.*, 2022).

Non-homologous end-joining (NHEJ): The term **"non-homologous end-joining" (NHEJ)** was coined in 1996 by Moore and Haber (Moore and Haber, 1996). NHEJ is a primary DNA repair mechanism responsible for repairing **double-strand breaks (DSBs)** throughout the cell cycle. Unlike homologous recombination, NHEJ does **not require a DNA template** to repair the break. The repair process begins when the **Ku protein complex** binds to the broken DNA ends. This binding protects the ends and recruits additional NHEJ proteins that process the DNA termini and facilitate **ligation**, effectively joining the broken DNA ends. Through this pathway, NHEJ rapidly repairs DSBs, although it is more error-prone compared to homologous recombination because small insertions or deletions may occur at the repair site.

Spontaneous Lesions

Spontaneous lesions are naturally occurring damage to the DNA. Two of the most frequent spontaneous lesions result from depurination and deamination. **Depurination.** In depurination, a purine base is lost from its position resulting in an empty site. The space formed after removal is called apurinic (AP) site. The more common of the two, involves the interruption of the glycosidic bond between the base and deoxyribose and the subsequent loss of a guanine or an adenine residue from the DNA. A mammalian cell spontaneously loses about 10,000 purines from its DNA during a 20-hour cell generation period at 37^0C. If these lesions were to persist, they would result in significant genetic damage because, during replication, the resulting apurinic sites cannot specify a base complementary to the original purine.

Deamination. Deamination is the removal of amino group from the base. Deamination of cytosine yields uracil, which pairs with adenine instead of guanine in replication, resulting in the conversion of a G-C pair into an A-T pair (a GC → AT transition). Another example for deamination is conversion of 5-methylcytosine to thymine (5-methyluracil). Deamination of guanine yields xanthine, but since this purine base has the same pairing properties as guanine, no transition mutation results. Deamination changes adenine to hypoxanthine.

Oxidative damaged bases. The DNA bases in the genome are susceptible to oxidation by reactive oxygen species (ROS). Oxidative DNA damage refers to change in DNA structure by reactive oxygen species (ROS), such as hydrogen peroxide (H_2O_2), superoxide ($O_2^{\cdot}$), hydroxyl radicle ($OH^{\cdot}$). Oxidative DNA damage is the most frequent type of damage encountered by aerobic cells and may play an important role in mutagenesis. Among the bases, guanine is the most oxidizable base. Oxidation of Guanine forms 8-Oxo-7-hydrodeoxyguanosine which base pairs with adenine during DNA replication.

Mutation Rate

The mutation rate is expressed as the number of mutations occurring in a single generation. Mutation rates vary substantially among taxa, and even among different parts of the genome in a single organism. Each gene has a characteristic mutation rate. Some genes undergo mutations more frequently than others in the same organism. Those with unusually high mutation rates are called unstable or mutable. But a wide range of mutation rates exists among genes that are considered stable. In Drosophila, for example, the spontaneous mutation rates are on the order of one mutation per gene per generation per 10^{-4} to 10^{-5} gametes. For eukaryotes in general, the spontaneous mutation rate is $\sim 10^{-4}$ to 10^{-6} and for bacteria and phages, the rate is 10^{-5} to 10^{-7}. In humans the rate varies between 10^{-4} and 4×10^{-6}. Spontaneous mutations are rare in terms of a human life.

Mutagenic Agents or Mutagens

A mutagen is a substance or effect that causes mutation at higher than spontaneous rate. Although many mutations are spontaneous some can be triggered by mutagens. Mutagens can alter DNA in many different ways, but such alterations are not mutations unless they can be inherited. Mutagens induce mutations by at least three different mechanisms. They can replace a base in the DNA, alter a base so that it specifically mispairs with another base, or damage a base so that it can no longer pair with any base under normal conditions. Although RNA and DNA polymerases make errors at about the same rate, RNA genomes typically accumulate mutations at much higher frequencies than DNA genomes.

Mutagens are invaluable tools not only for studying the process of mutation itself but also for increasing the yield of mutants for other genetic studies. A variety of agents have been shown to be mutagenic. There are three types of mutagens.

1. Biological mutagens

2. Chemical mutagens

3. Physical mutagens

Biological Mutagens

The biological mutagens can be divided into three groups: transposable genetic elements, viruses and bacteria.

Transposable Genetic Elements

Transposable genetic elements or transposons can insert themselves into genes. The insertion of a transposon into a gene will often render the gene nonfunctional. They are capable of causing rearrangements of nucleotide sequences, because their moving about may create new sequences via deletions, inversions, or the change to a new location. The rearrangement may also change the function of existing sequences by placing them under the control of a new regulatory sequence. Transposition occurs fairly infrequently, at the rate of 10^{-4} to 10^{-7} events per generation. Transposons can also cause damage to the gene by leaving a gap when it moves. This gap may not be repaired.

Viruses

Virus causes mutations when it inserts itself into the sequence of host genes. Examples include bacteriophage Mu, retroviruses. Mu is also a transposon and it can cause mutations when it transposes. Certain viruses, *e.g.,* Rous sarcoma virus, have been reported to induce cancer.

Bacteria

Toxin produced by bacteria can cause mutation. Certain inflammation-inducing bacteria like *Helicobacter pylori* produces reactive oxygen species that results in DNA damage and reduced DNA repair. This raises the likelihood of the mutation. Direct exposure of bacteria as a mutagen may increase the frequency of mutations in a population of cells from 10^{-3} to 10^{-6} fold.

Chemical Mutagens

The most common mutagens are toxic chemical substances. A wide variety of natural and synthetic, organic and inorganic, chemicals can react with DNA altering its structure and causing mutation. Most chemical mutagens are carcinogenic and cause cancer. They can be grouped into different classes based on their mechanism of action.

1. Base Analogs

Base analogs have molecular structures that are extremely similar to the normal bases of DNA. They are occasionally incorporated into the DNA in place of normal bases during replication. These compounds have base pairing properties different from the bases. They replace the bases and cause stable mutation. The most widely used base analogues are 5-bromouracil (5BU), 2-aminouracil (2AU), and 2-aminopurine (2AP). 5BU is an analogue of thymine. 5BU has a bromine residue instead of methyl group of thymine and will pair only with adenine. If cells are grown in a medium containing 5BU, a thymine is sometimes replaced by a 5BU in the replication of DNA. However, it can undergo a slight change in its structure called a tautomeric shift which causes it to base pair with guanine. In the next round of replication, the G-5BU base pair will be resolved into a G-C base pair. This process produces a transition mutation from A-T to G-C or from a G-C to an A-T.

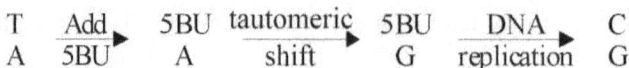

T	Add	5BU	tautomeric	5BU	DNA	C
A	5BU →	A	shift →	G	replication →	G

2-AP is widely used in research. It is an analog of adenine that normally pairs with thymine but it is able to form a single hydrogen bond with cytosine. Therefore, when 2-AP is incorporated into DNA by pairing with thymine, it can generate AT → GC transitions by mispairing with cytosine in subsequent replications. Or, if 2-AP is incorporated by mispairing with cytosine, then GC → AT transitions will result when it pairs with thymine. Genetic studies have shown that 2-AP, like 5-BU, is very specific for transitions.

2. Base-modifying Agents

These are chemicals that directly modify or change the chemical structure and properties of the bases in the DNA, causing mispairing. These mutagens are active on both replicating and nonreplicating DNA.

Alkylating agents

Most chemical mutagens are alkylating agents. Alkylating agents are chemicals that donate alkyl groups (usually methyl or ethyl) to the reactive sites on the bases, or phosphate groups of DNA. The introduction of the alkyl groups increases ionization that results in base-pairing errors and eventually inducing gaps in the DNA strand. Many carcinogens operate by alkylating the DNA. Some of the common alkylating agents are ethylene imine (EI), diethyl sulfate (DES), ethyl nitrosourethane (ENU), ethyl methane sulfonate (EMS), methyl methane sulfonate (MMS), dimethyl sulphate (DMS), nitrosoguanidine (NG), Di-(2-chloroethyl) sulfide (sulfur mustard), Di-(2-chloroethyl) methylamine (nitrogen mustard), ethylnitrosourea, vinyl chloride, busulfan, carmustine, lomustine, temozolomide, dacarbazine, ethyl ethane sulfate, and Thio-TEPA. They either alter the base-pairing properties, or cause structural distortion of the DNA molecule. When guanine is alkylated, producing 7-ethylguanine, it will base pair with thymine. Once again, if this occurs during replication, the wrong nucleotide can be inserted in the molecule being synthesized, leading to a mutation.

Among chemical mutagens, EMS is the most widely used alkylating agent in plants. It adds an ethyl group to bases in DNA and so changes their shape and their base-pairing properties. It alkylates guanine (G) to generate O-6-ethylguanine, which leads to mispairing with thymine (T). During replication, DNA polymerases preferentially mispair T with O-6-ethylguanine and

357

unrepaired alkylation damage results primarily in random G/C to A/T transition mutations. EMS also has the specificity to remove guanine and cytosine from the chain and results in gap formation. Any base (A, T, G, C) may be inserted in the gap. During replication chain without gap will result in normal DNA. In the second round of replication gap is filled by suitable base. If the correct base is inserted, normal DNA sequence will be produced. Insertion of incorrect bases results in transversion or transition mutation. Another example is methyl nitrosoguanidine that adds methyl group to guanine causing it to mispair with thymine. After subsequent replication, GC is converted into AT transition.

Nitrosoguanidine is among the most effective mutagens from this group.

Deaminating agents

They cause oxidative deamination of the amino group of adenine, guanine and cytosine. Nitrous acid (HNO_2) causes an oxidative deamination in which amino groups ($-NH_2$) are converted to keto groups ($=O$) and thus cytosine residues for example will be converted to uracil. Uracil will be capable of pairing with adenine, instead of guanine, producing a GC $\rightarrow$ AT transition. Similarly, deamination of adenine yields hypoxanthine, a base that pairs with cytosine rather than thymine, resulting in an AT $\rightarrow$ GC transition mutation. Deamination of guanine yields xanthine. Xanthine behaves like guanine because there is no change in pairing behaviour. Xanthine pairs with cytosine. Therefore, G-C pairing is replaced by X-C pairing. A nitrous acid induced mutation can be reverted by a second treatment with nitrous acid. Sodium bisulfite (bisulfite ions) specifically converts cytosine to uracil.

Hydroxylating agents

Hydroxylating agents such as hydroxylamine (NH_2OH) reacts specifically with cytosine, modifying it to hydroxylaminocytosine, which pairs only with adenine instead of with guanine, thus resulting in a CG $\rightarrow$ TA transition.

Intercalating agents

Intercalating agents are three ringed planar (flat) molecules of similar dimensions as those of purine pyrimidine pairs. In aqueous solution they can insert (intercalate) themselves in between the stacked nitrogenous bases at the core of the DNA double helix by a process called intercalation and hence distort the DNA. In this intercalated position, the agent can cause single-nucleotide-pair insertions or deletions during replication. Due to deletion or insertion of intercalating agents, there occur frameshift mutations. Examples include proflavin, acriflavin, acridine orange, acridine blue, actinomycin D, and ethidium bromide.

Metal ion

Metal ions like nickel, chromium, cobalt, cadmium, arsenic and iron, generate reactive oxygen species (ROS) that cause DNA hypermethylation, thereby promoting DNA damage and hindering the DNA repair process.

Physical Mutagens

Physical mutagens are in the form of physical substances. Physical mutagen causes physical damage to DNA. These include ionizing radiation, nonionizing radiation, and heat. Radiation is the most important among the physical mutagens.

Ionizing radiation is the electromagnetic radiation with a shorter wavelength and greater energy than visible radiation. X-rays, gamma rays (γ-rays) and cosmic rays are ionizing radiation which ionizes water and other molecules around the DNA of the cell and free radicals and ions (H^+, OH^-) are formed. Free radicals are molecular fragments having an unpaired electron that can attack DNA's structure. When free radicals collide with other molecules, they can create new free radicals. When they collide with the phosphate bonds in DNA, they can break them. A break in one strand of double-stranded DNA doesn't amount to much. When both strands break, repair enzymes often fail to restore the original base sequence. It is not the ionizing radiation itself that directly causes mutation. Rather, its agents are free radicals that it induces to form. The mutagenic effect of irradiation depends on doses, exposure time, cell cycle phase, and on the quality of repair mechanisms. Higher dose of X-rays can even cause death of an organism.

Ultra-violet (UV) light is a non-ionizing radiation. UV light has a shorter wavelength than visible light and has enough energy to damage DNA. It causes mutations because the bases in DNA absorb light very strongly in the ultraviolet range (254 to 260 nm). UV radiation can cause the formation of a covalent bond between adjacent pyrimidine residues in the same polynucleotide strand. These are called pyrimidine dimers, which prevents their participation in base pairing. The primary effect of UV on DNA is the creation of thymine dimers. These bonds distort the DNA conformation and inhibit DNA replication and transcription. Such mutations can lead to skin cancer. UV radiation is a strong mutagen for unicellular organisms; in multicellular organisms, it damages only their surface cells. The risk of UV radiation now increases as the ozone content in the upper atmosphere decreases.

Heat is also a significant environmental mutagen. It produces apurinic sites in the DNA polynucleotide which can cause point or deletion mutations when the DNA is replicated. DNA is sensitive to temperatures above 95°C. Above 95°C, the phosphodiester bonds break in DNA resulting in breakage of DNA strand.

The Relationship between Mutagens and Carcinogens

A mutagen is something that causes a mutation in an organism's genetic material. A carcinogen is a substance that induces unregulated growth processes in cells or tissues of multicellular animals, leading to the disease called cancer. Although mutagen and carcinogen are not synonymous terms, the ability of a substance to induce mutations and its ability to induce cancer are strongly correlated. Mutagenesis result in genetic change, and carcinogenesis may result from mutagenic events. A carcinogen causes a mutation that leads to cancer. So, a carcinogen is a type of mutagen that specifically leads to cancer. Approximately 90% known carcinogens are also mutagens. This is because cancer is generally caused by mutations in genes that control cell division.

Effects/Importance of Mutations

Mutations often result in the **loss of normal function** of the affected allele, either partially or completely. While some mutations do not produce any observable change in phenotype, most are **detrimental**. The phenotypic effects of mutations can vary widely, ranging from minor alterations detectable only through biochemical assays to severe disruptions of essential processes. In extreme cases, mutations can lead to **uncontrolled cell proliferation (cancer)** or the **death of a cell or organism**. Mutations play a significant role in the development of many diseases and can affect individuals in diverse ways. Among the major consequences of mutations are the following:

Role of Mutation in Genetic Diversity and Evolution: Mutations play a critical role in **increasing genetic diversity**, which is essential for evolution. Gene mutations can convert one allelic form of a gene into another, creating new alleles. Without mutations, all genes would exist in a single form, and allelic variation would not arise. Consequently, genetic analysis and inheritance studies would be impossible. **Germline mutations** are particularly important because they are heritable and constitute the ultimate source of all genetic variation in populations. This variation provides the raw material for natural selection and evolutionary change. Therefore, mutations are fundamental to the process of evolution, enabling species to adapt, survive, and diversify over time.

Change in a morphological trait: A mutation can lead to a noticeable change in an organism's **physical characteristics**. Such changes are referred to as alterations in **morphological traits**. Most mutant phenotypes studied in genetics courses fall into this category. For example, a mutation might result in **short plants instead of tall ones**, or altered flower color, leaf shape, or body structure. These visible changes provide a straightforward way to observe the effects of mutations.

Nutritional or biochemical variation: Mutations in genes that encode **enzymes of metabolic pathways** can lead to nutritional or biochemical changes in an organism. For example, if a mutation affects an enzyme required for the biosynthesis of an essential amino acid, the organism may no longer be able to produce that amino acid internally and must obtain it from its **dietary sources**. Such mutations alter the organism's metabolism and can be detected through changes in growth, nutrient requirements, or biochemical assays.

Change in behavior

Mutations can sometimes lead to changes in an organism's **behavior**, although these are often difficult to characterize and relatively few examples are known. One well-studied case involves **Drosophila (fruit flies)**: a mutation affected male mating behavior, causing mutant males to lose the ability to distinguish between males and females. As a result, they attempted to mate indiscriminately with any fly they encountered. Such examples illustrate that mutations can influence not only physical or biochemical traits but also behavioral patterns.

Changes in gene regulation: Mutations in genes that encode **transcription factors** can alter the expression of the genes they regulate. Such mutations may affect **when, where, or how strongly** target genes are expressed, potentially leading to developmental abnormalities, altered physiological responses, or disease. By changing gene regulation rather than the protein-coding sequence itself, these mutations can have wide-ranging effects on cellular function and organismal traits.

Lethality: Some mutations are **lethal**, meaning they cause the death of an organism either before birth or later in life. Examples include the **yellow coat color allele in mice**, which is lethal in the homozygous condition, and the **Huntington's disease allele in humans**, which leads to progressive neurodegeneration and eventual death. Lethal mutations illustrate how changes in genetic material can have severe consequences for survival.

CHAPTER TWELVE

CHROMOSOME MUTATION

Different cells of the same organism and different individuals of the same species have, as a rule; the same number of chromosomes, except that gametic cells have only half as many chromosomes as somatic cells. Homologous chromosomes are, as a rule, identical in number and in the order of genes they carry. Further, the organization and the number of genes on the chromosomes of an organism are the same from cell to cell. These rules have exceptions known as chromosomal mutations (aberrations). Chromosome mutations are variations from the wild type condition in either chromosome number or chromosome structure. In both prokaryotes and eukaryotes, chromosome mutations arise spontaneously, or can be induced by chemical or radiation mutagens. Sometimes chromosome mutations can be detected by microscopic examination (eukaryotes), sometimes by genetic analysis, that is, by changes in linkage arrangement of genes (both prokaryotes and eukaryotes), and sometimes by both.

Variations in Chromosome Structure

Variations in chromosome structure involve changes in parts of chromosomes rather than changes in the number of chromosomes, or sets of chromosomes in a genome. The four common types of changes in chromosome structure are deletions, duplications, inversions, and translocations. Duplication, inversion, and translocation mutations can revert to the wild type state by a reversal of the process by which they were formed. However, deletion mutations cannot revert because a whole segment of chromosome is missing.

All four classes of chromosomal structure mutations are initiated by one or more breaks in the chromosome. When breaks occur in chromosomes, any two broken chromosomal ends may reunite. Structural chromosome mutations are caused by the faulty repair of chromosome breaks or by recombination between homologous but nonallelic sites. The point at which fracture or recombination occurs is termed a chromosome breakpoint and structurally rearranged chromosomes are termed chromosome derivatives.

In discussing chromosomal structure mutations, it is convenient to use letters to represent different chromosome regions. These letters therefore represent large segments of DNA, each containing many genes. Structural mutations may be balanced or unbalanced.

Unbalanced Structural Mutations

Unbalanced structural mutations involve the loss or gain of genetic material. They are of two kinds: deletions and duplications. Both of these generally result from unequal crossing over during meiosis. As a result of unequal crossing over, one gamete receives a chromosome with a duplicated gene or genes while the other gamete receives a chromosome with a missing gene or genes.

Deletions or Deficiencies

A deletion is a chromosomal mutation involving the loss of a segment of a chromosome, which results in the absence of all of the genes in that particular region (Image 12.1). Two types of deletion are possible. The deletion where an internal part of the chromosome is missing is called an interstitial deletion. If the tip of the chromosome is lost then it is terminal deletion. Terminal deletions involve a single break, and then the fragment with a centromere obtains a new telomere. The acentric fragment will be lost at cell division. Two breaks can produce an interstitial deletion, followed by rejoining of the ends of the flanking pieces. Viral attacks, irradiation especially ionizing radiation, chemical, or other environmental factors may trigger a deletion.

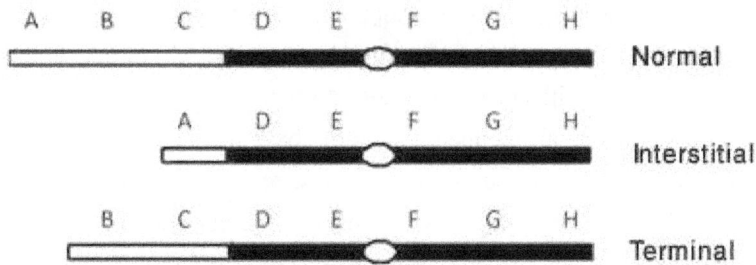

Image 12.1 A normal chromosome followed by two types of deletions.

Deletions are generally harmful to the organism, and the usual rule is the larger the deletion, the greater the harm. Very large deletions are usually lethal. Small deletions are often viable when they are heterozygous with a structurally normal homolog, because the normal homolog supplies gene products that are necessary for survival. The consequences of the deletion depend on the genes or parts of genes that have been removed. If the deletion involves the loss of the centromere, for example, the result is an acentric chromosome, which is usually lost during meiosis. This leads to the deletion of an entire chromosome from the genome. Depending on the organism, this chromosome loss may have very serious or lethal consequences. For example, there are no known cases of living humans who have one whole chromosome of a homologous pair of autosomes deleted from the genome. When deletions are homozygous, they are often lethal, because essential genes are missing. Even when heterozygous, lethals can cause abnormal development. A well-known example in humans is the deletion of a substantial part of the short arm of chromosome 5 (5p-), which when heterozygous causes *cri-du-chat* syndrome. Other examples of chromosomal deletion syndromes include Williams syndrome, Duchenne muscular dystrophy, Becker dystrophy, DiGeorge syndrome, Wolf-Hirschhorn syndrome (4p-), Prader-Willi syndrome, and Angelman syndrome.

Cri-du-Chat (Cat Cry) Syndrome

The most distinctive characteristic, and the one for which the syndrome was originally named is the cat-like mewing cry from small weak infants. The cry is caused by abnormal larynx development. Other characteristics are small head (microcephaly), an unusually round face, a small chin, saddle nose, widely spaced eyes with epicanthic folds, physical and mental retardation, and low IQ (20-40). A small number of children have heart defects, muscular or skeletal problems, hearing or sight problems, or poor muscle tone. As they grow, people with *cri-du-chat* usually have difficulty walking and talking. They may have behavior problems (such as hyperactivity or aggression) and severe intellectual disability. Generally, they die in infancy, or early childhood. Not every child will have every characteristic feature. Those only mildly affected may have very few.

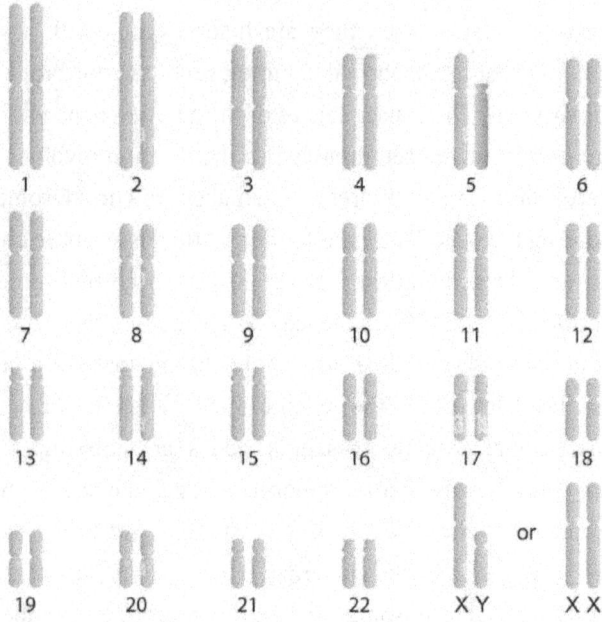

Image 12.2. *Cri-du-Chat* syndrome karyotype (source: https://www.alamy.com/cat-cry-syndrome-illustration-image613922740.html?imageid=40D74FCB-DA96-4791-89A3-64AE68C4CDB0&p=2085580&pn=1&searchId=cf3ee760a0114c34c29626 75d63a0d89&searchtype=0).

Cri-du-chat syndrome is caused by a deletion of the tip on the short arm of one of the homologs of chromosome 5, designated 5p- (Image 12.2). The specific bands detected in *cri-du-chat* syndrome are 5p15.2 and 5p15.3, the two most distal bands identified on 5p. In most cases, the deletion occurs while the sperm or egg cell is developing. Multiple genes are missing as a result of this deletion, and each may contribute to the symptoms of the disorder. In most cases the deletion is spontaneous and no specific cause can be identified. In 80 percent of the cases, the chromosome carrying the deletion comes from the father's sperm rather than the mother's egg. When this gamete is fertilized, the child will develop *cri-du-chat* syndrome. It is thought that more girls than boys are affected with this syndrome. The French geneticist Jerome Lejeune identified *cri-du-chat* syndrome in 1963. *Cri-du-chat* is one of the most common syndromes caused by a chromosomal deletion. It affects between 1 in 20,000 and 1 in 50,000 babies.

William's Syndrome

Williams syndrome, also known as Williams-Beuren syndrome, is a rare autosomal genetic disorder that affects a child's growth, physical appearance, and cognitive development. Williams syndrome is caused by a small deletion on chromosome 7 at location 7q11.23, which includes the elastin (ELN) gene. The elastin gene produces the elastin protein, which gives blood vessels the stretchiness and strength needed to withstand a lifetime of use. This protein is made primarily during embryonic development and childhood, when blood vessels are forming, and production largely stops afterward. In individuals with Williams syndrome, the absence or reduction of elastin leads to structural problems in the circulatory system, resulting in heart defects such as supravalvular aortic stenosis and other vascular disorders. Because the blood vessels lack sufficient elastin, they cannot stretch or handle normal blood pressure effectively, which underlies many of the cardiovascular complications characteristic of the syndrome. The most common symptoms of William's syndrome are intellectual disability, heart defects, and unusual facial features (small upturned nose, wide mouth, full lips, small chin, widely spaced teeth). Other symptoms include low birth weight, failure to gain weight appropriately, kidney abnormalities, low muscle tone, chronic ear infections and/or hearing loss, dental abnormalities, such as poor enamel and small or missing teeth, elevated calcium level in the blood, endocrine abnormalities, and sleep problems. People with this syndrome also exhibit characteristic behaviors, such as hypersensitivity to loud noises and an outgoing personality. Williams syndrome affects 1 in every 10,000 people worldwide. It affects boys and girls equally. William's syndrome is considered a microdeletion syndrome, because the deletion is too small to be seen with a microscope. Deletions that happen during egg and sperm formation are caused by unequal recombination. Unequal recombination occurs more often than usual at this location on chromosome 7, likely due to some highly repetitive DNA sequence that flanks the commonly deleted region. The syndrome was first described in 1961 by New Zealander John C. P. Williams.

Prader-Willi Syndrome

Prader-Willi syndrome (PWS) is a rare and complex genetic disorder that impacts the metabolic, endocrine, and neurologic systems and results in a number of physical, mental and behavioral problems, such as emotional outbursts and physical aggression. A key feature of Prader-Willi syndrome is a constant sense of hunger that usually begins at about 2 years of age. It occurs in approximately 01 out of every 15,000 births. PWS affects males and females with equal frequency and affects all races and ethnicities. PWS is recognized as the most common genetic cause of life-threatening childhood obesity. About 70% of Prader-Willi syndrome is caused by deletion of the proximal long arm of paternal chromosome 15 (15q11-q13 deletion), and hence referred to as a genomic imprinting disorder. These abnormalities usually result from random (sporadic) errors in egg or sperm development but are sometimes inherited.

Angelman Syndrome

Angelman syndrome (AS) is a rare complex neurogenetic disorder that primarily affects the nervous system and causes severe physical and learning disabilities. Characteristic features of this condition include severe mental retardation, delayed development, intellectual disability, severe speech impairment, specific electroencephalography (EEG) pattern, and problems with movement and balance. Most affected children also have recurrent seizures (epilepsy) and a small head size. Several distinctive behaviours are associated with Angelman syndrome including frequent and inappropriate laughter, being easily excitable, being restless, having a short attention span, trouble sleeping and needing less sleep than other children and a particular fascination with water. Angelman syndrome is most commonly caused by the deletion of a segment of the maternal chromosome 15 that contains the **UBE3A** gene, accounting for about 70% of cases. Since the paternal copy of this gene is typically inactive in certain brain regions due to **genomic imprinting**, the loss of the maternal copy leads to a complete lack of UBE3A function. In 10 to 20% of cases, the syndrome results from variants (mutations) in the maternal copy of UBE3A rather than deletion. The UBE3A gene encodes the **ubiquitin-protein ligase E3A (E6AP)**, an enzyme essential for regulating protein turnover inside cells. It works by attaching ubiquitin molecules to specific proteins, marking them for degradation. Disruption of this process affects normal cellular function, particularly in neurons, contributing to the developmental, neurological, and behavioral features of Angelman syndrome. It occurs in 1/10,000 to 1/20,000 live births. Angelman syndrome was first identified by Dr. Harry Angelman, an English paediatrician at Warrington General Hospital, UK in 1965.

DiGeorge Syndrome

DiGeorge syndrome (DGS) is a primary immunodeficiency disorder with a broad phenotypic presentation, which results predominantly from the microdeletion of chromosome 22 at a location known as 22q11.2. This mutation results in the failure of appropriate development of the pharyngeal pouches, which are responsible for the embryologic development of the middle and external ear, maxilla, mandible, palatine tonsils, thyroid, parathyroids, thymus, aortic arch, and cardiac outflow tract. Features of DGS include cardiac anomalies, thymic hypoplasia or aplasia, cleft palate, developmental delay, hypocalcemia, feeding problems, nasal sounding speech, hearing problems, an unusual shape or position of the eyes, ears, nose, mouth, and jaw, problems with the shape of the spine, low levels of some hormones, seizures, frequent infections, autism spectrum disorder, trouble interacting with peers, some types of mental health conditions, learning problems, vision problems, dental problems, *etc*. Features of DGS were first described in 1828 but properly reported by Dr. Angelo DiGeorge in 1965.

Wolf-Hirschhorn Syndrome (4p- Syndrome)

Wolf-Hirschhorn syndrome (WHS) is an extremely rare chromosomal disorder caused by terminal deletion in the short arm of chromosome 4, designated 4p- (4p16.3). NSD2 (Nuclear Receptor Binding SET Domain Protein 2), LETM1 (Leucine Zipper and EF-Hand Containing Transmembrane Protein 1), and MSX1 (Msh Homeobox 1) are the genes that are deleted in people with the typical signs and symptoms of this disorder. These genes play significant roles in early development, although many of their specific functions are unknown. This condition affects many parts of the body including heart and brain. The major features of this disorder include a characteristic facial feature (extremely wide-set eyes with a broad nose), a small head, prenatal and postnatal growth impairment, intellectual disability, language delay, severe delayed psychomotor development, low muscle tone (hypotonia), and seizures. Other features may include skeletal abnormalities, congenital heart defects, hearing loss, urinary tract malformations, and/or structural brain abnormalities. Wolf-Hirschhorn syndrome affects an estimated 1 out of every 50,000 births.

Duplications or Insertions or Repeat

The presence of two copies of a chromosomal region is called duplication, which results in the extra production of proteins from the repeated genes. We can categorize duplications on the basis of the position and order of the duplicated region ((Image 12.3). First, the duplication may be adjacent to the original chromosomal region. When this occurs, the order of gene in the duplicated segment may either be the same as the original order, called tandem duplication, or the opposite order, called reverse tandem duplication. Second, the duplicated segment may not be adjacent to the original segment, resulting in a displaced duplication. When the duplicated segment is at the end of the chromosome, the duplication is called terminal duplication.

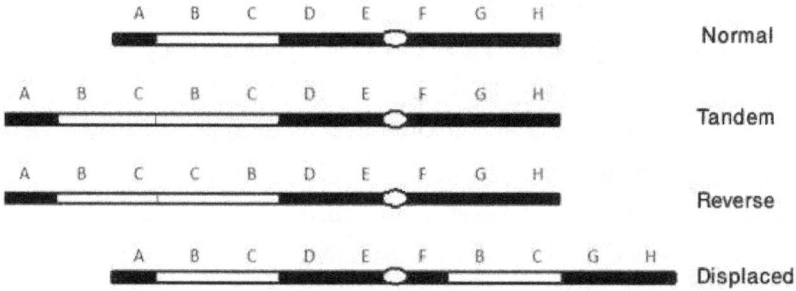

Image 12.3 A normal chromosome followed by different types of duplication.

The most frequent effect of duplication is a reduction in viability; in general, survival decreases with increasing size of the duplicated segments. Individuals that are either heterozygous or homozygous for small duplicated segments may be viable, although they often exhibit some phenotypic effects. If individuals are viable, there is a potential for further evolutionary change in these extra genes. Some cancers are caused due to duplication of genes. Duplication has played an important role in the evolution of multigene families. These genes may have descended from an ancestral gene that was duplicated, and then the duplicate copies diverged in their function. Duplications can be advantageous. Duplication is also one of the primary ways of increasing genome size. One example of a rare genetic disorder of duplication is called Pallister Killian syndrome, where part of the chromosome 12 is duplicated.

Pallister Killian Syndrome

Pallister-Killian syndrome (Pallister Mosaic Syndrome, Pallister-Killian Mosaic Syndrome, Killian Syndrome, Teschler-Nicola/Killian Syndrome,

Isochromosome 12p Syndrome) is a rare multi-system developmental disorder that is present at birth. The prevalence has been estimated to be 1 in 20,000. It is caused by the presence of at least four copies of the short arm of chromosome 12 instead of the normal two. It is a mosaic condition, meaning that not all cells in a particular tissue have these extra chromosomes and a percentage of cells are normal.

The signs and symptoms of Pallister-Killian syndrome vary from child to child and range in severity. Children with PKS may have extremely weak muscle tone, severe intellectual disability, distinctive facial features, sparse hair on the scalp, abnormally wide space between the eyes, extra skin folds over the inner corners of the eyes, high arched or cleft palate, pale areas on the skin, hearing loss, vision impairment, an extra nipple, genital abnormalities, heart defects, and skeletal abnormalities (*i.e.*, extra fingers/toes, and unusually short arms and legs). People with Pallister Killian have a shortened lifespan, but may live into their 40s. In 1977, the syndrome was reported independently by Pallister, and again in 1981 by Teschler-Nicola and Killian.

Isochromosomes

An isochromosome is an abnormal chromosome in which two identical arms are joined at the centromere, making it appear perfectly metacentric. Unlike normal chromosomes, which have one short arm (p) and one long arm (q), an isochromosome consists of either two p arms or two q arms. This structural abnormality results in the duplication of genetic material from one arm and the complete loss of genetic material from the other. Consequently, isochromosomes cause both a deficiency and an excess of genetic material, which can disrupt normal cellular and developmental processes. A well-known example of an isochromosome is **isochromosome Xq [i(Xq)]**, which is often associated with **Turner syndrome**. In this case, the chromosome has **two long arms (q arms) of the X chromosome** and lacks the short arms (p arms). Because the p arm of the X chromosome carries important genes involved in growth and development, its loss contributes to features of Turner syndrome such as short stature, gonadal dysgenesis, and infertility.

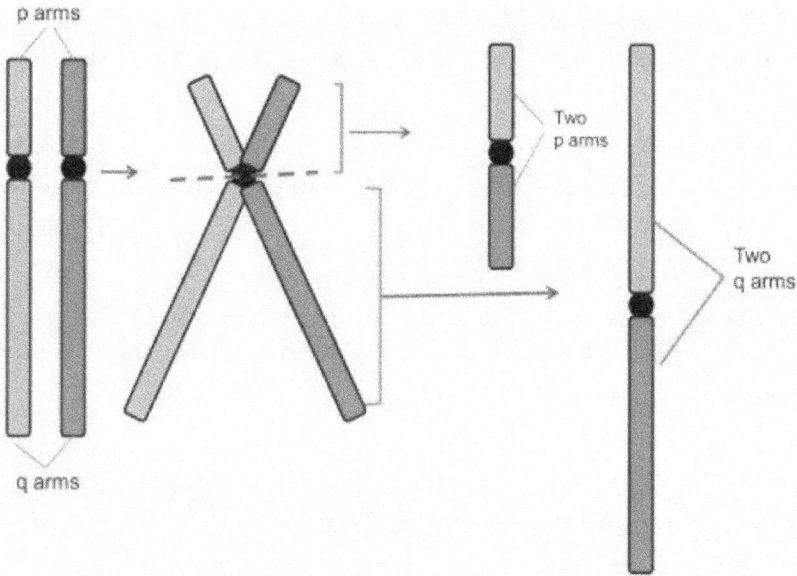

Image12.4 Isochromosome (source:
https://www.google.com/search?q=isochromosome+image&client=firefox-
b-).

Balanced Structural Mutation

Balanced structural mutations are known as chromosome rearrangements, and involve the rearrangement of genetic material on the chromosomes but no loss or gain of genetic material. They are of two kinds: inversion and translocations.

Inversions

Most of the homologous chromosomes in a population have genes in the same sequence. However, in some instances the sequence may differ on different chromosomes. An inversion is a chromosomal mutation where a segment of chromosome in which the order of the genes is the reverse of the normal orders. In this type of mutation, the part of the chromosome is inverted and then reinserted (Image 12.5). Inversions result when two breaks occur in a chromosome and the broken segment is rotated 180°. Inversions may be of two different kinds relative to the position of the centromere. When the inverted segment includes the centromere, the inversion is called a pericentric inversion. When the inverted segment occurs on one chromosome arm and does not include the centromere, the inversion is called a paracentric inversion. Inversions occur naturally, although rarely. More frequently they occur after exposure to X-ray. The X-ray irradiation creates hydrogen peroxide, superoxide and oxygen singlets that can react with and break the phosphodiester backbone of the DNA of the chromosome. Sometimes the pieces reattach correctly; but sometimes the pieces reattach in a reverse orientation.

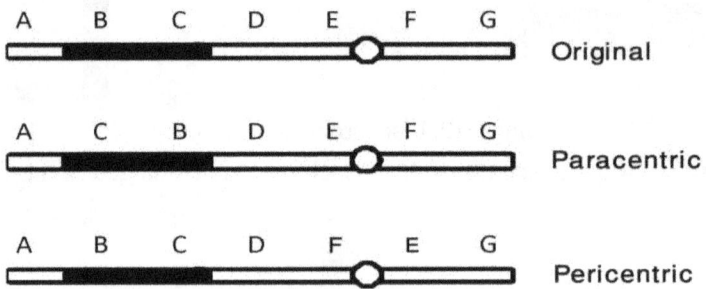

Image 12.5 Inversions.

Inversions have no phenotype unless the breakpoints interrupt a gene, or the rearrangement results in unfavorable position effects. All inversions cause abnormalities during meiotic chromosome synapsis in those heterozygous for them, and affect the viability of resulting gametes, embryos and progeny. Paracentric inversion heterozygotes have half the normal fertility.

Hemophilia A is caused due to inversion of the chromosomal region that involve a portion of the F8 (Factor VIII) gene (Xq28). The F8 gene encodes blood protein called factor VIII. Factor VIII is a clotting factor that are essential for proper clotting. This inversion (Inv1) completely splits the F8 gene into two parts and disrupts the F8 transcription, resulting in no factor VIII protein production.

Position Effect

In many cases, changes in location of a gene within a chromosome or genome as a result of inversion or translocation affect its level of expression or in some cases, its ability to function. Variation in the expression of a gene resulting from changes in its location within a chromosome or genome is called position effect. This has been well described in Drosophila with respect to eye colour and is known as position effect variegation (Weiler and Wakimoto, 1995). Position effect variegation (PEV) describes the mosaic expression of a phenotype in a cell population that is otherwise thought to be uniform. The phenotype is well characterised by unstable expression of a gene that results in the red eye coloration. In the mutant flies the eyes typically have a mottled appearance of white and red sectors.

Translocations

A translocation is a chromosomal mutation in which there is a change in position of chromosome segments and the gene sequences they contain (Image 12.6). Two simple kinds of translocations occur. One kind involves a change in position of a chromosome segment within the same chromosome: this is called an intrachromosomal translocation. The other kind involves the transfer of a chromosome segment from one chromosome into a nonhomologous chromosome: this is called an interchromosomal translocation. If this later translocation involves the transfer of a segment in one direction from one chromosome to another, it is a nonreciprocal (interstitial) translocation; if it involves the exchange of segments between the two chromosomes, it is a reciprocal translocation. Reciprocal translocations are the most common type of translocation.

Translocations are economically important. An organism that is heterozygous for a reciprocal translocation usually produces only about half as many offspring as normal – a condition that is called semisterility. In agriculture, translocation in certain crop strains can reduce yields considerably owing to the number of unbalanced zygotes that form. On the other hand, translocations are potentially useful: it has been proposed that insect pests could be controlled by introducing translocations into their wild populations. Most translocations are deleterious. Translocations can also cause several human diseases. Chromosomal translocations are frequently involved in the pathogenesis of leukemias, lymphomas and sarcomas. They can lead to aberrant expression of oncogenes or the generation of chimeric proteins.

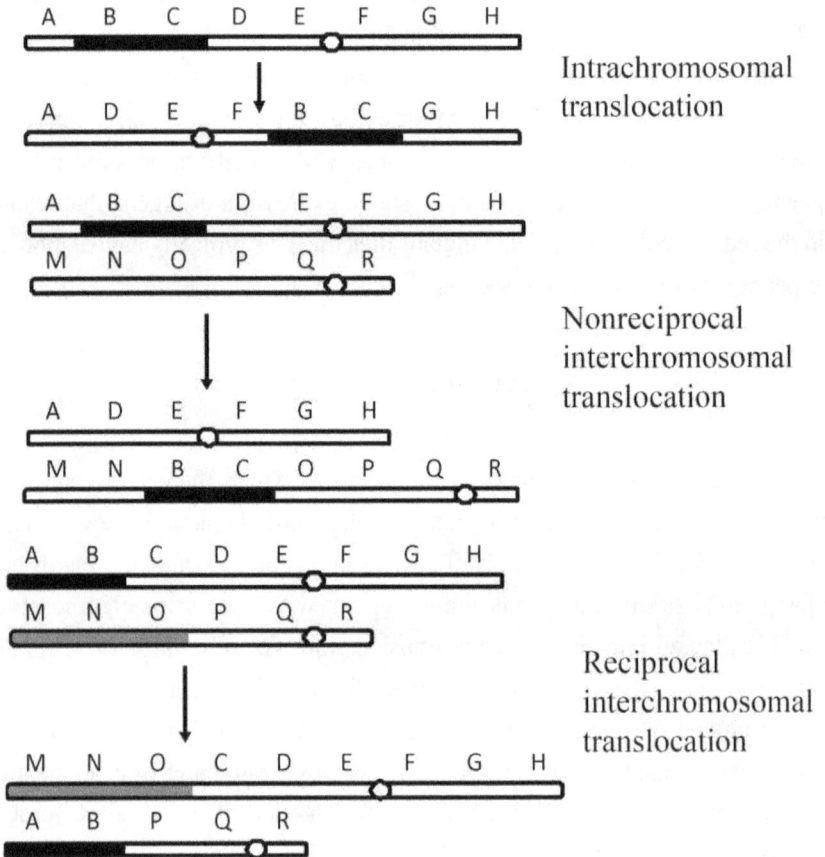

Image 12.6 Translocations.

Philadelphia Chromosome

The **Philadelphia chromosome** is the first tumor-specific chromosomal abnormality identified and is considered the **cytogenetic hallmark of chronic myelogenous leukemia (CML)**. It was discovered in 1960 by Nowell and Hungerford in Philadelphia. This abnormal chromosome 22 results from a **reciprocal translocation between the long arms of chromosomes 9 and 22**, designated as **t(9;22)(q34;q11)**. Through this rearrangement, the **ABL gene** from chromosome 9q34 fuses with the **BCR gene** on chromosome 22q11, creating the **BCR-ABL fusion oncogene**. This gene encodes a chimeric protein with constitutive tyrosine kinase activity, which drives uncontrolled proliferation of immature white blood cells, leading to their accumulation in the bone marrow and blood. The Philadelphia chromosome is detected in the bone marrow cells of nearly all individuals with CML and in some cases of acute lymphoblastic leukemia (ALL) and acute myelogenous leukemia (AML), making it a critical diagnostic and therapeutic marker in hematologic malignancies.

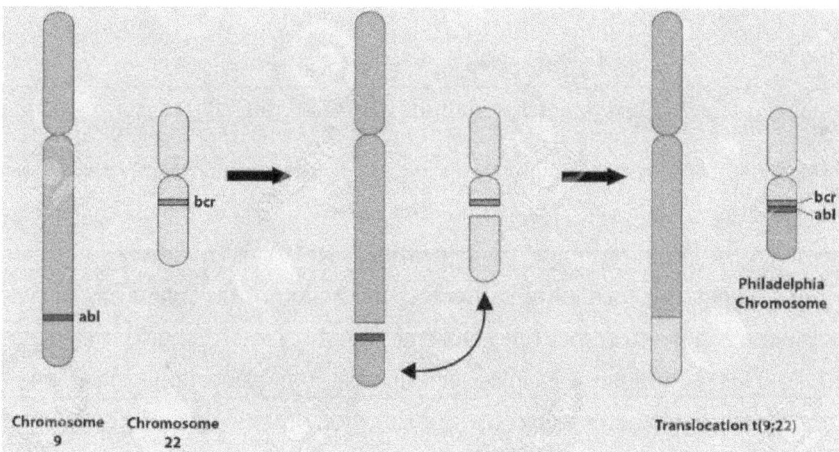

Image 12.7 Philadelphia Chromosome (source:
https://www.123rf.com/photo_36981787_philadelphia-chromosome.html).

Robertsonian Translocation

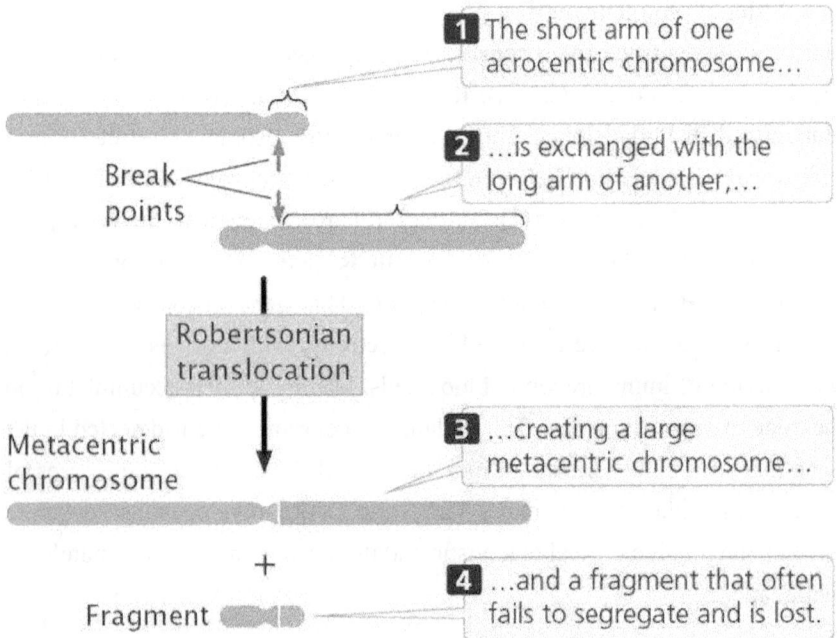

1. The short arm of one acrocentric chromosome...

2. ...is exchanged with the long arm of another,...

Break points

Robertsonian translocation

Metacentric chromosome

3. ...creating a large metacentric chromosome...

+

Fragment

4. ...and a fragment that often fails to segregate and is lost.

Image 12.8 Robertsonian translocation (source: https://www.pinterest.com/pin/29906785003207896/).

Translocations generally do not result in loss of genetic material. Robertsonian translocations, however, result in the loss of small parts of the chromosomes involved. A **Robertsonian translocation (ROB)** is a special type of chromosomal translocation that involves the fusion of the long arms of two **acrocentric chromosomes** (chromosomes 13, 14, 15, 21, 22, and Y) (Image 12.8). This is sometimes called centric-fusion translocation. This newly formed chromosome is called the translocation chromosome. Robertsonian translocations result in a reduction in the number of chromosomes. In this process, the two long (q) arms join at or near the centromere to form a single metacentric or submetacentric chromosome, while the two short (p) arms are lost. This fusion occurs because the short arms of the acrocentric chromosomes have repetitive DNA sequences that make them prone to joining together. Since the short arms of acrocentric chromosomes mainly contain repetitive DNA and ribosomal RNA genes (present in multiple copies elsewhere in the genome), their loss typically does not cause harm. As a result, carriers of Robertsonian translocations usually have no physical health problems. However, they may produce unbalanced gametes, increasing the risk of infertility, recurrent miscarriages, or chromosomally abnormal offspring. For example, a Robertsonian translocation between chromosomes 14 and 21 can give rise to a form of Down syndrome (translocation trisomy 21). Individuals with a ROB have a total of **45 chromosomes instead of 46**, because the fused q arms count as a single chromosome. Robertsonian translocations are the most common structural chromosomal rearrangements in humans, with an incidence of about 1-1.23 per 1000 live births, and the most frequent ones involve fusions such as 13;14, 13;21, or 21;22.

Disease Caused by Translocation Mutation

Chromosomal translocations have been linked to aneuploidy, infertility, mental retardation, cancer and other conditions:

Cancer and Chromosomal Translocation

Chromosomal translocations play a significant role in the development of many human cancers by creating fusion genes or activating oncogenes. In leukemias and lymphomas, specific translocations are often considered diagnostic markers because they are consistently associated with certain cancer types. For example, the **Philadelphia chromosome [t(9;22)]** produces the **BCR-ABL fusion oncogene**, which drives chronic myelogenous leukemia (CML). Similarly, the **Burkitt lymphoma translocation [t(8;14)]** brings the MYC oncogene under the control of immunoglobulin gene promoters, causing uncontrolled cell growth. Translocations are also seen in some cases of **prostate cancer**, where they can place genes under abnormal regulatory control. By altering gene expression or producing abnormal fusion proteins with oncogenic activity, chromosomal translocations disrupt normal cell cycle regulation, ultimately leading to malignant transformation.

Infertility and Chromosomal Translocation

Chromosomal translocations are known to negatively affect fertility, particularly in men. A common example is the **Robertsonian translocation**, which occurs in about **1 in 1,000 men**. In this condition, two acrocentric chromosomes fuse at their long arms, reducing the total chromosome number to 45. While carriers are usually physically healthy, the abnormal pairing and segregation of chromosomes during meiosis can produce **unbalanced gametes**, leading to reduced fertility, recurrent miscarriages, or offspring with chromosomal abnormalities. Despite this decreased fertility, natural conception is still possible for many men with Robertsonian translocations, although assisted reproductive techniques may sometimes be needed to improve chances of a successful pregnancy.

Translocation Down Syndrome

Most cases of Down syndrome are caused by **trisomy 21**, where an individual has three full copies of chromosome 21. However, in about **3-4% of cases**, Down syndrome results from a **chromosomal translocation**. In this form, part or all of chromosome 21 becomes fused with another chromosome, most commonly chromosome 14. As a result, the affected child inherits **two normal copies of chromosome 21 plus an extra long arm of chromosome 21 (21q) attached to chromosome 14**, giving them the effect of having three copies of chromosome 21 genetic material. This condition, known as **translocation Down syndrome**, produces the same characteristic features and developmental challenges as typical trisomy 21, but it is inherited differently. In some families, a parent may be a **balanced translocation carrier**, which increases the risk of passing the condition to offspring.

Lymphoma and Acute Myeloid Leukemia (AML)

Chromosomal translocations are a major cause of certain blood cancers, including **lymphomas** and **acute myeloid leukemia (AML)**. A well-documented example is the **translocation between chromosomes 8 and 11 [t(8;11)]**, which disrupts normal gene regulation and contributes to uncontrolled cell growth. This translocation is associated with **myeloproliferative disorders** that can progress to AML. Similar mechanisms are also involved in some **lymphomas**, where translocations place oncogenes under the control of highly active gene promoters, driving malignant transformation. These chromosomal rearrangements are clinically important, as they not only aid in diagnosis but also guide treatment strategies for affected patients.

Patau Syndrome (Trisomy 13)

Patau syndrome is a severe genetic disorder that most commonly arises from **trisomy 13**, where an individual has three full copies of chromosome 13 instead of the usual two. In some cases, the syndrome can also result from a **chromosomal translocation**, in which part or all of chromosome 13 fuses with another chromosome. This translocation leads to an extra copy of the genetic material from chromosome 13, producing the same clinical features as typical trisomy 13. Individuals with Patau syndrome often present with severe developmental abnormalities, including cleft lip and palate, heart defects, brain or spinal cord malformations, and extra fingers or toes, and the condition is associated with high infant mortality.

Variations in Chromosome Number

Ploidy refers to the number of sets of chromosomes in a cell or an organism. Humans normally have 2 sets of 23 chromosomes (total 46). When an organism or cell has one complete set of chromosomes, or an exact multiple of complete sets, that organism or cell is said to be euploid. Thus, eukaryotic organisms that are normally diploid (*e.g.*, mammals, birds, fruit flies) and eukaryotic organisms that are normally monoploid (*e.g.*, yeast) are all euploids. Variations in chromosome number involve departure from the normal diploid (or haploid) state of the organism. The number of chromosomes may vary in two basic ways: euploid variants (balanced), in which the number of chromosome sets differ, and aneuploid variants (unbalanced), in which the number of a particular chromosome differ. Occasionally, abnormal events occur before or during cell division, then gametes and new individuals end up with the wrong chromosome number.

Aneuploidy (Aneuploid Variants)

Aneuploidy is the second major category of chromosome mutations in which chromosome number is abnormal. In aneuploidy, one or several whole chromosomes are lost from or added to the normal set of chromosomes (Image 12.9). An **aneuploid** is an organism whose chromosome number differs from the wild type by part of a chromosome set. Having a missing or extra chromosome, resulting in incomplete sets, is referred to as a somy.

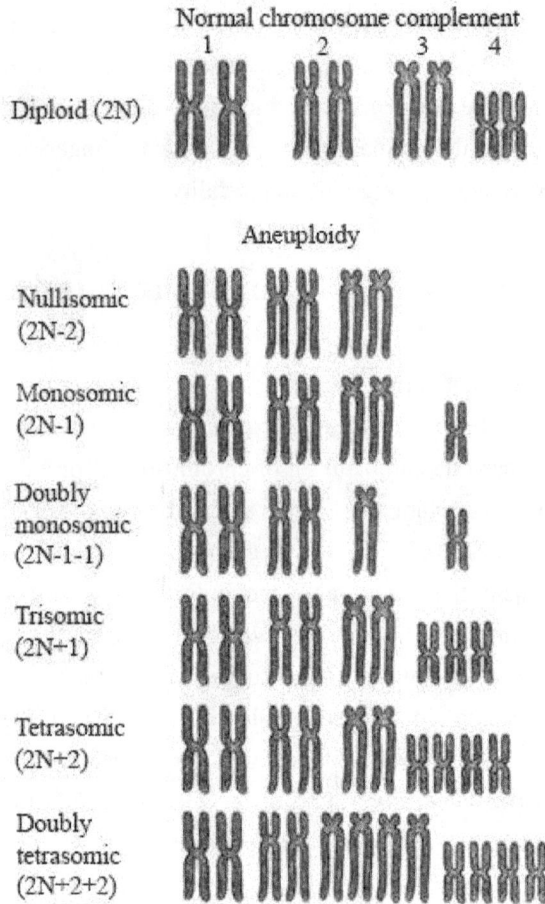

Image 12.9 Normal set of metaphase chromosomes and examples of aneuploidy.

In diploid organisms, aneuploid variations take four main forms:

1. Monosomy (a monosomic cell) involves a loss of a single chromosome; that is, the cell in $2N - 1$.

2. Nullisomy (a nullisomic cell) involves a loss of one homologous chromosome pair; that is, the cell is $2N - 2$. Generally, embryos that are nullisomic don't survive to be born.

3. Trisomy (a trisomic cell) involves a single extra chromosome; that is, the cell has three copies of one chromosome type and two copies of every other chromosome type. A trisomic cell is $2N + 1$.

4. Tetrasomy (a tetrasomic cell) involves an extra chromosome pair, resulting in the presence of four copies of one chromosome type and two copies of every other chromosome type. A tetrasomic cell is $2N + 2$. Tetrasomy is extremely rare.

Aneuploidy may involve the loss or the addition of more than one specific chromosome or chromosome pair. For example, a double monosomic has two chromosomes present in only one copy each; that is, it is $2N - 1 - 1$. A double tetrasomic has two chromosomes present in four copies each; that is, it is $2N + 2 + 2$.

Most chromosome conditions are referred to by category of aneuploidy followed by the number of the affected chromosome. For example, trisomy 13 means that three copies of chromosome 13 are present.

Disomy: When there is one chromosome more in a monoploid organism, *i.e.*, $N + 1$.

Aneuploidy Cause

Aneuploidy is caused due to non-disjunction during cell division. When sister chromatids fail to segregate during mitotic or meiotic cell division, it leads to one cell with an extra chromosome and another with one less chromosome. Interestingly, nondisjunction events seem to occur more frequently in older women.

Mosaic Variegated Aneuploidy

Mosaic variegated aneuploidy, also known as **mosaicism**, occurs when only some cells in the body have an abnormal number of chromosomes, while others remain normal. This condition arises when **nondisjunction** happens **after fertilization** during one of the early cell divisions in the developing zygote. As a result, the individual has a mixture of normal and aneuploid cells, which can lead to variable clinical features depending on the proportion and distribution of abnormal cells in different tissues.

Aneuploidy Disorders

Aneuploidy changes the number of chromosomes as well as the total amount of genetic material. Aneuploidy is deleterious in both plants and animals, although the effects in animals are more severe. Aneuploidy is usually lethal in animal, so in mammals it is detected mainly in aborted fetus, but tolerated to a greater extent in plants.

Aneuploidy is the most common chromosome abnormality in humans, and is the leading genetic cause of miscarriage and congenital birth defects. Human embryos that are missing a copy of any autosome fail to develop to birth. In other words, human autosomal monosomies are always lethal. Aneuploidy accounts for 1 in 160 live births. Trisomy is the most common aneuploidy disorder. Some of the common diseases are listed below:

- ❖ Down's syndrome – Trisomy of 21st chromosome
- ❖ Edward's syndrome – Trisomy of 18th chromosome
- ❖ Patau syndrome – Trisomy of 13th chromosome
- ❖ Klinefelter's syndrome – There is an additional number of the X chromosome, XXY.
- ❖ Turner's Syndrome – One X chromosome is less, XO

Down's Syndrome (Trisomy 21)

The most common form of aneuploidy is **Down syndrome**, first described in 1866 by **John Langdon Down**. It is caused by the presence of an **extra copy of chromosome 21**, making it a **trisomy 21** condition (Image 12.10). This chromosomal abnormality was identified in 1959 by the French geneticist **Jerome Lejeune**, who demonstrated that the syndrome results from the inheritance of three copies of chromosome 21 instead of the normal two. It affects 1 out of every 800 to 1,000 babies. The current nomenclature to indicate an individual with trisomy 21 is 47, +21, in which 47 indicates the total number of chromosomes and +21 indicates that there are three, rather than two, copies of chromosome 21. Having an extra copy of this chromosome means that individuals have three copies of each of its genes instead of two, making it difficult for cells to properly control how much protein is made. Producing too much or too little protein can have serious consequences. In 90% of Trisomy 21 cases, the additional chromosome comes from the mother's egg rather than the father's sperm.

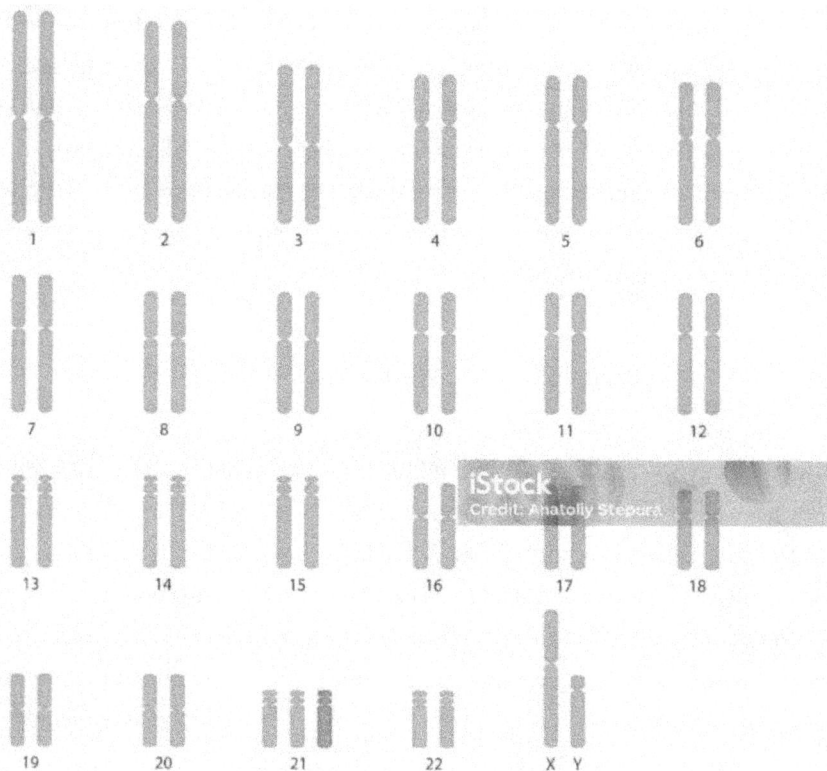

Image 12.10 Down syndrome karyotype (source: https://www.istockphoto.com/vector/down-syndrome-human-karyotype-gm1494398167-517751758?searchscope=image%2Cfilm).

Down syndrome is typically caused by nondisjunction. The incidence of Down syndrome and to some extent other aneuploidies increases with the age of the mother because increased age affects the meiosis of chromosomes adversely. The incidence of Down syndrome for mothers of age 45 is nearly 50-fold that for teenage mothers.

People with Down syndrome have distinct features: broad, flat face; small, flattened nose; abnormally shaped ears; upward-slanting eyes with epicanthic folds; short stature; short and broad hands; large, wrinkled protruding tongue; low muscle tone; excessive space between first and second toe; round head; open mouth; short neck; mental retardation; under-developed genitals and gonads. People with Down syndrome have an increased risk of developing a number of medically significant problems, including respiratory infections, gastrointestinal tract obstruction, leukemia, heart defects, hearing loss, hypothyroidism, and eye abnormalities. They also have moderate to severe intellectual disability. Children with Down syndrome usually develop more slowly than their peers and have trouble learning to walk, talk, and take care of them. Females may be fertile and may produce normal or trisomic progeny, but males have never reproduced. Mean life expectancy is about 17 years, and only 8 percent survive past age 40.

Klinefelter's Syndrome (47, XXY)

Klinefelter syndrome is a genetic condition that affects males and occurs when they have an **extra X chromosome**, resulting in a **47, XXY** karyotype (Image 12.11) instead of the normal **46, XY**. Individuals with this condition are genetically male. The syndrome arises due to **nondisjunction of the sex chromosomes**, which occurs when a pair of sex chromosomes fails to separate properly during the formation of an egg or sperm. If an egg or sperm carrying an extra X chromosome fuses with a normal sperm or egg, the resulting embryo will have **three sex chromosomes (XXY)**. During development, this extra chromosome is replicated in every cell. Klinefelter syndrome primarily affects **sexual development**, often leading to **small testes, reduced testosterone production, infertility, and delayed or incomplete puberty**.

The signs and symptoms of Klinefelter syndrome vary. In some cases, the features are so mild that the condition is not diagnosed until puberty or adulthood. A small percentage of affected individuals are born with undescended testes (cryptorchidism). As adults, nearly all XXY males are unable to make sperm and sterile. Many men discover their condition only after they seek medical help for infertility. The other physical changes associated with Klinefelter syndrome are usually subtle. Most commonly, affected individuals are taller than average, with proportionally longer arms and legs. Other features can include curved pinky fingers, flat feet, and less commonly abnormal fusion of certain bones in the forearm. Children with Klinefelter syndrome may have low muscle tone, difficulty coordinating movements, and mild delays of certain developmental skills, such as rolling over or walking. Affected children have an increased risk of mild delays in speech and language development. People with Klinefelter syndrome tends to have better receptive language skills than expressive language skills and may have difficulty communicating and expressing themselves. Affected individuals have an increased risk for learning disabilities, most commonly problems with reading and written expression. People with Klinefelter syndrome very rarely have intellectual disabilities.

Individuals with Klinefelter syndrome may have anxiety, depression, impaired social skills, or behavioral differences, such as emotional immaturity during childhood or difficulty with frustration. Affected individuals also have an increased risk for attention-deficit/hyperactivity disorder (ADHD), though they tend to have problems with attention and distractibility rather than hyperactivity. People with Klinefelter syndrome have an increased risk of developing metabolic syndrome including high blood glucose levels during prolonged periods without food (fasting), high blood pressure, increased belly fat, and high levels of fats (lipids) such as cholesterol and triglycerides in the blood. Compared with unaffected people, adults with Klinefelter syndrome also have an increased risk of developing involuntary trembling (tremors) in their arms or hands, breast cancer, and autoimmune disorders such as systemic lupus erythematosus and rheumatoid arthritis. Autoimmune disorders are a large group of conditions that occur when the immune system attacks the body's own tissues and organs. Children and adults may have less-muscular bodies, wider hips, narrower shoulders, or minor to moderate learning disabilities. Changes that appear at puberty can include low growth of facial and body hair, development of breast tissue, and small testes.

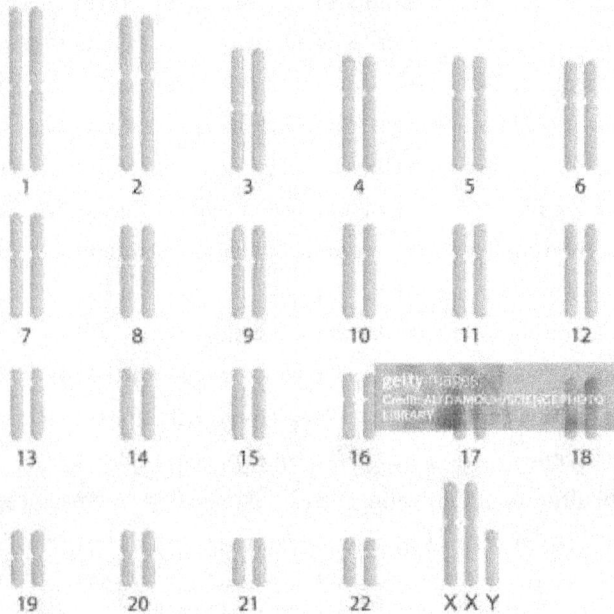

Image 12.11 Klinefelter syndrome karyotype (source:
https://media.gettyimages.com/id/2167925942/vector/klinefelter-syndrome-
illustration.jpg?s=2048x2048&w=gi&k=20&c=4vk8UUHNpqZEPVqMK6d
V5LusC2yCHLANr51qRS1FNBg=).

XXY individuals are also more likely to develop certain medical conditions, including osteoporosis, varicose veins, type 2 diabetes, and heart valve defects. Nondisjunction leading to XXY is equally likely to happen in the mother's egg and the father's sperm. Most of them develop as males, often not knowing they have an extra chromosome. A small proportion will develop as intersex or female. Injection of the hormone testosterone can reverse the feminizing traits but not the fertility. Except for their low fertility, many affected individuals show no outward symptoms at all.

Similar conditions are caused by additional X chromosomes (48, XXXY; 49, XXXXY), but they are much rarer. The more X chromosomes a person has, the stronger the physical characteristics and health problems tend to be, including intellectual disability. It is the most common genetic conditions, affecting about 1 in 660 genetic males. Klinefelter syndrome is named for Dr. Harry Klinefelter, who first reported its symptoms in 1942. The risk of a woman having a son with Klinefelter syndrome may be slightly higher if the mother or father are older.

Jacob's Syndrome (47, XYY)

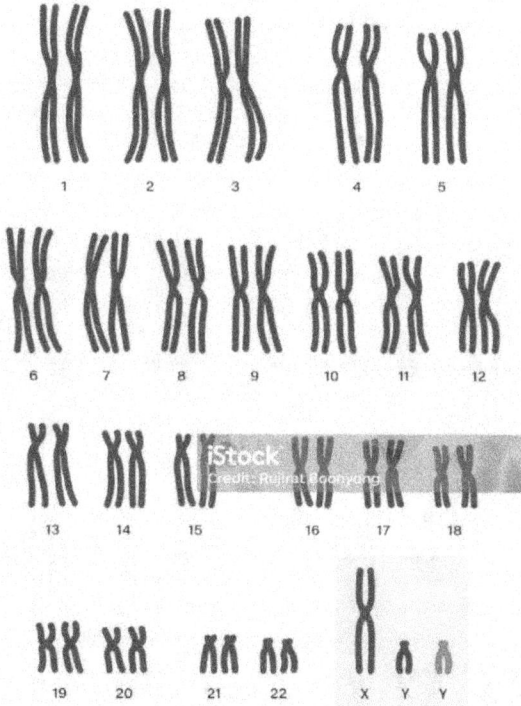

Image 12.12 Jacob's syndrome karyotype (source: https://www.istockphoto.com/vector/human-karyotype-of-double-y-syndrome-xyy-male-has-an-extra-y-chromosome-gm2135391165-568216667?searchscope=image%2Cfilm).

Jacob's syndrome is also sometimes called XYY karyotype, XYY syndrome, or YY syndrome. It is a sex chromosome aneuploidy where males receive an additional Y chromosome (47, XYY) (Image 12.12). The 47, XYY syndrome results from parental nondisjunction during meiosis, resulting in an extra Y chromosome. XYY syndrome has an incidence of 1 per 850-1000 live male births (Nielsen and Wohlert, 1991; Abramsky and Chapple, 1997). Typically, men with 47, XYY have normal internal and external genitalia, normal appearance, and normal sexual development, but are taller than average. Most XYY men are fertile; however, there is a subset of men who show severe oligospermia or azoospermia, have a variable risk of cognitive, language, and behavioral deficits. The syndrome is associated with an increased risk of social and emotional difficulties, learning disabilities, attention deficit, speech delay, hyperactive disorder and autistic spectrum disorder.

Triple X Syndrome (Trisomy X)

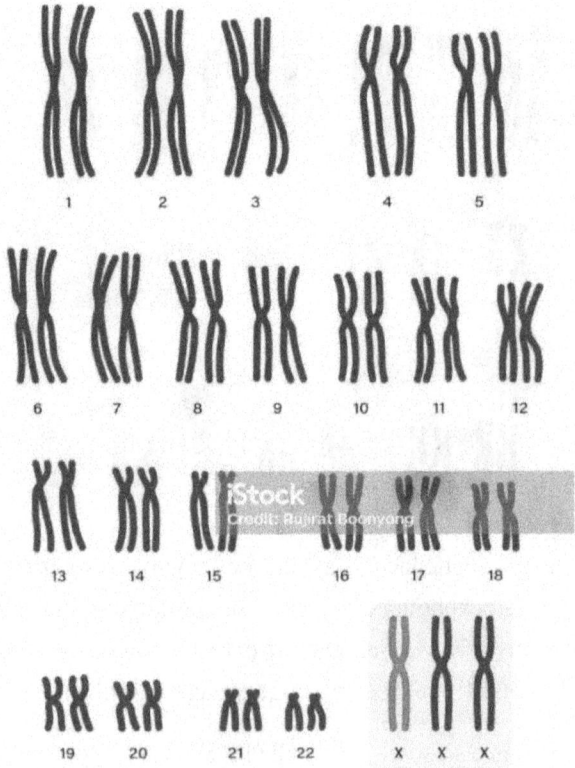

Image 12.13 Triple XXX syndrome karyotype (source: https://www.istockphoto.com/vector/human-karyotype-of-triple-x-syndrome-xxx-female-has-an-extra-y-chromosome-gm2135391162-568216665?searchscope=image%2Cfilm).

Trisomy X, also called triple X syndrome or 47, XXX or XXX syndrome. Triple X syndrome, is characterized by the presence of an additional X chromosome in each of a female's cells (Image 12.13). This disorder occurs in one in 900 to 1,000 females. Most girls and women with triple X syndrome don't experience symptoms or have only mild symptoms. They can grow up healthy, have normal sexual development and fertility, and lead productive lives. Occasionally, the disorder causes significant problems.

Girls with triple X syndrome can have some or all of these physical signs to some degree: taller than average height (usually, very long legs), low muscle tone, or muscle weakness (hypotonia), very curved pinky finger (clinodactyly), and widely spaced eyes (hypertelorism). Girls with triple X syndrome may experience delayed development in their social, language, and learning skills. This can manifest as difficulties with reading, understanding math, and mild delays in coordination. They might also face challenges with social interactions and emotional regulation. Some individuals may experience delays in motor skills development, making tasks like sitting and walking more challenging. Behavioral and emotional difficulties, including anxiety, depression, and Attention-deficit/hyperactivity disorder (ADHD), can also be associated with triple X syndrome. Seizures or kidney abnormalities occur in about 10 percent of affected females.

Turner's Syndrome (45, X)

Turner syndrome affects only females. It results when one of the X chromosomes is missing (monosomy of the X chromosome) or partially missing (Image 12.14). Turner syndrome is named for Dr. Henry Turner, who in 1938 published a report describing the disorder.

Turner syndrome is typically caused by nondisjunction and anaphase lag during meiosis. When sperm with no X chromosome unites with a normal egg to form an embryo, that embryo will have just one X chromosome (X rather than XX). At least 98% of all XO zygotes spontaneously abort early in pregnancy. In about 20 percent of Turner syndrome cases, both X chromosomes are present, but one is abnormal. It may be shaped like a ring or missing some genetic material. Turner syndrome can cause a variety of medical and developmental problems. Girls with this disorder are shorter than normal and may not start puberty when they should. This is because the ovaries don't develop properly. These adults have immature ovaries, have limited secondary sexual characteristics, and usually sterile. Affected females have short stature, low-set ears, a receding lower jaw, a short-webbed neck, a low hairline at the back of the neck, shield-shaped thorax, small uterus, and poorly developed breasts. Turner syndrome usually does not affect intelligence. Medical symptoms can include lymphedema (swelling of hands and feet), heart and/or kidney defects, high blood pressure, and infertility. This disorder is seen in 1 of every 2000 to 2500 baby girls.

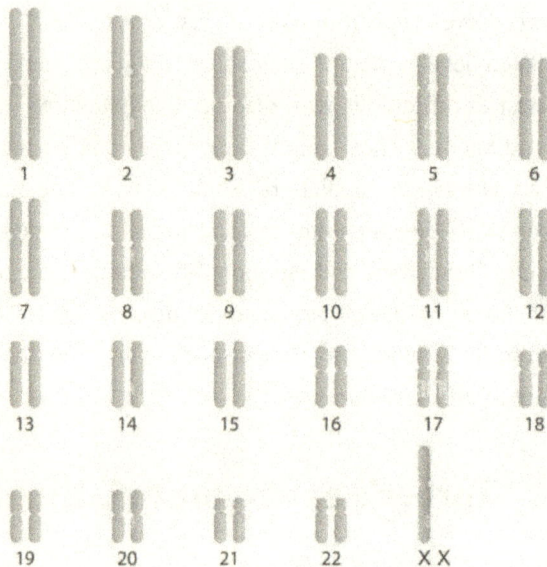

Image 12.14 Turner syndrome karyotype (source:
https://www.alamy.com/scientific-designing-of-turner-syndrome-
monosomy-x-karyotype-colorful-).

Turner syndrome can be diagnosed prenatally, during infancy, or in early childhood. In some females with mild signs and symptoms, the diagnosis may be delayed until the teenage years or young adulthood. Girls and women with Turner syndrome require ongoing medical care from a range of specialists. With regular checkups and appropriate treatment, most can lead healthy and independent lives.

Edward's Syndrome (Trisomy 18)

Children with Edward's syndrome have 3 copies of part or all of chromosome 18, instead of the usual 2 copies (Image 12.15). It is also called Trisomy 18. This can be caused by a mistake in the formation of the egg or sperm, or the problem can arise while the baby is developing in the womb. In 90% of the cases, the defect has been attributed to the egg. The likelihood that a parent will have a child with Edwards syndrome increases with maternal age at the time of pregnancy. It is much more common in girls.

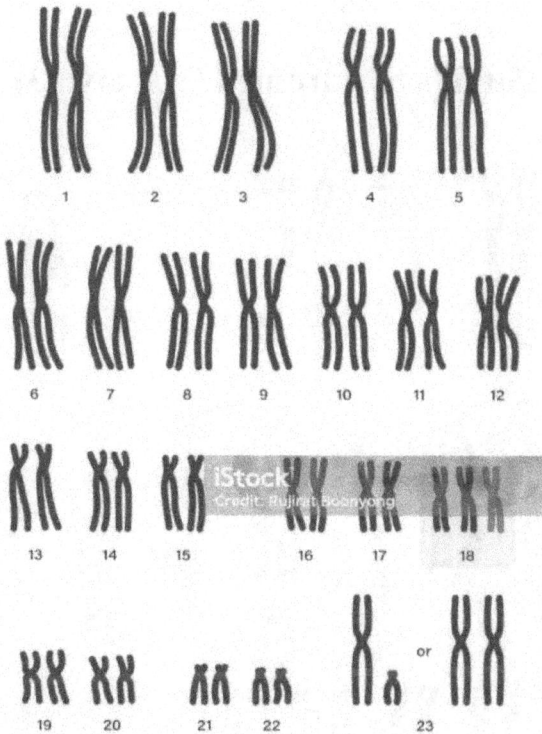

Image 12.15 Edward's syndrome karyotype (source:
https://www.istockphoto.com/vector/human-karyotype-of-edward-syndrome-autosomal-abnormalities-trisomy-18-human-gm2135384932-568216659?searchscope=image%2Cfilm).

Edward's syndrome is the most severe form of trisomy 18. It occurs in an estimated 1 out of every 5,000 to 6,000 live births. The condition is more common during pregnancy (1 out of every 2,500 pregnancies), but most (at least 95%) fetuses don't survive full term. Babies with Edwards syndrome are either miscarried, stillborn or born with severe physical abnormalities. Children have a low birth weight, small head, mouth and jaw, an unusual-looking face and head, unusual hands and feet with overlapping fingers and/or clubfeet, decreased muscle tone, low-set ears, weak cry and minimal response to sound, problems with feeding, breathing, seeing and hearing. It is very rare for a baby with Edwards syndrome to survive their first year of life, and most die within a week of birth. John Hilton Edwards *et al.* discovered Edwards syndrome in 1960.

Patau's Syndrome (Trisomy 13)

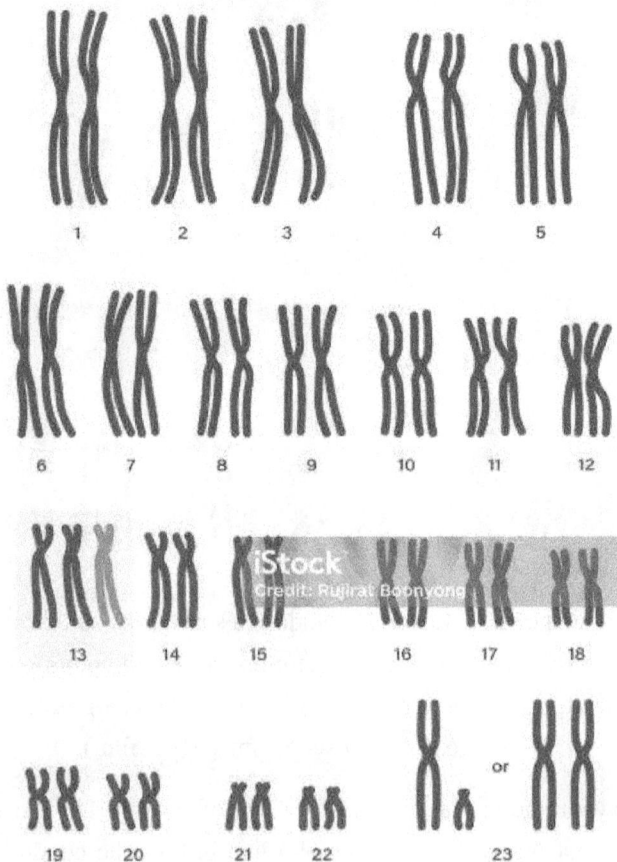

Patau's syndrome (also called trisomy 13) is a serious, rare genetic disorder caused by having an additional copy of chromosome 13 in some or all of the body's cells (Image 12.16). This severely disrupts normal development and, in many cases, results in miscarriage, stillbirth or the baby dying shortly after birth. Most babies born with it don't live past their first year, though some can survive much longer. Trisomy 13 occurs in an estimated 1 out of 10,000 to 20,000 live births and the incidence increases with increased maternal age.

Babies with Patau's syndrome grow slowly in the womb and have a low birth weight, along with a number of other serious medical problems that may include the following: cleft lip or palate, decreased muscle tone, extra fingers or toes, hernias, low-set ears, severe intellectual disability, small eyes, heart abnormalities, underdeveloped internal organs, small head and lower jaw, very small, close together or underdeveloped eyes, absent eyes, missing skin on the scalp, undescended testicle, *etc.*

Aberrant Euploidy (Euploid Variants)

Eukaryotes normally carry either one chromosome set (monoploid) or two sets (diploids). Both are normal cases of euploids. For diploids, two classes of euploid variants are observed: monoploidy and polyploidy. Monoploidy and polyploidy involve variations from the normal state in the number of complete sets of chromosomes. Euploidy is more tolerated in plants than in animals.

Monoploidy

The number of chromosomes in a basic set is called the monoploid number (x). In monoploidy an organism that is usually diploid has only one set of chromosomes (Image 12.17). Monoploidy is sometimes called haploidy, although haploidy is typically used to describe the chromosome complement of gametes. Monoploidy is typical of gametes, some fungi, and males of haplo/diplo species (bees, ants, wasps and sawflies). Monoploidy is seen only rarely in normally diploid organism. In most species monoploid individuals are abnormal, arising in natural populations as rare aberrations. The germ cells of a monoploid cannot proceed through meiosis normally because the chromosomes have no pairing partners. Thus, monoploids are characteristically sterile. Male bees, ants, wasps and sawflies bypass meiosis in forming gametes; here, mitosis produces the gametes.

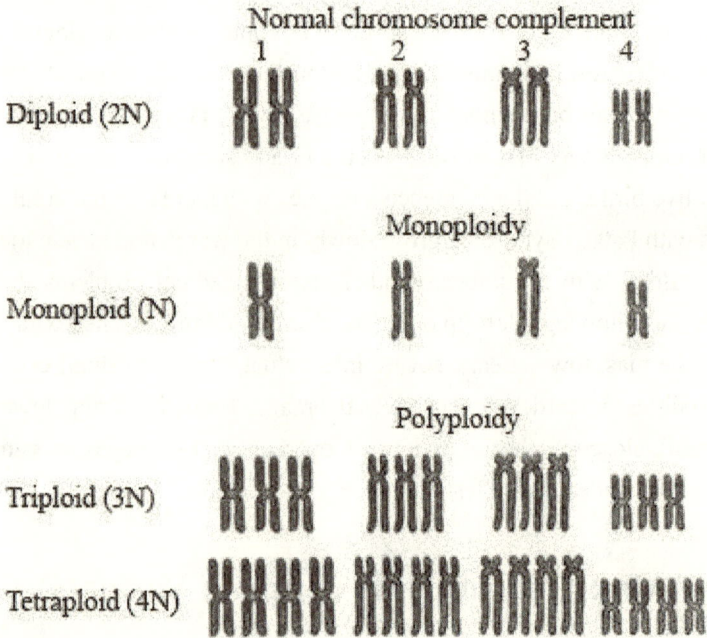

Image 12.17 Variations in number of complete chromosome sets.

Polyploidy

Polyploidy is the situation in which a cell or organism has one or more additional sets of chromosomes with respect to the number most frequently found in nature for a given species (Image 12.17). Organisms with additional sets of chromosomes are called polyploids. The polyploid types are named triploid ($3x$), tetraploid ($4x$), pentaploid ($5x$), hexaploid ($6x$), heptaploid ($7x$), octaploid ($8x$), nonaploid ($9x$), decaploid ($10x$), dodecaploid ($12x$), and so on. The effects of polyploidy on organisms can vary depending on the specific type and the organism itself. Polyploidy is relatively common in plants but rare in most animals. About two-thirds of all the grasses and half of all flowering plants are polyploids. All commercial grains, most crops, and many common commercial flowers are polyploids. In fact, polyploidy is the rule rather than the exception in plants. Polyploidy can happen due to errors during cell division or by merging chromosomes from different species.

The haploid number (n) refers strictly to the number of chromosomes in gametes. In most animals and many plants that we are familiar with, the haploid number and monoploid number are the same. Hence, n or x (or $2n$ or $2x$) can be used interchangeably. For organisms that are regularly polyploid, n and x are different. Thus, in a hexaploid organisms with 60 chromosomes, $6x = 2n = 60$, so that $x = 10$ and $n = 30$.

There are two general classes of polyploids: those that have an even number of chromosome sets and those that have an odd number of sets. If an organism has an even number of sets, the organism is often fertile. Odd number of sets usually results in sterility. An odd number of chromosome sets makes an organism sterile because there is not a partner for each chromosome at meiosis, whereas even number of sets can produce standard segregation ratios. Polyploid plants found in nature almost always have an even number of sets of chromosomes.

Two kinds of polyploids can be distinguished according to the origin of the chromosome sets: autopolyploids and allopolyploids. Autopolyploids receive all their chromosome sets from the same species. Allopolyploids obtain their chromosome sets from different species ("allo" means different). Many important crop plants, such as wheat, cotton, and potatoes, are allopolyploids. All the chromosome sets in an autopolyploid are homologous. But in allopolyploids, the different chromosome sets generally vary somewhat and are called homeologous, or partially homologous. All autopolyploid with odd number of sets of chromosomes ($3n$, $5n$, $7n$) are sterile due to abnormal meiosis. Organisms with even numbered sets ($4n$, $6n$, *etc*.) can have equal distribution of chromosomes during meiosis and may form some normal gametes. In agricultural setting, autopolyploidy (particularly, autotriploidy) is applied in producing seedlessness in watermelon and bananas. One way to produce an autotriploid is by mating a diploid with haploid (n) gamete and a tetraploid with $2n$ gamete. The offspring would have an unbalanced gametes and therefore possibly sterile. Thus, autotriploids are mostly propagated asexually. Other ways of producing autopolyploids are by the union of two diploid gametes, by somatic doubling, by fertilizing an egg with two sperms, and crossing a tetraploid with a diploid.

Cultivated banana is an example of triploid autopolyploid plant. Because it has an odd number of chromosome sets, the gametes have a variable number of chromosomes and few fertile seeds are set, thereby making most bananas seedless and highly palatable. In general, the development of 'seedless' fruits rely on 'odd number' polyploidy. Most naturally occurring polyploids are allopolyploids, and they may result in a new species. Cultivated bread wheat is an allohexaploid with 42 chromosomes. This plant species is descended from three distinct diploid species, each with a diploid set of 14 chromosomes. Meiosis is normal because only homologous chromosomes pair, so the plant is fertile. Polyploidy occurs when cytokinesis does not follow karyokinesis.

Polyploids occur naturally, but in very low frequency, when a cell undergoes abnormal mitosis or meiosis. For example, if during mitosis all chromosomes go to one pole, that cell has an autotetraploid chromosome number. Most changes in chromosome number arise as an outcome of nondisjunction during meiosis and gamete formation. If abnormal meiosis takes place, an abnormal gamete may result, having $2n$ chromosomes. However, in most situations this diploid gamete would combine with a normal haploid gamete and produce a triploid. Polyploids can also be produced artificially using colchicine, a chemical that interferes with the formation of spindle fibers. As a result of colchicine application, chromosomes do not move to the poles, and autotetraploids are often formed.

Some crops are found in a variety of ploidy. Apples, tulips and lilies are commonly found as both diploid and as triploid. Daylilies (*Hemerocallis*) cultivars are available as either diploid or tetraploid. Kinnows can be diploid, triploid or tetraploid.

Polyploidy in Plant Breeding

Polyploidy is a significant phenomenon in plants, playing a vital role in their evolution, diversity, and adaptation. It can occur naturally or be induced artificially, and it is particularly valuable in plant breeding for several reasons:

1. **Enhanced growth and yield:** Polyploidy often increases the size of individual cells, resulting in plants that are larger, more vigorous, and capable of producing bigger fruits and seeds. These plants typically grow faster and yield more biomass than their diploid counterparts.
2. **Economic benefits:** Polyploid plants often have higher production levels and are richer in nutritional and chemical content, making them economically valuable.
3. **Greater genetic diversity:** Polyploidy introduces additional genetic variation, allowing breeders to select plants with desirable traits more effectively.
4. **Maintaining crop purity:** Polyploid plants are often reproductively isolated from diploid plants, which helps prevent unwanted cross-pollination and maintain crop purity.
5. **Seedless fruit production:** Polyploidy can produce seedless varieties of fruits, such as watermelons and grapes.
6. **Stress tolerance:** Polyploid plants are generally more resilient to environmental stresses like drought, heat, and cold, making them valuable in breeding climate-resilient crops.

Polyploidy is often induced artificially using agents such as colchicine, which disrupts cell division to double chromosome numbers. This enables breeders to improve crop productivity, quality, and resilience. Polyploid plants are also better adapted to environmental fluctuations, further enhancing their agricultural value.

Beyond practical breeding, polyploidy plays a major role in genetics, evolution, and systematics. It contributes to speciation, promotes variation, and is considered a major evolutionary force in angiosperm development. Polyploid plants often exhibit a high degree of variation, which is beneficial both for natural evolution and for the development of improved crop varieties.

Examples of Polyploidy

Examples of Polyploidy in Plants

Triploid plants (3x): Seedless watermelon, blueberries, grapes, bananas, tulips, cassava, marigold, apples, ginger, some potatoes, and Winesap apples. These crops are typically propagated asexually.

Tetraploid plants (4x): Wheat, mustard, rapeseed, cotton, kiwi, potatoes, coffee, lucerne, some forage grasses, rye, macaroni, maize, cabbage, tobacco, kinnow, pelargonium, alfalfa, peanuts, and McIntosh apples. Cultivated potatoes (*Solanum tuberosum*) are natural autotetraploids (4x = 48).

Pentaploid plants (5x): Kenai Birch (*Betula papyrifera* var. *kenaica*).

Hexaploid plants (6x): Chrysanthemum, bread wheat, triticale, oat, kiwi, sweet potato, Prunus species, rose.

Heptaploid/septaploid plants (7x): Some cultured Siberian sturgeon (aquatic examples).

Octaploid plants (8x): Dahlia, pansies, sugarcane, oca, cultivated strawberry (8x = 56).

Decaploid plants (10x): Certain strawberries and roses.

Dodecaploid/duodecaploid plants (12x): *Celosia argentea* and *Spartina anglica*.

Tetratetracontaploid plants (44x): Black mulberry.

The ploidy level of wild rose species ranges widely from diploid (2n = 2x = 14) to decaploid (2n = 10x = 70), including nearly all even and odd levels of ploidy. Edible banana cultivars are propagated vegetatively and include diploid, triploid, and tetraploid hybrids. Sugarcane cultivars exhibit highly variable ploidy, often containing 10-13 sets of the basic 10 chromosomes.

Examples of Polyploidy in Animals

While some animal species are naturally polyploid, in most cases, monoploidy and polyploidy are lethal in animals. This is likely due to problems in gene expression when abnormal numbers of gene copies are present. A few examples of polyploid animals include Drosophila, flatworms, leeches, frogs, brine shrimp, goldfish, salmon, lizards, and salamanders. Polyploid animals are often sterile and typically reproduce by parthenogenesis, yet many are fit and well-adapted to their environments.

In mammals and birds, deviations from diploidy are poorly tolerated, often resulting in grossly abnormal offspring. Even in rare viable individuals, normal gamete formation is often disrupted due to segregation problems during meiosis. Polyploidy is lethal in humans, with over 99% of cases resulting in prenatal death, and the rare newborns usually die shortly after birth. Although rare polyploid mammals exist, most result in prenatal mortality. The only known exception is the Red Viscacha-Rat (*Tympanoctomys barrerae*), an octodontid rodent from Argentina's desert regions, with a diploid number of 102. This species is related to guinea pigs and chinchillas, not true rats.

Amphibians: The African clawed frog (*Xenopus laevis*) is tetraploid, a rare occurrence in vertebrates, often resulting from hybridization events.

Ants: Polyploidy is observed in certain ant species, such as the fire ant (*Solenopsis invicta*).

Polyploidy in Fungi and Crop Improvement

Fungi: Polyploidy occurs in fungi as well. An example is the button mushroom (*Agaricus bisporus*).

Plant Breeding and Polyploidy: Plant breeders use multiple techniques, including polyploidization, to modify crop traits and meet market demands. Polyploidy can create strong phenotypes and high vigor, making it one of the most effective methods for enhancing crops.

Interspecific Hybridization: Combining polyploidy with interspecific hybridization increases crop diversity and helps plants adapt to new environments. For instance, allopolyploid *Triticale* was developed by crossing hexaploid bread wheat with rye. This hybrid was created to achieve specific goals such as high yield, improved grain quality, disease resistance, and stress tolerance.

Polyploidy is common in newly domesticated crops and has played a significant role in crop speciation. Today, it is routinely used in plant breeding to develop new species and improve cultivated varieties.

Genetic Disorders

Genetic disorder is an inherited condition that sooner or later will cause mild to severe medical problems. A genetic disorder is caused by an altered gene or set of genes. Whether, when and to what extent a person with the genetic defect or abnormality will actually suffer from the disease is almost always affected by environmental factors in the person's development. Most genetic

disorders are quite rare and affect one person in every several thousands or millions. The three broad groups of genetic disorders are:
1. Single gene disorders.
2. Chromosomal disorders.
3. Multifactorial/polygenic/complex disorders

Chromosomal Disorders

A chromosome disorder means there is a change in either the structure or the number of chromosomes. Babies are rarely born with changes in chromosome numbers because most of these pregnancies end in miscarriage. Examples of this type of disorder are *cri-du-chat* syndrome, Down syndrome, Klinefelter syndrome, Turner syndrome, Williams syndrome, *etc.* A syndrome is a recognized set of symptoms that characterize a given disorder.

Multifactorial/Polygenic/Complex Disorders

Many disorders and traits result from the combined effects of multiple genes or from interactions between genes and the environment. These are referred to as **multifactorial, polygenic, or complex disorders**.
Examples in humans include:
- **Physical traits:** Skin color, eye color, hair color, body shape, height, weight.
- **Diseases and conditions:** Alzheimer's disease, breast and ovarian cancer, colon cancer, hypothyroidism, asthma, multiple sclerosis, various cancers, ciliopathies, cleft lip and palate, diabetes mellitus, congenital heart disease (CHD), hypertension, inflammatory bowel disease, intellectual disability, mood disorders, obesity, refractive errors, neural tube defects, pyloric stenosis, congenital dislocation of the hip, talipes (clubfoot), hypospadias, atherosclerosis, schizophrenia, epilepsy, nonspecific developmental delay, dementia, venous thrombosis, and infertility.

Example: Skin Color

Skin color is controlled by at least three major genes, with additional minor genes also influencing pigmentation. Each gene has two alleles located on different chromosomes. The darkest skin color corresponds to the presence of three dominant alleles (AABBCC), while the lightest skin color corresponds to the recessive alleles (aabbcc). Skin color is primarily determined by the amount of melanin pigment produced in the skin.

Mitochondrial Disorders

Some genetic material in humans is contained in the mitochondrial genome, and some diseases result from mutations in mitochondrial genes. Only females can transmit mitochondrial diseases to their children because only egg cells contribute mitochondria to the developing embryo. Mitochondrial disorders can appear in every generation of a family and can affect both males and females, but fathers do not pass these disorders onto their children *e.g.*, Leber's hereditary optic neuropathy (LHON), Kearns-Sayre syndrome, *etc.*

Uniparental Disomy

Uniparental disomy (UPD) occurs when a person receives two copies of a chromosome from one parent and no copies from the other parent. UPD can occur as a random event during the formation of egg or sperm cells or may happen in early fetal development. In many cases, UPD likely has no effect on health or development. In some cases, however, it does make a difference whether a gene is inherited from a person's mother or father. A person with UPD may lack any active copies of essential genes that undergo genomic imprinting. This loss of gene function can lead to delayed development, intellectual disability, or other health problems.

Several genetic disorders can result from UPD. The most well-known conditions include Prader-Willi syndrome and Angelman syndrome. Both of these disorders can be caused by UPD or genomic imprinting involving genes on the long arm of chromosome 15. Other conditions, such as Beckwith-

Wiedemann syndrome (a disorder characterized by accelerated growth and an increased risk of cancerous tumors), are associated with abnormalities of imprinted genes on the short arm of chromosome 11.

Genomic Imprinting

People inherit two copies of all genes – one from their mother and one from their father. Usually both copies of each gene are active, or 'turned on', in cells. In some cases, however, only one of the two copies is turned off in a parent-of-origin dependent manner: either only from the allele inherited from the mother (*e.g.*, H19 or CDKN1C), or from that inherited from the father (*e.g.*, IGF2). These genes are called 'imprinted' or 'silenced' and the phenomenon is known as genomic imprinting. The most well-known epigenetic effect in animal breeding is genomic imprinting. Genomic imprinting affects several dozen mammalian genes and results in the expression of those genes from only one of the two parental chromosomes.

Imprinted genes follow a non-Mendelian pattern of inheritance. The phenotype of the progeny is based on both the individual's genotype and the paternal or maternal origin of the genotype. This is termed monoallelic expression. Genomic imprinting involves methylation and histone modifications to achieve monoallelic gene expression without altering the DNA sequence that it is established in the germline and maintained throughout all somatic cells of an organism. In mammals, about 1% of genes are imprinted, depending upon its parental origin. Genomic imprinting occurs during gamete formation before fertilization, in either the egg or the sperm. Examples of genomic imprinting are Prader-Willi syndrome, Angelman syndrome, Beckwith-Wiedemann syndrome, and Silver syndrome. Prader-Willi syndrome and Angelman syndrome were the first imprinting diseases discovered in humans. The symptoms of these two disorders are very different, but both conditions are caused by indistinguishable deletions of certain critical genes on the long (q) arm of chromosome 15 in the region 15q11 to 15q13. What distinguishes these disorders is the parental origin of the affected chromosome.

Prader-Willi Syndrome (PWS)

Prader-Willi syndrome (PWS) was first described by John Langdon Down in 1887, and later reported by Andrea Prader, Alexis Labhart, and Heinrich Willi in 1956. It is a complex genetic condition that affects many parts of the body. It occurs in males and females equally and in all races. Prader-Willi syndrome affects an estimated 1 in 10,000 to 30,000 people worldwide. Affected individuals are small and weak at birth. In infancy, this condition is characterized by weak muscle tone (hypotonia), feeding difficulties, poor growth, and delayed development. Beginning in childhood, affected individuals may develop an insatiable appetite, which leads to chronic overeating (hyperplasia), obesity, and type 2 diabetes mellitus. People with Prader-Willi syndrome typically have mild to moderate intellectual impairment and learning disabilities. Behavioral problems such as temper tantrums, stubbornness, and compulsive behavior are common. Many have sleep abnormalities. Individuals with PWS have distinctive facial features, short stature, and small hands and feet. Some have fair skin and light-colored hair. Both males and females have underdeveloped genitals. Puberty is delayed or incomplete, and most are infertile.

Most cases of Prader-Willi syndrome (about 70 percent) is caused by the deletion of certain critical genes on the long (q) arm of chromosome 15 in the region 15q11 to 15q13 received from the father. The genes on the maternal copy are turned off (inactive) by genomic imprinting, even if it is 'normal' copy. Thus, the child must inherit the active gene from the father in order to be free of the disease. Failure to receive the gene can be due to a *de novo* deletion, or it can be the result of uniparental disomy.

Angelman Syndrome (AS)

Angelman syndrome is named after the physician Harry Angelman who first delineated the syndrome in 1965. Angelman syndrome is a complex genetic disorder that primarily affects the nervous system. Angelman syndrome affects an estimated 1 in 12,000 to 20,000 people. Characteristic features of this condition include delayed development, intellectual disability, severe speech impairment, and problems with movement and balance (ataxia). Physical features include deep-set eyes, a flattened back of the head and a wide ever-smiling mouth. Most affected children also have recurrent seizures (epilepsy) and a small head (microcephaly). Delayed development becomes noticeable by the age of 6 to 12 months, and other common signs and symptoms usually appear in early childhood. Children with Angelman syndrome typically have a happy, excitable demeanor with frequent smiling, laughter, and hand-flapping movements. Hyperactivity, a short attention span, and a fascination with water are common. Most affected children also have difficulty sleeping and need less sleep than usual. With age, people with Angelman syndrome become less excitable, and the sleeping problems tend to improve. However, affected individuals continue to have intellectual disability, severe speech impairment, and seizures throughout their lives. Other common features include unusually fair skin with light-colored hair and an abnormal side-to-side curvature of the spine (scoliosis). The life expectancy of people with this condition appears to be nearly normal.

In most cases, Angelman syndrome is caused by the deletion of certain critical genes on the long (q) arm of chromosome 15 in the region 15q11 to 15q13 inherited from the mother. The genes on the paternal copy are turned off (inactive) by genomic imprinting, even if it is 'normal' copy.

Mutational Heterogeneity

Mutational heterogeneity refers to the phenomenon in which different mutations occur at the same genetic locus, leading to the same or similar phenotype. In reality, most autosomal recessive disorders are caused by mutational heterogeneity, while a small proportion may be truly homogeneous, especially in consanguineous families.

Examples:
- Sickle cell anemia
- Cystic fibrosis
- Tay-Sachs disease

Locus Heterogeneity

Locus heterogeneity occurs when the same disease phenotype can be caused by mutations at different genetic loci. In other words, mutations in different genes can lead to the same clinical condition. One example is Osteogenesis imperfecta: Mutations in any one of three genes – two located on Chromosome 17 and one on Chromosome 7 – can cause the disorder.

CHAPTER THIRTEEN
EXTRANUCLEAR GENETICS

Most phenotypes of eukaryotes are determined by genes contained in the nucleus. However, the nucleus is not the only source of genetic material in the cell. All available evidences indicate that outside the nucleus genes are found in the DNA present in two principal self-replicating cytoplasmic organelles, the mitochondria and the chloroplasts. Genes contained in mitochondrial and chloroplasts DNA control several important phenotypes of eukaryotes. The genes in these mitochondrial and chloroplast genomes have become known as extrachromosomal genes, non-Mendelian genes, organellar genes (plastid genes and mitochondrial genes), extragenomic genes, plasma genes, cytoplasmic genes, cytogenes, or extranuclear genes. The genes contained in mitochondrial and chloroplasts DNA are not present in nuclear DNA. The DNA is replicated within the organelles and transmitted to daughter organelles as they form. The organelle genetic system is separate from that of the nucleus, and traits determined by organelle genes show a pattern of inheritance very different from the familiar Mendelian inheritance determined entirely by nuclear genes. Extranuclear inheritance refers to inheritance patterns involving genetic material outside the nucleus. The examples of such a role were so few that Morgan in 1926 concluded that the cytoplasm may be ignored genetically.

Criteria of Extranuclear Inheritance

Four main characteristics of extranuclear inheritance are summarized below:
1. Typical Mendelian ratios do not occur, because meiosis-based segregation is not involved.

2. Reciprocal crosses show marked differences for the characters governed by extranuclear genes. Reciprocal crosses usually show uniparental inheritance, with all progeny (both males and females) having the phenotype of one parent. Typically, for higher eukaryotes it is the mother's phenotype that is expressed exclusively, because the zygote receives nearly all of its cytoplasm (including organelles) from the ovum.

3. Extranuclear genes cannot be mapped to the chromosomes in the nucleus. If a new mutation does not show linkage to any of the nuclear genes, it is probably an allele of an extranuclear gene.

4. Extranuclear inheritance is not affected by substituting a nucleus with a different genotype. When a particular phenotype persists after the nucleus is replaced with one with a different genotype, this indicates that the phenotype is likely to be controlled by an extranuclear genome.

Organelle Genomes

Organelle genomes differ fundamentally from nuclear genomes in terms of their structure, organization, stability, and mechanisms of gene expression and regulation. The DNA of these organelles is not associated with histones. Several copies of the genome are present in each organelle. There are typically from 2 to 10 copies in mitochondria and from 20 to 100 copies in chloroplasts. In addition, most cells contain multiple copies of each organelle. The genomes of both organelles are naked.

Organelle genes code for DNA polymerases that replicate the organelle DNA. They also code for other components essential to their function or replication, but many organelle components are determined by nuclear genes. These components are synthesized in the cytoplasm and transported into the organelle. Both mitochondria and chloroplasts carry their own ribosomal RNA and transfer RNA genes, and their ribosomes are similar in size to prokaryotic ribosomes. Some of the proteins found in these organelles are synthesized on their own ribosomes from mRNA transcribed from the organelle genome, but most proteins are coded by nuclear genes and are imported from the cytoplasm.

The Genetic Codes of Organelles

The **genetic codes of organelles** refer mainly to the codes used by **mitochondria** and **chloroplasts**, which differ slightly from the "universal" genetic code found in the nuclear genome. The universal genetic code is used without changes in chloroplasts and in mitochondria of green plants (Jukes and Osawa, 1990). In some green algae chloroplasts, **UGA** can sometimes code for **Tryptophan**, similar to mitochondria. Non-plant mitochondria often use genetic codes different from the standard genetic code. Table 13.1 shows the variations in codon assignment. Vertebrate mitochondria depart from the standard code in three principal ways:

1. UGA is not a stop codon but encodes tryptophan.
2. AGA and AGG codons are stop codons rather than arginine codons.
3. AUA encodes methionine rather than isoleucine.

Table 13.1 Deviations from the 'universal' genetic code

Source	Codon	Usual Meaning	New Meaning
Fruit fly mitochondria	UGA	Stop	Tryptophan
	AGA and AGG	Arginine	Serine
	AUA	Isoleucine	Methionine
Human and Vertebrate mitochondria	UGA	Stop	Tryptophan
	AUA	Isoleucine	Methionine
	AGA and AGG	Arginine	Stop
Yeast mitochondria	UGA	Stop	Tryptophan
	AUA	Isoleucine	Methionine
	CUN*	Leucine	Threonine
Higher plant mitochondria	UGA	Stop	Tryptophan
	CGG	Argine	Tryptophan
Protozoa nuclei	UAA and UAG	Stop	Glutamine
Mycoplasma	UGA	Stop	Tryptophan

*N = Any base

Maternal (Organelle) Inheritance

In many higher eukaryotes, the mother provides all the cytoplasm in the egg whereas the father provides only the paternal nucleus. The female cytoplasm contributes the mitochondria for all species as well as the chloroplast for plant species. These two organelles contain DNA and control certain traits in the offspring. Those phenotypes controlled by organelle genes exhibit maternal inheritance. Maternal inheritance is the transmission of genes, and the traits they control, solely through the maternal line. In maternal inheritance, the hereditary determinants of a trait are extranuclear, and genetic transmission is only through the maternal cytoplasm. In maternal inheritance the progeny always have the maternal phenotype. The phenotype of the male parent is irrelevant, and its contribution to the progeny appears to be zero.

Mitochondrial Genome

A mitochondrion is an important cell organelle responsible for producing ATP (adenosine triphosphate), which is essential for proper cellular functioning. Besides energy production, mitochondria help in calcium storage, metabolism regulation, cell death control, cell signaling, and various other functions. Each cell contains hundreds to thousands of mitochondria, which are located in the cytoplasm of all aerobic animal and plant cells. It is the "power plants" of the eukaryotic cell. Although most of a cell's DNA is in the nucleus, the mitochondria also have a small but crucial piece of their own DNA. This genetic material is known as mitochondrial DNA or mtDNA. mtDNA is considered as one of the most valuable tools in population genetics and molecular phylogenetics.

Mitochondria contain the enzymes of the Krebs cycle, carry out oxidative phosphorylation (OXPHOS), and are involved in fatty acid biosynthesis. The DNA in most species' mitochondria (mtDNA) is circular, double-stranded and supercoiled. mtDNA of some protozoa and some fungi is linear. Mitochondria contain multiple copies of mtDNA. The gene content is very similar among mitochondrial genomes from different species in both number and function.

411

Mitochondrial genomes are usually smaller. The sizes of mitochondrial DNAs vary widely from one class of organisms to another. For example, the mitochondrial genome in mammals is about 16.5 kb, that in *Drosophila melanogaster* is about 18.5 kb. Plant mitochondrial genomes are much larger than those found in metazoans ranging from 200 to 2400 kilobase pairs (kbp). The mitochondrial genomes of higher plants are exceptional in consisting of two or more circular DNA molecules of different sizes.

Mammalian mitochondrial genome contains 37 genes, all of which are essential for normal mitochondrial function. Thirteen of these genes provide instructions for making enzymes involved in OXPHOS. OXPHOS is a process in which the cells' main energy source, ATP, is produced. Twenty-two genes encode tRNAs, two genes specify rRNA, all of which are indispensable for mitochondrial function.

The mitochondrial genome has two main functions: (1) it encodes some proteins that are actually in or associated with the electron transport chain, and (2) it encodes some proteins, all the tRNAs, and rRNAs necessary for mitochondrial protein synthesis. The remaining necessary components for both these functions are encoded by nuclear genes.

Replication of mtDNA is semi-conservative, uses mitochondrial DNA polymerases and RNA primers, and involves no proofreading. The mutation rate in mtDNA is ten times higher than in nuclear DNA because mtDNA are subject to damage from reactive oxygen molecules released as a byproduct during OXPHOS. In addition, the mtDNA also lacks the DNA repair mechanisms found in the nucleus. Replication of mtDNA occurs throughout the cell cycle (in contrast with nuclear DNA, which replicates only in S phase). Both strands of mtDNA in most animals replicate in a continuous manner, with replication of one strand initiating well before the other.

A number of mitochondrial genetic diseases have been identified. Some of these are quite serious and involve the central nervous system as well as muscle, heart, liver and kidney. Some examples of mitochondrial diseases include: Leber's hereditary optic neuropathy (LHON), aminoglycoside-induced deafness, Kearns-Sayre syndrome (KSS), mitochondrial encephalopathy with lactic acidosis and stroke-like episodes (MELAS), myoclonic epilepsy with ragged-red fibres (MERRF), maternally inherited diabetes and deafness (MIDD), leigh syndrome (a progressive brain disorder), age-related hearing loss, progressive external ophthalmoplegia, and neuropathy, ataxia and retinitis pigmentosa (NARP).

Leber's Hereditary Optic Neuropathy

The first human disease that was associated with a mutation in mitochondrial DNA is called Leber's Hereditary Optic Neuropathy (LHON), also known as Leber Optic Atrophy. It was named after Doctor Theodore Leber, who described a characteristic pattern of sudden vision loss in young men with family history of blindness in 1871. LHON is a rare maternally inherited mitochondrial disorder that can cause loss of vision quickly and unexpectedly. Mutations in any of the four mitochondrial genes, MT-ND1 (Mitochondrially Encoded NADH-Ubiquinone Oxidoreductase Core Subunit 1), MT-ND4 (Mitochondrially Encoded NADH-Ubiquinone Oxidoreductase Core Subunit 4), MT-ND4L (Mitochondrially Encoded NADH-Ubiquinone Oxidoreductase Core Subunit 4L), and MT-ND6 (Mitochondrially Encoded NADH-Ubiquinone Oxidoreductase Core Subunit 6), can cause LHON. These genes encode protein subunits of **Complex I (NADH:ubiquinone oxidoreductase)**, which is the first enzyme of the mitochondrial respiratory chain and plays a critical role in oxidative phosphorylation (OXPHOS). Complex I is responsible for transferring electrons from NADH to ubiquinone while pumping protons across the inner mitochondrial membrane to help establish the electrochemical gradient required for ATP synthesis. Mutations in these genes disrupt Complex I activity, leading to impaired electron transport, reduced ATP production, and increased generation of reactive oxygen species (ROS). The resulting oxidative stress causes selective degeneration of retinal ganglion cells, leading to the hallmark clinical feature of LHON − painless, bilateral, central vision loss. Vision loss is typically severe and irreversible. Vision loss may start in one eye and then progress to both eyes several months later. It usually starts in late childhood to early adulthood. It leads to blindness typically between 12 and 30 years of age. Males are approximately four times more likely to be affected than females. Males will not pass the gene to any of their children, but females with the mutation will pass it to all of their children. Many factors affect the development of LHON. Both alcohol and tobacco use are important environmental factors associated with an increased risk for blindness in carriers of the mutation.

Heteroplasmy

Homoplasmy refers to the condition in which all copies of mitochondrial DNA (mtDNA) within a cell or an individual are identical, meaning they are either entirely normal or entirely mutant. In contrast, heteroplasmy describes the presence of a mixture of more than one type of mtDNA variant within the same cell or individual. Heteroplasmy of mitochondria is unusual among animals. For instance, in humans, a typical cell contains between 1,000 and 10,000 mitochondria, and under normal circumstances, all of them carry genetically identical mtDNA. However, when heteroplasmy does occur, the proportion of normal to mutant mtDNA is critical, as it can determine whether mitochondrial dysfunction is sufficient to cause disease.

Cytoplasmic Male Sterility in Maize

Male sterility is a condition that results from several reasons such as failure to produce functional pollens or male gametes. Male sterility is a desirable trait for the production of hybrids in crop plants. Male sterility may result from mutations in nuclear genes (genetic male sterility – GMS) or in mitochondrial genes (cytoplasmic male sterility – CMS). Cytoplasmic male sterility (CMS) is the maternally inherited inability of a plant to produce or shed functional pollen, which prevents them from fertilizing female flowers. A mitochondrial mutation is transmitted to all progeny of a cross, which allows the breeder to maintain the male sterility trait without the inconvenience of gene segregation each generation. It has been documented in at least 150 plant species, including rice, wheat, pearl maize, mustard, sorghum, onion, sugar beet, carrot, sunflower, *etc.*, and results from either spontaneous or artificial mutations Male sterility found in pearl maize (*Sorgum vulgare*) is the best example for mitochondrial inheritance. It is called cytoplasmic male sterility. It is transmitted only through the female and never by the pollen. The gene for cytoplasmic male sterility is found in the mitochondrial DNA. When all of the chromosomes of the male sterile line were replaced with chromosomes of normal plants, the line still remained male sterile, showing thereby that male sterility in controlled by some agency in the cytoplasm. There are commonly two types of cytoplasm N (normal) and S (sterile). CMS is governed by plasmogenes located in the mt-DNA, which causes pollen abortion in higher plants.

Chloroplast Genome

Chloroplasts are cellular organelles found only in green plants, photosynthetic protists, and cyanobacteria (blue-green algae), and are the sites of photosynthesis in the cells containing them. Like mitochondria, chloroplasts contain their own genomes. In all cases the DNA is circular, double-stranded, and supercoiled. Chloroplast DNA (cpDNA) is larger and more complex than mtDNA, with a size between 80 kb and 600 kb and contains more genes. Chloroplasts contain multiple copies of cpDNA that contains genes for photosynthesis, electron transport, and chloroplast protein synthesis. It encodes proteins and structural rRNAs required for chloroplast gene expression, including tRNAs sufficient for all codons, rRNA, some ribosomal proteins, and RNA polymerase. Additionally, the chloroplast encodes products with a direct function in photosynthesis, including components of photosynthesis II and I. Most polypeptides used in the chloroplast, however, are encoded by the nuclear genome and imported. Mutations in cpDNA typically produce photosynthetic defects or drug resistance. The inheritance pattern of plastid characters due to genes located in plastid is known as plastid inheritance.

Maternal Inheritance (Plastid Inheritance) in Four o'clock Plant

Plastid inheritance was the first case of cytoplasmic inheritance discovered and described simultaneously by Carl Erich Correns in *Mirabilis jalapa* and Erwin Baur in *Pelargonium zonale* in 1908. Both of these geneticists studied the inheritance of variegation in the respective plant species.

Correns studied the variegated strain of the four o'clock plant (*Mirabilis jalapa*) in which the chloroplasts are inherited exclusively from the female parent. Variegation refers to appearance of white or yellow spots of variable size on the green back ground of leaves and stems of plants. Variegation may be produced by some environmental factors, some nuclear genes and in some cases, plasma genes. On the variegated plants, some branches have completely green, some completely white, and others variegated leaves. Branches of all three types produce flowers, and these flowers can be used to perform nine possible crosses. From each cross, the seeds are collected and planted, and the phenotypes of the progeny are examined. The results of the crosses are summarized in Table 13.2. In each case, type of offspring was depended only upon the phenotype of the branch from which the ovule was derived – not the pollen.

Although the white progeny plants do not live long because they lack chlorophyll, the other progeny types do survive and can be used in further generations of crosses. In these subsequent generations, the patterns of transmission are identical with those of the original crosses. These observations suggest direct transmission of the trait through the mother, or maternal inheritance.

The results can be explained in the following manner. All of the organelle DNA that is found in an embryo is from the female. The egg cell is many times larger than the pollen cells, and contains both mitochondria and chloroplasts. Pollen is small and is essentially devoid of organelles, and thus organelle DNA. So, any trait that is encoded by the organelle DNA will be contributed by the female. In the case of the four o'clock plant, the different color of the leaves, like green, white, or variegated, are indeed due to the presence or absence of chlorophyll within the chloroplasts. This trait is controlled by the chloroplast DNA (cpDNA). Thus, green shoots contain chloroplasts that have chlorophyll, the chloroplasts in the white shoots contain no chlorophyll, and the variegated shoots contain some chloroplasts with chlorophyll and some without chlorophyll. Thus, depending upon the location in the plant where the flower comes from, the egg can have chloroplast with chlorophyll, without chlorophyll, or a mixture of the two types of chloroplasts. This is the biological basis of maternal inheritance. It is clear that variegation is determined by agencies transmitted through the female and that it is not influenced by the type of pollen used. These agencies are the chloroplast. They are capable of self-duplication and are transmitted from generation to generation through the cytoplasm of the egg. Variegation is thus a heredity character determined by stable, self-duplicating, extra nuclear particles called plastids. Neither the nucleus of the female gamete nor the male gamete is involved in the control of this type of heredity character.

Table 13.2 Results of crosses of variegated Four o'clock plants

Phenotype of branch bearing egg parent	Phenotype of branch bearing pollen parent	Phenotype of progeny
White	White	White
White	Green	White
White	Variegated	White
Green	White	Green
Green	Green	Green
Green	Variegated	Green
Variegated	White	Variegated
Variegated	Green	Variegated
Variegated	Variegated	Variegated

Maternal Effect (Maternal Influence)

Maternal effect refers to an inheritance pattern for certain nuclear genes in which the nuclear genotype of the mother directly determines the phenotype of her progeny; *i.e.*, the phenotypes of the progeny are determined by the genotype of the mother only. Surprisingly, the genotype of the father and the genotype of the affected individual have no effect on the phenotype. It is one kind of non-Mendelian pattern of inheritance. The hereditary determinants are nuclear genes transmitted by both sexes, and in suitable crosses the trait undergoes Mendelian segregation. In the F_1 generation, this can appear to be maternal inheritance, but subsequent generations reveal a Mendelian pattern of inheritance.

The embryo is formed when a female gamete unites with a male gamete. In the vast majority of species, the female gamete is physically larger than the male gamete and provides the cytoplasm for the developing embryo. Within this cytoplasm there are factors that were encoded by the nuclear genes of the female. Those factors may have specific effects upon the developing embryo. Maternal effect is due to the nuclear gene products of mother present in the egg cytoplasm. Maternal effect genes (MEGs) encode factors (*e.g.*, RNA and proteins) that are present in the oocyte prior to fertilization and play important roles in early steps of embryogenesis, *e.g.*, cell division, cleavage pattern, body axis orientation, *etc.*

A classic example of a maternal effect is **gestational diabetes mellitus (GDM),** in which the mother's metabolic condition directly influences the phenotype of the offspring. Infants born to mothers with gestational diabetes are more likely to experience dangerously low blood glucose levels immediately after birth due to their prenatal exposure to high maternal blood sugar levels. Beyond these early complications, such individuals also face long-term health risks, including an increased likelihood of developing obesity, cardiovascular disease, and type 2 diabetes later in life. This illustrates how the maternal environment, independent of the offspring's own genotype, can have profound and lasting effects on development and disease susceptibility.

Maternal Effect in Snail Shell-Coiling

The classic phenotype which exhibits maternal effects is shell-coiling direction of the water snail *Limnaea peregra*. The shell-coiling trait is determined by a single pair of nuclear alleles, D for coiling to the right (dextral coiling) and d for coiling to the left (sinistral coiling). The D allele is completely dominant to the d allele and the shell-coiling phenotype is always determined by the genotype (not phenotype) of the mother rather than by the genes of the developing snail. Reciprocal crosses between homozygous strains give the following results:

Dextral♀ × Sinistral ♂→ all F_1 Dextral

Sinistral ♀ × Dextral ♂→ all F_1 Sinistral

In these crosses, all the F_1s have the same genotype (Dd), but the phenotype is different in that the direction of coiling is the same as that of the mother. This result is typical of traits with maternal inheritance. All F_2 progeny from both crosses exhibit dextral coiling. The result does not sound like maternal inheritance; neither does it sound like Mendelian inheritance. Finally, in the F_3 generation, we see a Mendelian pattern.

In the cross of a dextral (DD) female with a sinistral (dd) male (Image 13.1), the F_1s are all Dd in genotype and dextral in phenotype. Selfing the F_1 (the snail is hermaphroditic) produces F_2s with a 1 : 2 : 1 ratio of DD, Dd, and dd genotypes. All of the F_2s are dextral, even the dd snails whose genotype would seem to indicate sinistral phenotype. The dd snails have a coiling phenotype not specified by the genotype they have, but one specified by the genotype of their mother (Dd). Selfing the F_2 snails give F_3 progeny, $^3/_4$ of which are dextral and $^1/_4$ of which are sinistral. The later are the dd progeny of the F_2 dd snails; these F_3 snails are sinistral because their phenotype reflects their mother's (the F_2) genotype.

Similar results are seen in the reciprocal cross of a sinistral (dd) female with a dextral (DD) male. The F_1s are all Dd in genotype yet they are sinistral in phenotype because the mother was dd in genotype. Selfing the F_1 produces F_2s all of which are dextral for the same reason as the reciprocal cross already described. The genotypes and the phenotypes of the F_2 and F_3 generations are the same as for the reciprocal cross and for the same reason. This means that the direction of shell-coiling is not a case of maternal inheritance but rather a case of maternal effect. Cytological analysis of developing eggs has provided their explanation: the genotype of the mother determines the orientation of the spindle in the initial mitotic division after fertilization, and this in turn controls the direction of shell-coiling of the offspring. This classic example of maternal effect indicates that more than a single generation of crosses is needed to provide conclusive evidence of extranuclear inheritance of a trait.

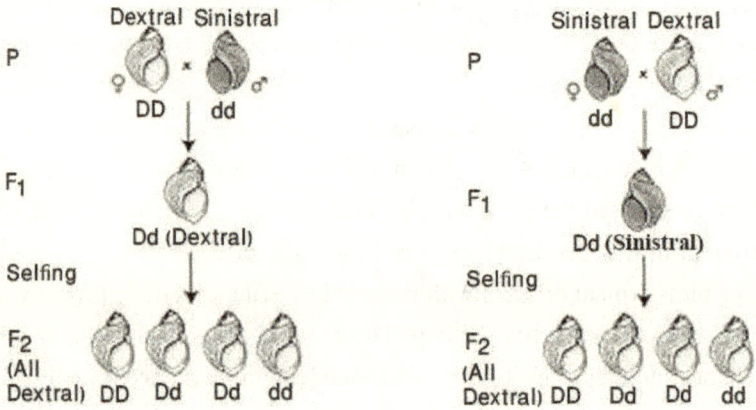

Image 13.1 Inheritance of shell-coiling direction in the snail *Limnaea peregra.*

Progeny resembling the mother is characteristic of both maternal inheritance and maternal effects.

CHAPTER FOURTEEN   THE GENETICS OF BACTERIA AND VIRUSES

―――――――――⇒ ⇒⇒ ⋅⋅⟫⋅⋅⧈⋅⋅⟪⋅⋅ ⇐ ⇐――――――――

A large part of the history of genetics and current genetic analysis is concerned with prokaryotic organisms and with viruses. Advances in microbial genetics within the past few decades have provided the foundation for recent advances in molecular biology. The role of DNA in heredity was first discovered by studying bacteria and bacteriophage. Genetic studies of bacteria and viruses are so productive that they have shaped our concepts of the gene and gene activity. Since all organisms from bacteria to humans have DNA as their genetic material, many of these concepts are directly applicable to higher organisms, or at least have provided the foundation for the study of the genetics of higher organisms. Besides providing this theoretical framework, bacteria have given us the tools of recombinant DNA technology that have been crucial for investigating the genetics of all organisms.

Merits of Bacteria and Viruses as the Materials for Genetic Experimentation

There are several advantages in the genetic study of bacteria and viruses over multicellular plants and animals:

1. They are haploid, so dominance and recessiveness of alleles is not a complication in identifying genotype. In this respect, the study of microorganisms is relatively simple because, barring interaction, every gene that the organism possesses can be expressed in its

phenotype.

2. The most important qualification of bacteria for genetic studies is their extremely rapid rate of growth, their simple growth requirements, and their small size. They grow on simple, cheap medium and often give rise to large populations in a matter of 24 hours. Under favorable conditions, a new generation is produced in minutes rather than weeks or months, which vastly increases the rate of accumulation of data. Given ideal growth conditions, *E. coli* cells divide every twenty minutes. A single *E. coli* cell will grow overnight into a visible colony containing millions of cells, even under relatively poor growth conditions. Thus, genetic experiments on *E. coli* usually last one day, whereas experiments on corn, for example, take months. It is easy to isolate their genomic material, manipulate it in the test tube and then place it back into the microbe.

3. The individual members of these large populations are genetically identical; that is, each laboratory population is a clone of genetically identical cells.

4. They are easy to grow in enormous numbers under controlled laboratory conditions, which facilitates molecular studies and the analysis of rare genetic events.

5. Microbes provide tools for use in molecular biology. These tools have allowed scientists to make rapid progress in investigating many types of microorganisms.

Working with Bacteria

Bacteria can be grown in a liquid medium or on a solid surface, such as an agar gel, as long as basic nutritive ingredients are supplied. In a liquid medium, the bacteria divide by binary fission: they multiply geometrically until the nutrients are exhausted, or until toxic factors (waste products) accumulate to levels that halt the population growth. A small amount of such a liquid culture can be pipetted onto a petri plate containing an agar medium and spread evenly on the surface with a sterile spreader, in a process called plating. Each cell then reproduces by binary fission. Because the cells are immobilized in the gel, all the daughter cells remain together in a clump. When this mass reaches more than 10^7 cells, it becomes visible to naked eye as a colony. Overnight, the single cells on the plate give rise to colonies. If the initially plated sample contains very few cells, then each distinct colony on the plate will be derived from a single original cell. Members of a colony that share a single genetic ancestor are known as clones.

Genetic analysis of bacteria and other microorganisms typically involves studies of mutants; with bacteria, three types are particularly useful:

Antibiotic-resistant mutants

Antibiotics are substances produced by living organisms that are toxic to other living organisms. There are about 258 antibiotics known. Alexander Fleming discovered the first antibiotic, penicillin in 1928, while a recent example of a last-line antibiotic is gepotidacin, approved by The Food and Drug Administration (FDA), USA in March 2025. Antibiotic-resistant mutants are able to grow in the presence of an antibiotic, such as streptomycin (Str) or tetracycline (Tet). For example, streptomycin-sensitive (Str^s) cells have the wild type phenotype and fail to form colonies on medium that contains streptomycin, but streptomycin-resistant (Str^r) mutants can form colonies on such medium.

Nutritional mutants

Each bacterial species (or any other microorganisms) has a characteristic minimal medium on which it will grow. A minimal medium contains only inorganic salts, a carbon source for energy, and water. Wild type bacteria can grow on minimal medium and can synthesize all essential nutrients. Such organisms are called prototrophs. Mutant cells are unable to synthesize one or more essential nutrients and cannot grow unless the required nutrients are supplied in the medium. Such a mutant bacterium is said to be an auxotroph for the particular nutrients. For example, a *bio* mutant of *E. coli* will die unless it is supplied with the vitamin biotin, while a *leu* mutant will die without the amino acid leucine.

Carbon-source mutants

Such mutant cannot utilize particular substances as sources of carbon atoms or of energy. For example, Lac⁻ mutants cannot utilize the sugar lactose for growth and are unable to form colonies on minimal medium that contains only lactose as the carbon source.

A medium on which all wild type cells form colonies is called a nonselective medium. If the medium allows growth of only one type of cell (either wild type or mutant), then it is said to be selective medium. For example, a medium containing streptomycin is selective for the Str^r phenotype and selective against the Str^s phenotype; similarly, minimal medium containing lactose as the sole carbon source is selective for Lac$^+$ cells and selective against Lac$^-$ cells.

Table 14.1 Some genotypic and phenotypic symbols used in bacterial genetics

Genotype	Phenotype	Description of phenotype
lac$^-$	Lac$^-$	Cannot utilize lactose as a carbon source
mal$^-$	Mal$^-$	Cannot utilize maltose as a carbon source
gal$^-$	Gal$^-$	Cannot utilize galactose as a carbon source
ara$^-$	Ara$^-$	Cannot utilize arabinose as a carbon source
trp$^-$	Trp$^-$	Unable to make the amino acid tryptophan
pro$^-$	Pro$^-$	Unable to make the amino acid proline
leu$^-$	Leu$^-$	Unable to make the amino acid leucine
met$^-$	Met$^-$	Unable to make the amino acid methionine
bio$^-$	Bio$^-$	Unable to make the amino acid biotin
arg$^-$	Arg$^-$	Unable to make the amino acid arginine
ton^r	Ton^r	Resistant to the phage T_1
ton^s	Ton^s	Infected by the phage T_1
str^r	Str^r	Resistant to the antibiotic streptomycin
str^s	Str^s	Sensitive to the antibiotic streptomycin
tet^r	Tet^r	Resistant to the antibiotic tetracycline
tet^s	Tet^s	Sensitive to the antibiotic tetracycline

In bacterial genetics, phenotype and genotype are designated in the following way. A phenotype is designated by three letters, the first of which is capitalized, with a superscript + to denote the wild type, or a - to denote the mutant; and with *s* or *r* for sensitivity or resistance. A genotype is designated by lowercase italicized letters. Thus, a cell unable to grow without a supplement of leucine (a leucine auxotroph) has a Leu$^-$ phenotype, and *leu*$^-$ genotype. Often the -

superscript is omitted, but using it prevents ambiguity. Table 14.1 lists some often seen bacterial genotypes and phenotypes.

Transfer of Genetic Material in Bacteria (Methods of Genetic Exchange in Bacteria)

Bacteria do not show Mendelian inheritance, because they are haploid and lack meiosis. However, they can undergo genetic recombination, or transfer of genes between individuals. Genetic material can be transferred from organism to organism through vertical transfer, or through horizontal transfer methods. Vertical gene transfer is a process in which genes are passed from parent to its offspring when the cell divides through binary fission. Horizontal gene transfer, also known as lateral gene transfer, is a process in which an organism transfers genetic material to another organism that is not its offspring. Vertical gene transfer uses reproduction as a means of gene transfer through generations, whereas horizontal gene transfer uses non-reproductive methods of gene transfer.

Bacterial genes are usually transferred to members of the same species but occasionally transfer to other species can also occur. There are three mechanisms of horizontal gene transfer in bacteria: transformation, conjugation, and transduction. Transformation involves acquisition of DNA from the environment, conjugation involves acquisition of DNA directly from another bacterium, and transduction involves acquisition of bacterial DNA via a bacteriophage intermediate. In each case: (1) transfer is unidirectional; (2) unlike eukaryotes, no true diploid zygote is formed; and (3) only genes included in the circular chromosome will be inherited stably. The most obvious difference between these three processes is the mode of transfer of DNA from one cell to another. In each process, there is a donor cell and a recipient cell.

Bacterial Transformation

Bacterial cells take up DNA from the environment and integrate it into the chromosome, if homologous regions exist between the chromosomal DNA and the DNA taken up. Transformation is a process in which recipient cells acquire free or naked chromosomal or plasmid DNA from the surrounding medium. Transformation was discovered in *Streptococcus pneumoniae* in 1928 by Frederick Griffith. Transformation was the first experimental proof that DNA is the genetic material. The uptake of naked viral genomic DNA/RNA is termed transfection. Transformation is a naturally occurring but rare event in which DNA can be transferred into bacteria. Transformation can occur in two ways: natural transformation and artificial transformation.

In natural transformation bacteria are naturally able to take up naked DNA from the cell's natural environment. Transformation begins with uptake of a DNA fragment from the surrounding medium by a recipient cell and terminates with one strand of donor DNA replacing the homologous segment in the recipient DNA. During natural transformation, a DNA fragment (usually about 10 genes long) in the surrounding medium is bound to the surface of a competent living recipient bacterium by a protein receptor. After binding, the DNA is transported across the cell membrane by the transformation machinery. Uptake of DNA by Gram-positive and Gram-negative bacteria differs. In Gram positive bacteria the DNA is taken up as a single stranded molecule and the complementary strand is made in the recipient. In contrast, Gram negative bacteria take up double stranded DNA. Depending on the bacterium, either both strands of DNA penetrate the recipient, or a nuclease degrades one strand of the fragment and the remaining DNA strand enters the recipient. This DNA fragment from the donor is then exchanged for a piece of the recipient's DNA by means of RecA proteins. This involves breakage and reunion of paired DNA segments. The replaced segment of host DNA will be degraded. If the DNA taken up is not homologous to genes already present in the cell, the DNA is usually broken down and the nucleotides released are used to synthesize new DNA during normal replication. Transformation usually involves only homologous recombination. Typically, this involves similar bacterial strains or strains of the same bacterial species. Few species such as *Bacillus subtilis, Haemophilus influenzae, Neisseria gonorrhoeae, Neisseria meningitidis, Legionella pneomophila, Streptococcus pneumoniae,* and *Helicobacter pylori* are naturally competent for transformation (able to be transformed). However, these bacteria only take up DNA at a particular time in their growth cycle (log phase) when they produce a specific protein called a competence factor. Only a small proportion of the cells involved in transformation will actually take up DNA. Those recipients whose phenotypes are changed by a recombination event are called transformants. In most species, with suitable recipient cells and excess external DNA, transformation takes place at a frequency of about 1 cell per 1000 cells.

In artificial transformation bacteria have been altered to enable them to take up exogenous DNA. The most common method of artificial transformation of bacteria involves use of divalent cations (*e.g.*, $CaCl_2$) to increase the permeability of the bacterium's membrane, making them chemically competent, thereby increasing the likelihood of DNA acquisition. Another artificial method of transformation is electroporation, in which cells are shocked with an electric current, to create holes in the bacterial membrane. In cloning protocols, artificial transformation is used to introduce recombinant DNA into host bacteria (*E. coli*).

Transformation leads to genetic diversity and increased virulence. Bacterial transformation is generally used to map genes in bacterial species in which mapping by other methods (*i.e.*, conjugation or transduction) is not possible. Transformation is used for cloning or to move DNA molecules around between strains.

Plasmids

The term plasmid was first introduced by the American molecular biologist Joshua Lederberg in 1952. Plasmids are naturally occurring, small, self-replicating, supercoiled, double-stranded DNA molecules present in almost all bacteria that are distinct from, and additional to, the cell's chromosomal DNA (Image 14.1). Usually, they are covalently closed circular molecules; however, they occur as linear molecule in *Borrelia burgdorferi*. They are most commonly found in bacteria; however, plasmids are sometimes present in archaea and eukaryotic organisms. In prokaryotes they are found in *Escherichia coli, Pseudomonas* species, *Agrobacterium* species, *etc.* In eukaryotes they are mainly found in *Saccharomyces cerevisiae*.

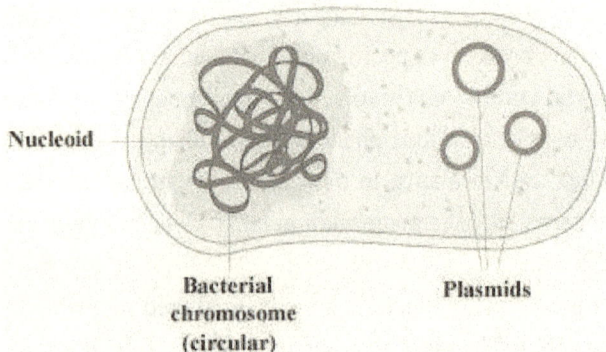

Image 14.1 Plasmid (source: https://www.sciencelearn.org.nz/resources/1900-bacterial-dna-the-role-of-plasmids).

A bacterium can have no plasmids at all, or have many plasmids (10-700), or multiple copies of a plasmid. Although a cell can contain more than one plasmid, they cannot be closely related genetically. In natural populations of bacteria, the amount of plasmid DNA may approach 1 or 2 percent of the total amount of cellular DNA. The distribution of any one plasmid within a species is generally sporadic; some cells have the plasmids whereas others do not. Plasmids range in size from a few kilobases to a few hundred kilobases. Plasmids carry genetic information and replicate independently of the main chromosome to yield daughter plasmids, which pass into the daughter cells at cell division. For this replication to occur, the plasmids must contain genes that control their replication. Plasmids are considered replicons, capable of autonomous replication within a suitable host. All plasmids possess an origin of replication which helps them to multiply within the cell independently of the main bacterial chromosome. A few plasmids, known as integrative plasmids or episomes are able to replicate by inserting themselves into the bacterial chromosome. Plasmids carry genes not found on the bacterial chromosome, which often confer useful properties, such as resistance to antibiotics to the bacteria. In many cases plasmids can move from one bacterium to another, and even between bacteria of different species. The transfer process is directed by genes carried on the plasmids. The ability of plasmids to move between bacteria can be utilized to study transfer of genes from the bacterial chromosome.

Plasmids usually contain between 5 and 100 genes. Plasmids are not essential for normal bacterial growth and bacteria may lose or gain them without harm. Plasmids can be easily separated from host bacteria. Plasmids code for synthesis of a few proteins not encoded by the chromosomal DNA. For example, some exotoxins, such as the tetanus exotoxin and *Escherichia coli* enterotoxin are encoded by plasmids.

Depending on the phenotype many different types of plasmids have been found in bacteria; *e.g.*, fertility (F) plasmids, resistance (R) plasmids, Col plasmids, virulence plasmids, degradative plasmids, and tumor-inducing plasmids. Fertility or F plasmids are conjugative plasmid found in F^+ bacterium with higher frequency of conjugation. F plasmid carries transfer gene (*tra*) and has the ability to form conjugation bridge (F pilus) with F^- bacterium. For example, F plasmid of *E. coli*. Col plasmids have genes that code for colicins, proteins that kill other bacteria. For example, ColE1 of *E. coli*. Degradative plasmids allow the host bacterium to metabolize unusual molecules such as toluene and salicylic acid. For example, TOL of *Pseudomonas putida*. Virulence plasmids confer pathogenicity on the host bacterium. For example, Ti plasmids of *Agrobacterium tumefaciens*, which induce crown gall disease on dicotyledonous plants. Cryptic plasmids do not have any apparent effect on the phenotype of the cell harboring them. They just code for enzymes required for their replication and maintenance in the host cell.

Conjugation

In 1946 Joshua Lederberg and Edward L. Tatum discovered that some bacteria can transfer genetic information to other bacteria through a process known as conjugation. Conjugation is the process by which one bacterium transfers genetic material to another through direct cell-to-cell contact. Joshua

Lederberg was just 33 years old when he won the 1958 Nobel Prize for discovering that bacteria can mate and exchange genes. Conjugation in bacteria is a one-way transfer of genetic information. The most common mechanism for horizontal gene transfer among bacteria is conjugation. While only very small fragments of the bacterial chromosome are transferred in transduction and transformation, in conjugation it is possible for large segments of the chromosome, and in special cases the entire chromosome, to be transferred. Although bacteria can acquire new genes through transformation and transduction, this is usually a rarer transfer among bacteria of the same species or closely related species. During conjugation, one bacterium serves as the donor of the genetic material, and the other serves as the recipient. The recipients receiving donor DNA are called transconjugants. Only certain strains of a bacterium can act as donor cells. The steps of bacterial conjugation are: mating pair formation, conjugal DNA synthesis, DNA transfer, and maturation. Conjugation involves the transfer of plasmids from donor bacterium to recipient bacterium. Some plasmids are designated as F factor (F plasmid, conjugative plasmid, fertility factor, or sex factor). The main structure of the F factor that allows mating pair formation is the F pilus or sex pilus. F plasmid is approximately 100 kb in length and contains many genes for its maintenance in the cell and its transmission between cells. These genes code for the production of a thin, tubelike structure called the sex/conjugation pilus (structure formed between the bacteria through which F can be transferred) and enzymes necessary for conjugation. Cells possessing the F factor are designated F^+ (male) and act as donors. Those cells lacking it are F^- (female) and act as recipient. Each Gram-negative F^+ bacterium has 1 to 3 sex pili that bind to cell-surface receptors on recipient bacteria to initiate mating. Plasmid-mediated conjugation occurs in *Bacillus subtilis, Streptococcus lactis*, and *Enterococcus faecalis* but is not found as commonly in the Gram-positive bacteria as compared to the Gram-negative bacteria. Plasmid transfer in Gram-negative bacteria occurs only between strains of the same species or closely related species.

Conjugation (Image 14.2) begins with physical contact between a donor cell (F^+ or Hfr) and a recipient cell (F^-). Conjugation pilus (plural: pili) on the donor bacterium binds to a recipient bacterium. A series of membrane proteins encoded by the conjugative plasmid then forms a bridge and an opening between the two bacteria, now called a mating pair. One strand of the F^+ plasmid is broken with a nuclease at the origin of transfer (*oriT*) site of the plasmid. The nicked strand enters the recipient bacterium progressing in the 5' to 3' direction. The other strand remains behind in the donor cell. Both the donor and the recipient plasmid strands then make a complementary copy of themselves. During conjugation, no cytoplasm or cell material except DNA passes from donor to recipient. Following successful conjugation, the recipient becomes F^+ and the donor remains F^+.

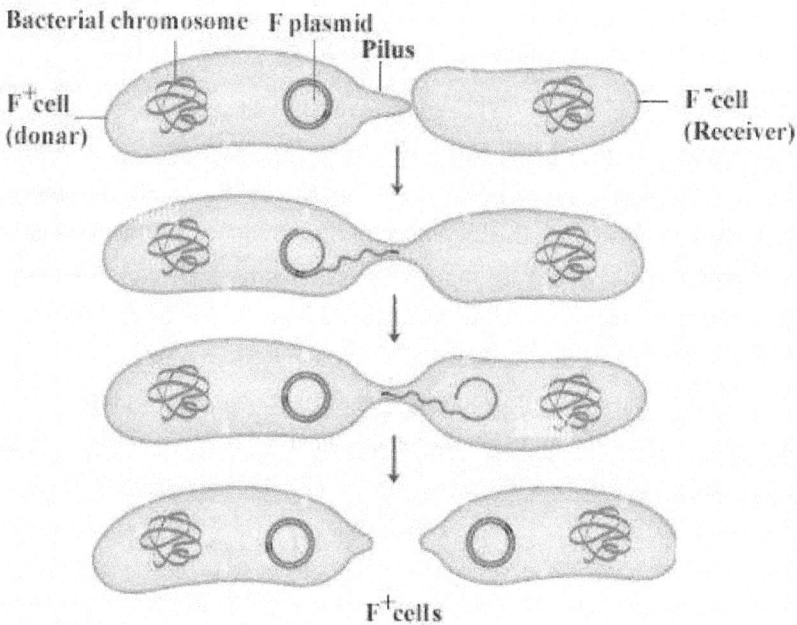

Image 14.2 Diagram showing how an F^- cell acquires an F factor from an F^+ cell.

Among the Gram-negative bacteria this is the major way that bacterial genes are transferred. Transfer can occur between different species of bacteria. Transfer of multiple antibiotic resistant genes by conjugation has become a major problem in the treatment of certain bacterial diseases. Since the recipient cell becomes a donor after transfer of a plasmid, an antibiotic resistant gene carried on a plasmid can quickly convert a sensitive population of cells to a resistant one.

431

Creation of an Hfr Cell

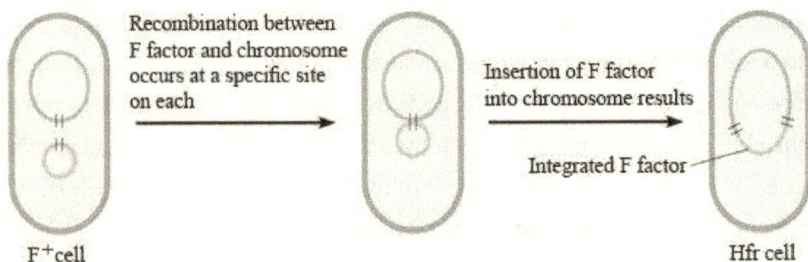

Image 14.3 Creation of an Hfr cell.

Sometimes the F factor will spontaneously integrate itself into the chromosome of an F^+ cell and replicates as part of that chromosome. When an F factor becomes integrated into the chromosome of an F^+ cell, it makes the cell an Hfr cell (Image 14.3). Hfr stands for high frequency of recombination, which refers to the relatively high frequency with which donor genes are transferred to the recipient. It still retains its ability to transfer itself to F^- cells but now carries the rest of chromosome with it. F is an example of episome, a term that refers to any genetic element that can exist either as a plasmid in the cytoplasm, or as an integrated part of a chromosome.

Antimicrobial Resistance (AMR)

Antimicrobial resistance (AMR) is the ability of microorganisms such as bacteria, viruses, fungi, and parasites to resist the action of antimicrobial drugs that were previously effective against them. This phenomenon has become one of the most serious global health challenges because it renders standard treatments ineffective, leading to persistent infections, longer illness, and higher risk of death. AMR arises mainly due to the overuse and misuse of antimicrobials in humans, animals, and agriculture, as well as incomplete treatment courses that allow partially resistant microbes to survive and multiply. In addition, poor infection control practices, lack of hygiene, and the slow pace of new drug development contribute significantly to the problem.

432

Microbes develop resistance through several mechanisms, such as producing enzymes that destroy or inactivate drugs (*e.g.*, β-lactamase in bacteria), altering the drug's target site, reducing cell permeability to drugs, pumping drugs out of the cell through efflux mechanisms, and forming biofilms that protect microbial colonies. Common examples of resistant organisms include *Staphylococcus aureus* (MRSA), multidrug-resistant *Mycobacterium tuberculosis* (MDR-TB), drug-resistant malaria parasites, and extended-spectrum β-lactamase (ESBL)-producing bacteria. The consequences of AMR are severe: treatments become less effective or fail completely, infections last longer, healthcare costs increase, and medical procedures like surgeries, organ transplants, and cancer therapy – which depend on effective antimicrobials – are put at risk. Preventing and controlling AMR requires rational use of antimicrobials, strict infection prevention and control measures, vaccination, improved sanitation, effective surveillance of resistant strains, and public awareness programs. Global cooperation is essential, and initiatives such as the World Health Organization's Global Action Plan on AMR emphasize research, stewardship, and innovation in developing new drugs and alternative therapies.

R Plasmids and Antibiotic Resistance

The F plasmid is not the only plasmid capable of moving between bacteria. Some plasmids can carry multiple antibiotic resistant genes between bacteria by conjugation and are called R plasmids. **R plasmids** play a critical role in the spread of antibiotic resistance among bacteria. These plasmids are capable of transferring between bacterial cells by **conjugation** and often carry multiple genes that confer resistance to antimicrobial substances and even heavy metals, such as chloramphenicol, ampicillin, and mercury. For instance, the **RP4 plasmid**, commonly found in *Pseudomonas* and many other bacteria, provides resistance to several antibiotics while also encoding the machinery for pilus formation. Antibiotics kill sensitive bacteria, but not bacteria with specific R plasmids. A bacterium with an R plasmid for penicillin resistance is able to survive treatment by that antibiotic. Over time, natural selection increases the proportion of resistant bacteria in the population, which poses a major challenge in clinical medicine. The widespread emergence of antibiotic-resistant strains is largely due to the mobility of these plasmids. It is believed that R plasmids originated when **transposons**, carrying resistance genes, became integrated into self-transmissible plasmids. Because of their ability to spread resistance genes rapidly, R plasmids have profound consequences in the treatment of bacterial infections and represent a serious global health concern.

Viruses

Viruses are the simplest forms of life which have sub-microscopic entities. They can only be seen at magnifications provided by the electron microscope. They consist of a single molecule of nucleic acid surrounded by a protective shell or capsid made up of protein molecules and, in some cases, a membranous envelope. However, their size, shape, molecular constituents, and structural complexity vary greatly. Some viruses contain only one kind of protein in their capsid (*e.g.*, the tobacco mosaic virus). Other viruses contain dozens or hundreds of different kinds of proteins.

A virus particle by itself is essentially a package of genes. Viruses exist in two states. Outside the host cells, viruses are simply nonliving particles called virions, which are regular in size, shape, and composition and can be crystallized. Once a virus or its nucleic acid component gains entry into a specific host cell, it seems to come to life, becomes an intracellular parasite. Suddenly the host cell begins making viral proteins. Then the viral genes are replicated and the newly made genes, together with viral coat proteins, assemble into progeny viral particles. Viruses use enzymes, ribosomes, and small molecules of host cells to synthesize progeny viruses. As a result, hundreds of progeny viruses may arise from the single virion that infected the host cell. Because of their behavior as inert particle outside, but lifelike agents inside their hosts, viruses resist classification. Many biologists regard viruses as distinct entities that in some sense are not fully alive. Some scientists refer to them as 'living things' or even organisms.

A different type of response results from some viral infection, in which viral DNA becomes integrated into the host's chromosome and is replicated with the host's own genes. Integrated viral genome have little or no effect on the host's survival, but they often cause profound changes in the host cell's appearance and activity.

All viruses are absolutely dependent on living, functioning cells for reproduction. They are incapable of producing ATP or proteins. Viruses are able to infect all cellular life forms. But each virus is host specific, and some, such as HIV and polio, can be tissue specific as well.

Retroviruses

The retroviruses are a large family of positive single-stranded RNA viruses characterised by encoding the enzyme reverse transcriptase and by their ability to integrate the viral genome into the host genome which replicate their genome through a double-stranded DNA intermediate in the nucleus of the host cell. The unique feature of the retroviruses is their ability to produce double-stranded DNA from their RNA template. It is the reverse (retro) of the normal transcription process. For example, Rous sarcoma virus (RSV), feline leukemia virus, human immunodeficiency virus-1 (HIV-1), HIV-1, HIV-2, Human T-Lymphotropic Virus type I (HTLV-I), HTLV-II and mouse mammary tumor virus (MMTV). This is in contrast to all other RNA viruses that replicate their genomes through double-stranded RNA intermediates almost always in the cytoplasm of host cells. Most retroviruses contain an RNA genome of 9 to 10 kilobases in length. Retroviruses are named for the enzyme reverse transcriptase, which was discovered independently in 1971 by American virologists Howard Temin and David Baltimore. Retroviruses are the cause of various cancers, leukemias, and immunodeficiencies in a wide variety of animals. Some retroviruses, such as RSV and MMTV, are responsible for the induction of cancerous tumors.

Virus replication

Viruses are obligate intracellular pathogens, meaning they are entirely dependent on the host cell's machinery and metabolic processes for replication.

While the details of the viral life cycle vary widely among different virus species and categories, all viruses undergo six fundamental stages that are essential for successful replication (Image 14.4). These stages ensure that the viral genome is delivered into the host, replicated, packaged into new virions, and released to infect other cells.

Attachment: The first stage of viral replication is **attachment**, during which viral proteins on the capsid or phospholipid envelope bind to specific receptors on the surface of the host cell. This interaction is highly specific and determines the **host range** or **tropism** of the virus, meaning which cell types or species the virus can infect.

Penetration: The second stage is **penetration**, in which binding to a specific host receptor triggers conformational changes in the viral capsid proteins or lipid envelope, allowing the viral and cellular membranes to fuse. In some DNA viruses, entry into the host cell occurs instead via **receptor-mediated endocytosis**, enabling the viral genome to gain access to the intracellular environment for subsequent replication.

Uncoating: The third stage is **uncoating**, during which the viral capsid is removed and degraded by either viral or host enzymes. This process releases the viral genomic nucleic acid into the host cell, making it accessible for replication and transcription.

Replication: The fourth stage is **replication**, which begins once the viral genome has been uncoated. During this stage, the viral genome undergoes transcription and translation to produce viral proteins. The exact mechanisms of replication vary significantly between DNA and RNA viruses, as well as between viruses with different nucleic acid polarities. This stage ultimately results in the **de novo synthesis of viral proteins and genomes**, providing the components needed for assembly of new virions.

Assembly: The fifth stage is **assembly**, also called **maturation**, during which newly synthesized viral proteins and replicated genomes are packaged into complete virions. Some viral proteins may undergo post-translational modifications before packaging, ensuring that the new virions are fully functional and ready for release from the host cell.

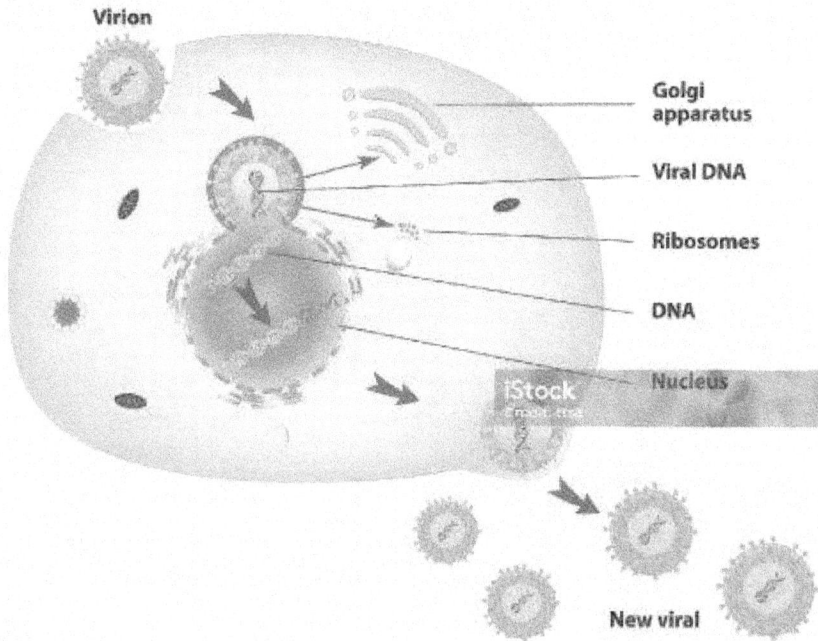

Image 14.4 Virus replication (source: https://www.istockphoto.com/vector/virus-replication-vector-illustration-gm508170355-45813938?searchscope=image%2Cfilm).

Virion release: The sixth and final stage of viral replication is **virion release**, which occurs by either **lysis** or **budding**. In lytic release, the host cell is destroyed, and the virus is termed **cytolytic**; an example of this is *variola major* (smallpox). In contrast, enveloped viruses, such as **influenza A virus**, are typically released by **budding**, a process in which the virus acquires its phospholipid envelope from the host cell membrane. Budding usually does not kill the host cell, and such viruses are called **cytopathic viruses**. After virion release, some viral proteins remain embedded in the host cell membrane, serving as potential targets for circulating antibodies. Additionally, viral proteins retained within the cytoplasm can be processed and presented on **MHC class I molecules**, where they are recognized by T cells, contributing to the host immune response.

Bacteriophages

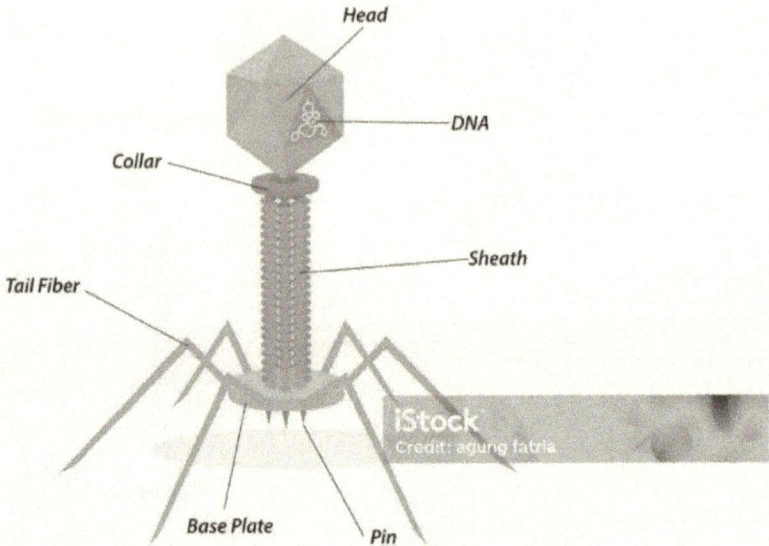

Image 14.5 Bacteriophage (source:
https://www.istockphoto.com/vector/virus-replication-vector-illustration-
gm508170355-45813938?searchscope=image%2Cfilm).

Viruses that specifically infect bacteria and replicate within them are called
bacteriophages or simply **phages** (Image 14.5). Most bacteria are susceptible
to attack by these phages, which show great diversity in their structure.
Typically, a bacteriophage consists of a head, which can vary in size and shape,
and is enclosed by a protein coat called the **capsid**. This capsid is made up of
many copies of one or more proteins and protects the phage's nucleic acid,
which may be either DNA or RNA. The head is connected to a tail structure
through a collar. The **tail**, which functions as a hollow tube, allows the transfer
of nucleic acid into the bacterial cell during infection. The tail length can vary
widely, and some phages do not even have a tail structure. Beyond their role
as bacterial parasites, many bacteriophages are important in **gene transfer**
processes, influencing bacterial evolution and diversity. Their replication can
occur via two major life cycles: the **lytic cycle**, where the phage multiplies
rapidly, lyses, and kills the host bacterium, and the **lysogenic cycle**, where the

phage genome integrates into the bacterial chromosome and remains dormant until conditions trigger lytic growth.

Lytic cycle

During infection, a phage attaches by tail fibers to a susceptible bacterium. Phage lysozyme opens cell wall and the bacteriophage genome enters the bacterium. The phage genetic information then takes over the machinery of the bacterial cell by turning off the synthesis of bacterial components and redirecting the bacterium's metabolic machinery to manufacture bacteriophage components and enzymes. Phage DNA replicates, and is transcribed into RNA, then translated into new phage proteins. New phage particles are assembled and ultimately, many phage descendants are released when the bacterial cell wall breaks open, the bacterium is killed. This breaking-open process is called lysis. All phage species can undergo a lytic cycle. After lysis, the progeny phages infect neighboring bacteria. This is an exponentially explosive phenomenon. Within 15 hours after the start of an experiment of this type, the effects are visible to the naked eye: a clear area, or plaque, is present on the opaque lawn of bacteria on the surface of a plate of solid medium. Depending on the phage genome, such plaques can be large or small, fuzzy or sharp, and so forth. Thus, plaque morphology is a phage character that can be analyzed.

Lysogenic cycle

In lysogenic cycle, the viral DNA is incorporated into a specific region of the bacterial DNA where it is passively replicated as part of the bacterial DNA. In this integrated state the phage DNA is called a prophage, which is in dormant state, and surviving cell is called a lysogen. The bacterium that contains a phage in the prophage state is said to be lysogenic for that phage; the phenomenon of the insertion of a phage DNA into a bacterial DNA is called lysogeny. In this situation the bacterium metabolizes and reproduces normally, the viral DNA being transmitted to each daughter cell through all successive generations. The bacterium is unaffected but is immune to further infection (superinfection) with the same strain of phage. The lysogenic cell can replicate nearly indefinitely without the release of phage progeny. Occasionally with low frequency the prophage loses its dormancy, the viral DNA removes from the host's DNA and converts to the lytic cycle, replicating phages and lysing the bacterium. This phenomenon is called prophage induction, and it is initiated by damage to the bacterial DNA. Later studies showed that a variety of agents, such as ultraviolet light or certain chemicals, could activate the prophage, inducing lysis and infective progeny phage (lysate) release in a large fraction of a population of lysogenic bacteria.

Phages can be categorized into two types: virulent or lytic, and avirulent or temperate. A phage capable only of lytic growth causes lysis and death of the bacterial cell is called virulent phage *e.g.*, T4. Temperate phages usually initiate a lysogenic cycle, in which the phage exists as a prophage within the bacterial DNA and is replicated with it and after some cycle become virulent cause lysis. Temperate phages also caused lysis when the prophage is induced or activated.

Transduction

Transduction is the process by which DNA is transferred from one bacterium to another through a bacteriophage. This process can occur via either the lytic cycle or the lysogenic cycle. It was first discovered in 1956 by Joshua Lederberg and Norton Zinder. Transduction has been found to occur in a variety of prokaryotes, including bacterial genera such as *Desulfovibrio*, *Escherichia*, *Pseudomonas*, *Rhodococcus*, *Rhodobacter*, *Salmonella*, *Staphylococcus*, and *Xanthobacter*, as well as in the archaeon *Methanobacterium thermoautotrophicum*. There are two main types of transduction: generalized and specialized. In generalized transduction, any gene from the bacterial chromosome can be randomly transferred, although the efficiency of this process is generally low. In specialized transduction, only specific genes from restricted regions of the bacterial chromosome are transferred.

Generalized transduction

Generalized transduction occurs during the replication of both lytic and temperate bacteriophages. In this process, the bacterial DNA is fragmented into small pieces. Occasionally, by mistake, a phage capsid assembles around a fragment of bacterial DNA instead of a phage genome. When the bacteriophages are released, a very small proportion of the progeny phages carry bacterial DNA rather than viral DNA. These rare phages, present at frequencies as low as 10^{-6} in a phage stock, are called *transducing phages* or *transducing particles*. They retain the ability to infect bacteria because of their intact phage coat proteins, but since they carry only bacterial DNA, they function merely as vehicles for DNA transfer between bacteria.

When a transducing particle infects another bacterium, the bacterial DNA packaged inside the phage head is injected into the recipient cell, where it may recombine with homologous regions of the host DNA. The bacterium that acquires new genetic material through this process is called a *transductant*. Because any random fragment of the bacterial chromosome can be transferred in this way, the phenomenon is referred to as generalized transduction.

Generalized transduction has been observed in several bacterial genera, including *Staphylococcus*, *Escherichia*, *Salmonella*, and *Pseudomonas*. Interestingly, plasmids, such as the penicillinase plasmid of *Staphylococcus aureus*, can also be transferred from one bacterium to another through this mechanism.

Specialized Transduction

Specialized transduction occurs during the lysogenic life cycle of temperate bacteriophages. In this process, the phage genome is integrated at a specific site on the bacterial chromosome. When a lysogenic prophage loses its dormant state and begins to replicate in the host bacterium, the prophage is usually excised from the host DNA precisely. However, in rare cases – about 1 in 10^6 to 10^7 cells – an excision error occurs. In such cases, a portion of the phage DNA is left in the host DNA and a reciprocal portion of the bacterial DNA is excised with the phage DNA.

Such 'mistakes' during prophage excision are responsible for the formation of *specialized transducing particles*. These particles are defective in that some phage genes have been left behind in the host. As the bacteriophage replicates,

441

the segment of bacterial DNA replicates as part of the phage's genome. Every phage now carries that segment of bacterial DNA. They cannot carry out their full life cycle, but like the transducing particles of the virulent phages they can transfer genes between bacteria. Only host genes located close to the site of prophage insertion can be excised with the phage DNA and packaged in 'phage' particles. Thus, specialized transduction is restricted to the transfer of genes located within a short distance on each side of the prophage attachment site.

Specialized transduction is widely used in genetics to transfer specific bacterial genes and to construct strains with desired genotypes. A well-known example is bacteriophage λ, which infects *Escherichia coli*. Phage λ integrates between the *gal* (galactose metabolism) genes and the *bio* (biotin synthesis) genes in the *E. coli* chromosome, making these genes potential candidates for transfer by specialized transduction.

Viroids

Viroids are plant pathogens composed of short, circular, naked single-stranded RNA molecules that can replicate only within a host cell. Unlike viruses, viroids **do not code for any proteins**. Their RNA is highly resistant to enzymatic degradation due to the absence of free ends and the formation of a compact, stable secondary structure. Replication occurs in the nucleus of infected cells through a **double-stranded RNA intermediate**, using host enzymes. Viroid genomes are extremely small, ranging from **246 to 467 nucleotides**, which is far smaller than the genomes of the smallest viruses capable of independent infection, which are around 2,000 nucleobases. Viroids differ from viruses not only in genome size but also in structure, function, and evolutionary origin, highlighting their unique position among infectious agents.

CHAPTER FIFTEEN   GENETIC ENGINEERING

The term "genetic engineering", also called "genetic modifications", "gene cloning", "DNA cloning", "molecular cloning", "gene splicing", "gene manipulation", "new genetics", or "recombinant DNA technology" is used to describe the process by which the genome of an organism can be altered to get the desired heritable characteristics. This involves the use of laboratory tools to insert, delete, or alter DNA sequence that contain one or more genes of interest. Genetic engineering typically changes an organism in a way that would not occur naturally. With conventional breeding, there is little or no guarantee of obtaining any particular gene combination and can take a long time. Undesirable genes can be transferred along with desirable genes; or, while one desirable gene is gained, another is lost. These problems limit the improvements that breeders can achieve. In contrast, genetic engineering allows the direct transfer of one or more genes of interest, between either closely or distantly related organisms to obtain the desired trait. Recombinant DNA (rDNA) technology is the main pillar of genetic engineering that allows scientists to **combine DNA from different sources** – which can be different species or even different individuals of the same species either within cells or *in vitro* – to create hybrid DNA. This hybrid DNA can then be **introduced into an organism** where it can replicate and sometimes express a new trait. In this process, DNA from one species or individual is isolated and cut at specific sequences using restriction endonucleases. These DNA fragments are then joined with DNA from another source using DNA ligase, forming a recombinant molecule. The resulting rDNA can be inserted into a host organism – such as bacteria, yeast, plants, or animals – where it can replicate and sometimes express new traits or proteins. This technology enables the creation of genes with novel functions or the combination of DNA from different species, such as mouse and human DNA, human and bacterial DNA, viral and bacterial DNA, or even DNA from different humans. The organism, from which the candidate DNA is isolated, is called **donor organism**. The organism which will accept the foreign gene is called **host organism**. The recombinant DNA technology was developed in 1973 by Boyer and Cohen and resulted in the appearance of the biotechnology industry.

Tools of Recombinant DNA Technology

The various biological tools used in recombinant DNA technology are:
1. Restriction enzymes
2. Cloning vectors
3. DNA polymerase
4. Host organism
5. DNA ligases
6. Target DNA
7. Selectable markers

Restriction Enzymes: Molecular Scissors

The possibility for recombinant DNA technology emerged with the discovery of restriction enzymes, also called restriction endonucleases. It is the most widely recognized enzymes used in molecular genetics that cut the long DNA molecule into fragments. These enzymes were discovered in 1970. Swiss microbiologist Werner Arber, and American microbiologists Hamilton Othanel Smith and Daniel Nathans discovered restriction enzymes, for which the three men shared the 1978 Nobel Prize for Physiology or Medicine. A **restriction enzyme (RE)** is a specialized protein that recognizes specific DNA sequences, typically 4- or 6-base long, and cleaves the phosphodiester bonds within those sequences wherever they occur. The DNA sequence that a restriction enzyme recognizes for cleavage is called the **restriction site** (also referred to as the restriction, recognition, cleavage, or target sequence). Importantly, the enzyme recognizes this site **regardless of the source of the**

DNA. The enzyme binds to DNA at the restriction site and makes precise cuts in each strand of the DNA molecule. Unlike most enzymatic, chemical, or physical methods, which break DNA randomly, **restriction enzymes cut DNA into specific, unique fragments in a sequence-specific manner.** The DNA fragment produced by a pair of adjacent cuts in a DNA molecule is called a restriction fragment. Different restriction enzymes recognize different sequences, but the sequences recognized share one property – they are all palindromes (sequences read the same from either direction). For example, the palindromic sequence recognized by the restriction enzyme *Eco*RI is GAATTC (Image 15.1). When *Eco*RI encounters its recognition sequence, it cuts the phosphodiester bond between the G and the first A (G↓A) on both sequences. This causes the two segments of DNA to separate. Each fragment of DNA has a four-base single-stranded 'tail' protruding from one end. Such a single-stranded projection is often called a 'sticky or cohesive end'. 'Sticky' means a restriction fragments single-stranded tail that has a capacity to base pair with a complementary tail of any other DNA fragment or molecule cut by the same restriction enzyme.

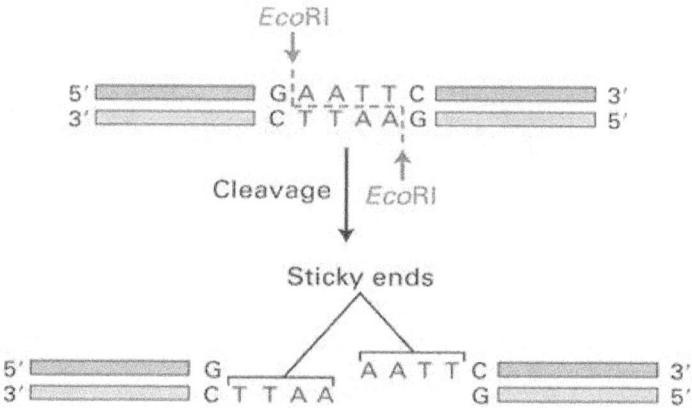

Image 15.1 The palindromic sequence recognized by the restriction enzyme EcoRI.

DNA fragments cut by the same restriction enzyme when mixed together, the sticky ends of any two fragments having the complementary base sequences will base pair and form a recombinant molecule (Image 15.2). Any piece of DNA cut with a certain restriction enzyme will base pair to any other piece

445

cut with that same RE, even if they come from different sources. *Eco*RI makes tails four bases long (AATT). Tails made by *Taq*I are two bases long (CG).

Since each restriction enzyme cuts DNA at an enzyme specific sequence, the number of cuts the enzyme makes in a particular DNA molecule depends on the number of times the particular recognition sequence occurs in the DNA. In some cases, restriction enzymes isolated from different bacteria recognize and cleave the same DNA sequences; these enzymes are called isoschizomers. One of the first restriction enzymes isolated was *Eco*RI. Restriction enzymes are named according to the genus and species of the bacteria they are isolated from. The first letter of the name comes from the genus of the bacterium, the next two letters come from the species, and any subsequent designations are the strain of bacterium and the order in which the enzyme was discovered in that bacterium. For example, the enzyme *Eco*RI was discovered in *Eschericia coli*, strain RY13, and it was the first restriction enzyme identified in that bacterium (hence the designation I).

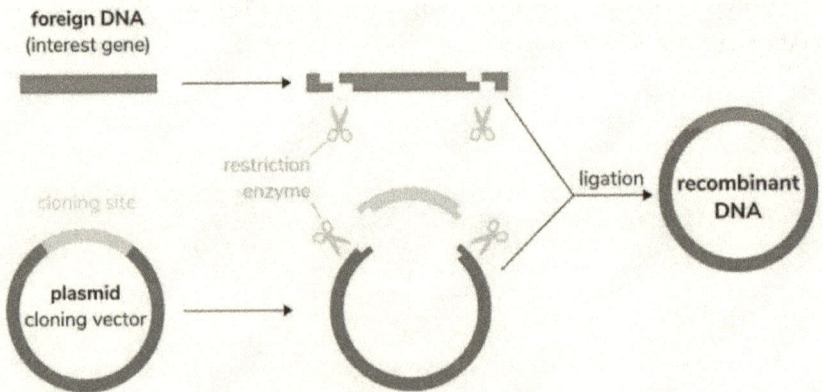

Image 15.2 Recombinant DNA (adapted from: https://stock.adobe.com/ Image ID: 488363444).

There are four different categories of restriction enzymes. Type I restriction enzymes cut DNA at random locations far from their recognition sequence, type III cut outside of their recognition sequence, and type IV typically recognize a modified recognition sequence. Types I, type III, and IV restriction endonucleases are not useful for gene cloning because they cleave DNA at sites other than the recognition sites and thus cause random, unpredictable cleavage patterns. Type II restriction enzymes cut within, or near to, their recognition sequence to give rise to discrete DNA fragments of defined length and sequence. Scientists use them to cut DNA molecules at interesting specific locations and then reattach different DNA sequences to each other using an enzyme called DNA ligase, creating new, recombined DNA sequences, or essentially new DNA molecules. This powerful approach to cutting and pasting DNA molecules is known as DNA cloning or recombinant DNA technology. More than 3500 different types of such enzymes have been characterized. Restriction enzymes are very powerful and useful tools, because they allow researchers to cut large DNA molecules into smaller, more easily manipulated fragments in a very predictable way.

Table 15.1 Some of the restriction sites of some restriction endonucleases

Restriction endonuclease	Recognition site
Eco RI	5'-G*AATTC-3' 3'-CTTAA*G-5'
Bam HI	5'-G*GATCC-3' 3'-CCTAG*G-5'
SAl I	5'-G*TCGAC-3' 3'-CAGCT*C-5'
Hae III	5'-GG*CC-3' 3'-CC*GG-5'
Hind III	5'-A*AGCTT-3' 3'-TTCGA*A-5'
Sma I	5'-CCC*GGG-3' 3'-GGG*CCC-5'

* indicates the cutting site.

Restriction enzymes are naturally found in bacteria and archaea. Bacteria use these enzymes to destroy viruses and foreign DNA. When the bacteriophage injects its DNA, the enzymes cut the DNA and prevent replication of the phage. Bacterium protects its own DNA, not by being free of these restriction sites but usually by methylating its own DNA in these sites. The same sequences that are attacked by the endonucleases in the unmethylated condition are protected when methylated. In *E. coli*, the restriction enzyme *Eco*RI site is protected by the action of the enzyme *Eco*RI methylase which adds a methyl group to the 3'-adenine. Invading viral DNA will not be methylated and can be cut by the restriction enzyme. Thus, it is said that foreign DNA proliferation is restricted in the cell by the restriction enzyme, and bacterial DNA is modified by the methylase to prevent cleavage by the restriction enzyme. Some of the restriction sites of some REs are given in Table 15.1.

Today, researchers rely on restriction enzymes to perform virtually any process that involves manipulating, analyzing, and creating new combinations of DNA sequences. Among the many new combinations are DNA cloning, hereditary disease diagnosis, paternity testing, forensics, genomics, epigenetics, genetically modified organisms, and biotechnology.

Cloning Vectors/Vehicles

The molecular analysis of DNA has been made possible by the cloning of DNA. Gene cloning is the process of creating genetically identical copies of a specific gene. In order to clone a gene, its DNA sequence must be attached to some kind of carrier, also made of DNA, which can take it into the cell. Biologists call these carriers vectors. Vector is the central component of a gene cloning experiment. A vector is an autonomously replicating DNA molecule that carries foreign DNA into a host cell, facilitates stable integration and replication inside the host cell and produces many copies of itself and the foreign DNA. It serves as a taxi for delivering foreign DNA into a bacterium, yeast, or some other cell that starts a 'cloning factory'. If it is used for reproducing the DNA fragment, it is called a cloning vector. If it is used for expressing a certain gene in the DNA fragment, it is called an expression vector. An ideal vector should have the following properties/characteristics:

1. A vactor should be small in size.
2. It should be easily introduced into the host cell.
3. It should be stably maintained inside the host cell.
4. Should have an **origin of replication (*ori*)**, so that the vector DNA can be replicated independently inside the host.
5. Contain at least one **selectable marker** (usually in the form of antibiotic resistance genes or genes for enzymes missing in the host cell) to distinguish host cells that carry vectors from host cells that do not contain a vector. Often, the marker is a gene that provides the bacteria with resistance to a particular antibiotic. If a culture of bacteria is treated with the antibiotic, those with the vector will live, and those without it will die. The result is a pure culture of vector-containing bacteria.
6. Presence of unique **restriction enzyme recognition site**, so that the vector can be cleaved only once, instead of being cut into multiple pieces, which would make the vector useless.
7. Relatively easy to recover from the host cell.

Types of Cloning Vectors

There are many types of vectors, including plasmid, bacteriophage, cosmid, bacterial artificial chromosome (BAC), yeast artificial chromosome (YAC), yeast 2-micron plasmid, retrovirus, baculovirus vector, *etc*. Each type of vector has been specially constructed in the laboratory to possess the features necessary to make it an efficient cloning vector.

Plasmid Vectors

Most recombinant DNA technology involves the insertion of foreign genes into the plasmids. Plasmids have several properties that make them extremely useful as cloning vectors. They exist as single or multiple copies within the bacterium and replicate independently from the bacterial DNA. Plasmids of different sizes and possessing different copy number are present. The complete sequence of many plasmids is known; hence, the precise location of restriction enzyme cleavage sites for inserting the foreign DNA is available. Plasmids are smaller than the host chromosome and are therefore easily separated from the later, and the desired plasmid-inserted DNA can be readily removed by cutting the plasmid with the enzyme specific for the restriction site into which the original piece of DNA was inserted. Plasmids range in size from about 1.0 kb to over 250 kb.

Plasmids encode only few proteins required for their own replication and these proteins encoding genes are located very close to the *ori*. All the other proteins required for replication are provided by the host cell. Thus, only a small region surrounding the *ori* site is required for replication. Other parts of the plasmid can be deleted and foreign sequences can be added to the plasmid without compromising replication.

Minimum Requirements for Plasmids in Recombinant DNA Technology
For plasmids to be useful in recombination techniques, they must possess:

1. **Origin of replication (ori):** Allows the plasmid to replicate independently within a host cell.
2. **Selectable marker:** Provides a way to identify cells that have successfully taken up the plasmid (e.g., antibiotic resistance genes).
3. **Restriction enzyme sites:** Specific sequences where DNA can be cut and inserted.

Additional Desirable Features of Plasmids
Several properties make plasmids highly convenient and versatile tools in genetic engineering:

1. **Small size:** Typically 1,000-20,000 base pairs, making them easier to isolate, manipulate, and introduce into host cells.
2. **Multiple restriction enzyme sites:** Offers flexibility in cloning strategies and allows for **directional cloning**.
3. **Multiple ORIs:** When inserting two genes into the same plasmid, different ORIs are needed to ensure proper replication.
4. **Stability:** Plasmids remain stable over time, both as purified DNA and when stored in bacteria (e.g., as glycerol stocks).
5. **Broad functionality:** Many plasmids are engineered to drive gene expression in diverse organisms, ranging from bacteria to plants, animals, and even cultured human cells.

The plasmids are most convenient for cloning relatively small fragments of DNA (5-10kb).

Phages

Phases often have double-stranded, linear DNA molecules into which foreign DNA can be inserted at several restriction sites without disrupting its life cycle. The chimeric DNA is collected after the phage proceeds through its lytic cycle and produces mature, infective phage particles. A major advantage of phage vector is that they can accept DNA fragments of 10 to 20 kb long. Amongst various bacteriophages available such as λ, T4, T5, and T7 phages; the λ phage gained favourable attention due to its unique life cycle.

Lambda (λ) Phage

Bacteriophage lambda (λ) has been widely used in recombinant DNA since engineering of the first viral cloning vector in 1974. Phage λ vectors are particularly useful for preparing genomic libraries, because they can hold a larger piece of DNA than a plasmid vector. Bacteriophage λ contains ~49kb of DNA and has a very efficient mechanism for delivering its genome into a bacterium. One-third λ genome is nonessential and could be replaced with foreign DNA. The λ genome is linear, but in the ends has 12-nucleotides overhangs, termed as *cos* sites, which are complementary to each other. A λ cloning vector can be circularized using *cos* site which can be manipulated and replicated inside *E. coli* via the process of transfection. Bacteriophage (λ) has the ability to infect *E. coli* and can integrate up to 15-16 kilobases of the DNA segment (Nicholl, 2008).

Cosmids

A **cosmid** is an artificial hybrid vector that combines features of both plasmids and bacteriophage λ. It is essentially a plasmid carrying a λ *cos* site (cohesive end site), which enables the recombinant DNA to be packaged into a phage head for efficient delivery into host cells. Unlike phages, however, cosmids replicate as plasmids once inside the host and do not lyse the bacterial cell. The main advantage of cosmids is their ability to clone large DNA fragments, up to about 50 kilobases, which makes them particularly useful for cloning large eukaryotic genes that cannot be maintained in ordinary plasmid or phage vectors. Structurally, cosmids contain *cos* sites that allow circularization of DNA after infection, a plasmid origin of replication to ensure independent replication in the host, and a selectable marker such as an antibiotic resistance gene for identifying transformed cells. Since much of the unnecessary λ DNA has been removed, more foreign DNA can be inserted, making cosmids highly efficient tools. In essence, cosmids combine the large insert capacity of phage vectors with the replication stability of plasmids, providing a versatile system for genetic engineering and molecular cloning.

Bacterial artificial chromosomes (BACs)

Bacterial artificial chromosomes (BACs) are artificially constructed circular DNA molecules derived from the **F plasmid** of bacteria, designed to accommodate very large DNA fragments – up to about **350 kilobases**. They were developed to overcome the limited insert size of plasmids, allowing researchers to clone much larger segments of DNA. BAC vectors contain a **single-copy F-plasmid origin of replication (ori)**, which ensures stable maintenance of large inserts within the host cell. The natural F plasmid had already been studied in detail and was shown to carry up to a million base pairs of DNA, making it an ideal backbone for vector development. In 1992, Hiroaki Shizuya and colleagues engineered a streamlined version of the F plasmid by retaining only the essential components, producing an efficient cloning vector. BACs are equipped with all the necessary features of a functional vector, including a replication origin, selectable antibiotic resistance genes, and multiple cloning sites for DNA insertion. Their ability to stably maintain large inserts makes them invaluable for studying **large genes, multiple genes simultaneously, or even entire viral genomes**, and they played a key role in large-scale genome projects such as the Human Genome Project.

Yeast artificial chromosomes (YACs)

Yeast artificial chromosomes (YACs) are specially constructed linear DNA molecules designed to function like yeast chromosomes and to accommodate very large DNA fragments – up to **1 million base pairs**. Although yeast is a eukaryote, it is a small, single-celled organism that can be manipulated and grown in the laboratory much like bacteria, making it an excellent model system. YACs combine components of both yeast and bacterial plasmids. Normally, plasmids introduced into yeast are often lost from cells, but this limitation has been overcome by constructing plasmids containing a **yeast centromere (CEN)** and a **yeast origin of replication (ARS, autonomously replicating sequence)**, ensuring stable inheritance from one generation to the next. To create a YAC, plasmids can be linearized and supplied with **telomeric sequences**, allowing them to mimic natural yeast chromosomes. As a result, YACs are capable of carrying extremely large DNA fragments, far exceeding the capacity of plasmids or cosmids. They have been particularly valuable in **cloning very large segments of DNA**, such as those from the **human genome**, and have greatly advanced recombinant DNA studies in eukaryotic gene regulation.

Yeast artificial chromosomes (YACs) are engineered with several essential features that allow them to function like natural yeast chromosomes while carrying very large DNA fragments. Each YAC contains a **yeast telomere at both ends**, which seals the chromosome ends and provides stability. A **yeast centromere (CEN)** ensures proper segregation of the artificial chromosome into daughter cells during cell division. To facilitate identification, each arm of the YAC carries a **selectable marker**, allowing detection and maintenance of the plasmid in yeast. In addition, YACs contain an **origin of replication (ARS, autonomously replicating sequence)**, enabling independent replication within the yeast cell. Finally, they are equipped with **unique restriction sites** that provide convenient points for the insertion of foreign DNA. Together, these features make YACs powerful vectors for cloning and studying very large DNA fragments, including those from complex eukaryotic genomes.

Retroviral vectors

Retroviral vectors are engineered from retroviruses, which naturally integrate their genetic material into the host cell's DNA and are stably propagated as the host cell divides. This property makes them highly useful tools in genetic engineering and gene therapy, as they can deliver and permanently integrate a **foreign gene (transgene)** into the host genome. Retroviral vectors can typically accommodate **7-8 kilobases** of foreign DNA, and once integrated, the transgene is efficiently expressed and inherited by daughter cells during cell division.

Adenoviral vectors (AdVs)

Adenoviral vectors (AdVs) are non-enveloped, double-stranded DNA viruses that have been widely engineered as tools for gene transfer. They are particularly valuable in research because they can be **easily produced in large**

quantities and are capable of driving **very high levels of transgene expression** in host cells, both *in vitro* and in small animal models. Adenoviral vectors can typically accommodate up to **10 kilobases of foreign DNA**. To make them safe and suitable for use as vectors, key **structural genes such as** *gag*, *pol*, **and** *env* – which in retroviruses encode proteins necessary for viral particle assembly – are deleted. This modification prevents the production of infectious viral particles while still allowing efficient delivery and expression of the desired transgene.

Choice of vector depends on nature of protocol or experiment and type of host cell (prokaryotic/eukaryotic) to accommodate recombinant DNA.

Expression of Foreign DNA in Eukaryotic Cells

Foreign DNA can be introduced into eukaryotic cells in methods similar to bacterial transformation, but it is called transfection. Eukaryotic organisms that take up foreign DNA are referred to as transgenic. The term transformation in eukaryotes refers to the changing of a normal cell into a rapidly growing cancerous one.

DNA Extraction, Purification and Quantification

DNA extraction is the process of isolating deoxyribonucleic acid (DNA) from the cells or viruses in which it naturally occurs. The first successful DNA isolation was carried out in 1869 by Friedrich Miescher. Today, DNA extraction has become a routine technique in molecular biology laboratories and forensic analyses. The process involves breaking open cells to release DNA and then separating it from proteins and other cellular components. The main purpose of DNA extraction is to obtain DNA in a relatively purified form, which can then be used for downstream applications such as polymerase chain reaction (PCR), sequencing, cloning, or genetic analysis. The procedure typically involves three fundamental steps: **lysis** (breaking the cell membrane to release DNA), **precipitation** (separating DNA from proteins and other cellular debris), and **purification** (cleaning the DNA to obtain it in usable form).

Step 1: Lysis

The first step of DNA extraction is breaking open the cells or viral particles to release the DNA. This process can be achieved by physical methods such as sonication, bead beating, or vortexing, sometimes combined with heating. Chemical agents are also used: detergents like sodium dodecyl sulfate (SDS) disrupt lipid membranes, while phenol treatment helps dissolve proteinaceous cell walls or viral capsids. To remove proteins bound to DNA, proteases are added to degrade them. Protein precipitation can be further aided by salts such as ammonium acetate or sodium acetate. In the phenol-chloroform extraction method, centrifugation separates the sample into phases: proteins remain in the organic phase, while DNA is located at the aqueous-organic interface, from which it can be carefully recovered..

Step 2: Precipitation

After lysis, the DNA is released but remains mixed with fragmented cellular material and proteins. To separate it from this debris, the DNA is precipitated using cold ethanol or isopropanol. Because DNA is insoluble in alcohol, it comes out of solution and forms visible strands or a pellet upon centrifugation. The DNA pellet is then washed with cold alcohol, which helps remove residual salts and other impurities introduced during the lysis step. Finally, centrifugation allows the DNA pellet to be collected for further processing.

Step 3: Purification

Once the DNA has been separated and collected as a pellet, it undergoes purification to remove any remaining contaminants. The pellet is rinsed with alcohol to wash away residual proteins, salts, and cellular debris. After drying, the purified DNA is re-suspended in a suitable buffer, commonly Tris or TE (Tris-EDTA), which stabilizes the DNA and makes it ready for downstream applications such as PCR, sequencing, or cloning.

Step 4: DNA Quantification

After extraction and purification, the concentration and quality of DNA can be assessed using several methods. One common approach is measuring UV absorbance, typically at 260 nm, which provides an estimate of DNA concentration based on optical density. Agarose gel electrophoresis can also be used to visualize DNA and estimate its quantity and integrity. Additionally, fluorescent DNA-binding dyes offer sensitive detection and quantification, often with minimal sample requirements. Each method varies in terms of equipment needed, ease of use, and the type of calculations required, but all are effective for evaluating the extracted DNA before downstream applications.

Host Organism

In recombinant DNA (rDNA) technology, the host organism is the final recipient of the engineered DNA and serves as the tool for expressing the desired phenotype. It is a crucial component of rDNA experiments, as it takes up the vector carrying the inserted gene with the help of specific enzymes and then replicates or expresses the recombinant DNA. The choice of host organism determines the efficiency of gene expression, replication, and production of the target protein or trait.

Steps in Recombinant DNA Technology

Inserting a desired gene into the genome of a host organism is a complex process that requires careful planning and execution. Recombinant DNA technology involves multiple sequential steps to achieve the intended outcome. First, the gene of interest must be identified and isolated. Next, an appropriate vector is selected, into which the gene is integrated to form recombinant DNA. This recombinant DNA is then introduced into the host organism. Finally, the recombinant DNA must be maintained within the host and properly replicated or passed on to its offspring to ensure stable expression of the desired trait.

Isolation of Desired DNA (Gene of Interest)

The first step in recombinant DNA (rDNA) technology is to identify and isolate the gene of interest in its purest form, free from other cellular macromolecules. Within cells, DNA exists alongside RNA, proteins, polysaccharides, and lipids, all of which must be removed to obtain pure DNA. This is achieved using specific enzymes: lysozymes break down bacterial cell walls, cellulases digest plant cell walls, and chitinases degrade fungal cell walls. Ribonucleases remove RNA, while proteases eliminate proteins such as histones that are associated with DNA. Additional treatments remove other cellular contaminants, and finally, the purified DNA is precipitated using chilled ethanol. The DNA forms fine threads that can be spooled out, yielding high-quality DNA ready for further manipulations. The source of the DNA may be bacterial, plant, or animal cells, depending on the gene of interest.

Restriction Enzyme Digestions

Once the desired DNA has been isolated and purified, it is treated with selected restriction enzymes that recognize and cut DNA at specific sequences. These enzymes generate DNA fragments with "sticky ends," which facilitate later recombination. The digestion is performed under conditions optimized for the particular enzyme being used. To ensure compatibility, both the purified DNA fragment and the chosen vector are cut with the same restriction enzyme, producing complementary sticky ends. As a result, the DNA fragment and the vector are both "opened" and ready for the next step – ligation.

Isolation of Desired DNA Fragment

After restriction enzyme digestion, the DNA fragments can be visualized and separated using agarose gel electrophoresis. Because DNA molecules carry a negative charge, they migrate toward the positive electrode (anode) when an electric field is applied. Smaller fragments move faster than larger ones, allowing separation based on size. The fragment of interest is then carefully excised from the gel for further use. The vector DNA undergoes the same procedure to ensure that both the DNA fragment and the vector are properly prepared for subsequent ligation.

Amplification of Gene of Interest using Polymerase Chain Reaction

When the number of copies of a specific gene is insufficient for cloning or further manipulation, the gene can be amplified using the polymerase chain reaction (PCR). This technique rapidly produces millions of copies of the target DNA sequence, providing an adequate amount for subsequent recombinant DNA experiments. PCR is highly specific, efficient, and widely used in molecular biology to ensure that even small amounts of DNA can be reliably multiplied for downstream applications.

Ligation (Joining/Splicing) of DNA Molecules with a Suitable Vector

In recombinant DNA technology, plasmids are the most commonly used vectors. To create a recombinant DNA molecule, the vector DNA, the foreign DNA fragment, and the enzyme DNA ligase are combined at appropriate concentrations. The sticky ends of the DNA fragment and the vector align through complementary base pairing, forming temporary hydrogen bonds. However, these hydrogen bonds are not strong enough to maintain the connection permanently. DNA ligase catalyzes the formation of covalent phosphodiester bonds between the DNA backbones, permanently joining the fragments. The resulting recombinant DNA molecule – also called a recombinant plasmid, hybrid plasmid, chimera, or hybrid vector – is now ready for either in vivo gene transfer (gene cloning in a host organism) or further amplification via PCR.

Insertion of rDNA into the Host Cell

Once a recombinant DNA (rDNA) molecule has been constructed in vitro, it must be introduced into host cells to enable gene cloning – the process of producing many identical copies of a foreign gene (insert) within a living organism. *Escherichia coli* is commonly used as a host for general cloning purposes. Since bacterial cells do not naturally take up foreign DNA efficiently, they are first treated to become "competent." The recombinant DNA can then be delivered into host cells using various methods, including transformation, transduction, transfection, electroporation, microinjection, biolistics (gene gun), thermal shock, alternated cooling and heating, calcium ion treatment, virus-mediated transfer, liposome-mediated transfer, silicon carbide fiber-mediated transfer, polyethylene glycol-mediated transfer, or dextran-mediated transfer. Once inside the host, the recombinant DNA can replicate and, if designed properly, express the desired gene.

Screening of Transformed Host Cells

The process of transformation produces a mixed population of host cells – some that have successfully taken up the recombinant DNA and others that have not. Only the transformed cells, which carry the recombinant gene, are able to pass it on to their offspring. To identify these cells, a **selection procedure** is used, often involving a selectable marker such as an antibiotic resistance gene carried by the vector. For example, if host cells are grown on a medium containing ampicillin, only cells transformed with an ampicillin-resistance gene will survive, while non-transformed cells die. In this case, the antibiotic resistance gene functions as the selectable marker. Additional methods for screening transformed cells include immunological assays, nucleic acid hybridization, blue-white screening, and insertional inactivation. To confirm that the cells have incorporated the desired transgene, **transgenic markers** are also used and tested using specific screening techniques.

Validation of Recombinant DNA Integration

After introducing recombinant DNA into host cells, it is essential to confirm that the DNA has integrated correctly and remains intact. Various techniques are employed for this purpose, including polymerase chain reaction (PCR), DNA sequencing, restriction fragment analysis, nucleic acid hybridization, and immunological assays. Methods such as blue-white screening can also be used to verify successful insertion. These approaches allow researchers to ensure that the recombinant DNA is present, correctly oriented, and capable of proper expression within the host genome.

Multiplication or Expression of the Introduced Gene in the Host

The primary goal of recombinant DNA technology is either to produce multiple copies of a specific gene or to synthesize the corresponding gene product within a host organism. Once the recombinant DNA is successfully introduced into the host cell, it replicates as the host multiplies in culture, generating numerous copies of the foreign DNA. Under optimal conditions, the gene is also transcribed and translated, producing the desired protein – now called a **recombinant protein**. This approach enables bacteria like *Escherichia coli* to produce medically and commercially important proteins such as human growth hormone (somatotropin), bovine somatotropin (BST), human insulin, and many others.

Obtaining Foreign Gene Product

Small-scale cultures of host cells typically produce only limited amounts of recombinant protein. To generate sufficient quantities for practical or commercial use, large-scale production is required. This is achieved using **bioreactors**, which are specialized vessels designed for continuous culture. In a bioreactor, fresh growth medium is supplied continuously while spent medium is removed, allowing cells to grow and produce proteins efficiently.

Bioreactors can process volumes ranging from 100 to 1,000 liters or more, providing optimal conditions such as temperature, oxygen levels, pH, and nutrients. Under these controlled conditions, host cells convert raw materials into the desired recombinant proteins, enzymes, or other gene products, enabling large-scale production for medical, industrial, or research applications.

Downstream Processing

Before a recombinant protein can be marketed as a final product, it undergoes downstream processing, which ensures its purity, safety, and efficacy. This process includes the separation and purification of the protein from host cells and culture medium, formulation with suitable preservatives or stabilizers, and rigorous quality control tests. Additionally, clinical trials are conducted to evaluate the protein's safety and effectiveness in humans. Downstream processing is a critical step in translating recombinant DNA technology into commercially and medically useful products.

Application of Recombinant DNA Technology

Recombinant DNA technology (RDT) has revolutionized biological research and continues to influence many aspects of everyday life. It is widely applied across medicine, agriculture, industry, and environmental science. Molecular cloning, as a central tool of RDT, enables the creation of recombinant cells and organisms with applications in both basic and applied research.

1. Production of recombinant proteins:

Genes encoding proteins of diagnostic, therapeutic, or commercial value can be cloned, expressed, and purified in large quantities. Examples include **insulin** for diabetes, **growth hormone**, **interferon** for immune disorders, **plasma protein α1-antitrypsin**, and **blood-clotting factors VIII and IX**. Recombinant proteins are also used in laboratory experiments and to generate antibody probes for studying protein synthesis.

2. Gene therapy:

RDT is used to treat genetic disorders by repairing or replacing defective genes. This approach shows promise for diseases such as **cystic fibrosis, muscular dystrophy, and certain cancers**.

3. Recombinant vaccines:

Recombinant DNA methods have enabled the development of vaccines for diseases such as **hepatitis B, herpes, and rabies**, using vectors like bacteria, yeast, and viruses.

4. Immunotherapy and diagnostics:

RDT contributes to advanced immunotherapies, such as **T-cell therapy**, which enhances the immune system's ability to target cancer cells. It is also widely used in diagnostics, for example in **ELISA tests**.

5. Agricultural advancements:

Recombinant DNA has transformed agriculture through the development of **genetically modified (GM) crops** with improved traits such as pest resistance, drought tolerance, and enhanced nutrition. Examples include **Flavr Savr tomatoes, golden rice, and Bt-cotton**.

6. Bioremediation and environmental protection:

Engineered microbes such as **bacteria, yeast, and fungi** can be used to clean up pollutants. For instance, *Escherichia coli* strain JM109 has the ability to remove **mercury** from contaminated soil or water.

7. Targeted drug delivery:

RDT enables the design of **drug delivery systems** that direct therapeutic agents to specific cells or tissues, increasing treatment effectiveness and reducing side effects.

8. Industrial applications:

Recombinant microbes are used for the large-scale production of **enzymes, biofuels, food products, and fine chemicals**. Examples include enzymes for **cheese production, sugar processing, and biofuel generation**.

9. Gene identification and functional studies:

RDT allows scientists to **identify, map, and sequence genes**, and to study their functions in development, physiology, and disease.

Genetically Modified Organisms (GMO)

A genetically modified organism (GMO) is a living creature (plant, animal, or microbe) whose genome has been altered using genetic engineering techniques in order to change its function or behavior. Genetic engineering typically changes an organism in a way that would not occur naturally. Any living organism that has had its DNA engineered by humans is a GMO. GMOs are taken as one of the revolutionary technological advancements in the field of biology (genetics/biotechnology). Recently, a wide variety of plants including crops, flowers, and other medicinal and decorative plant species, animals, and microorganisms are genetically modified. Foods produced from or using GM organisms are often referred to as GM foods. Some common examples of foods that are genetically modified include cantaloupe, corn, papaya, potatoes, soybeans, strawberries, tomatoes, and much more. The first GMO food produced was a GM tomato in 1983 and GMO crops were first approved by The Food and Drug Administration (FDA), USA in 1994. GMOs may be used for a variety of purposes, such as making human insulin, producing fermented beverages and developing pesticide resistance in crop plants. Pigs, chickens, cows, fish, and mosquitoes are some of the genetically modified animals produced by scientists. Today, approximately 90 percent of the corn, soybeans, and sugar beets on the market are GMOs. Genetically engineered crops produce higher yields, have a longer shelf life, are resistant to diseases and pests, and even taste better.

The first GMO was produced in 1973 and was a bacteria species resistant to the kanamycin antibiotic and the first GM animal was a mouse, which was created the following year. The field has grown by leaps and bounds in the nearly half century since then. With the advent of clustered regularly interspaced short palindromic repeats **(CRISPR)**, the field of genetic engineering was revolutionized, simplifying the creation of GMOs. While it is relatively easy to modify the genetics of bacteria, the process becomes more complex and difficult with larger animals, such as humans.

Process of Development of GMOs

Genetically modified organisms (GMOs) are created by introducing **foreign genes** that confer valuable traits into the genome of a target organism. This process allows specific genes to be transferred not only between closely related species but also across unrelated species. The gene of interest is first identified and prepared using gene-editing technologies such as **recombinant DNA technology, transcription activator-like effector nucleases (TALENs), zinc-finger nucleases (ZFNs),** and the widely used **CRISPR-Cas9 system**. Once prepared, the gene is incorporated into vectors such as **bacterial plasmids, cosmids,** or **yeast artificial chromosomes (YACs)**. These vectors are then introduced into host cells using methods such as **heat shock, electroporation, viral delivery, gene gun (particle bombardment), microinjection,** or **liposome-mediated transfer**.

The development of GMOs is a **multi-step process** that includes: selecting a trait of interest, isolating the gene responsible for that trait, tagging it with regulatory elements such as promoters, markers, and terminators, and finally integrating it into the host genome.

There are several strategies for transferring genes into target organisms:

1. **Direct transfer:** DNA fragments are directly delivered into gametes, seeds, or embryos. Methods include electroporation, heat shock, microinjection, or particle bombardment. Although relatively simple, this approach often has a low efficiency.

2. **Microinjection:** Using a fine glass micropipette, the gene of interest is directly injected into the nucleus or cytoplasm of a living cell.

3. **Particle bombardment (gene gun):** DNA-coated microscopic particles (usually tungsten or gold) are shot into cells under high pressure, enabling DNA entry even through rigid cell walls.

4. **Vector-mediated transfer:** Bacteria (*e.g.,* *Agrobacterium tumefaciens*) or viruses act as natural carriers to introduce the desired gene into the host genome.

5. **Direct genome editing:** Modern tools like **CRISPR-Cas9, TALENs,** and **ZFNs** allow scientists to precisely edit or modify the organism's own DNA without necessarily introducing a vector.

Through these techniques, GMOs are developed to express traits such as pest resistance, drought tolerance, improved nutritional content, or production of medically valuable proteins.

Transgenic Organism and GMO

A **transgenic organism** is a type of genetically modified organism (GMO) whose genome has been deliberately altered through the introduction of one or more genes from another species using genetic engineering techniques. Transgenic organisms are typically produced in the laboratory for research purposes, and genes can be **added, enhanced, or deleted** within a species, between different species, or even across kingdoms. While the terms "GMO" and "transgenic" are often used interchangeably, they are not identical: a **GMO** refers broadly to any organism whose DNA has been modified, without specifying the source of the genetic change, whereas a **transgenic organism** specifically contains DNA from a sexually incompatible organism. In other words, **all transgenic organisms are GMOs, but not all GMOs are transgenic**. Examples of transgenic animals include **transgenic fish, chickens, rabbits, cows, sheep,** and **mice**, which have been engineered for research, agricultural, or biomedical purposes.

Gene knockout or Gene Replacement

A gene knockout is a genetic technique in which a specific gene is deliberately mutated to permanently stop its expression. This method is one of the most powerful tools for studying gene function and for permanently altering the phenotypic characteristics **of cells or organisms**. Gene knockouts can be achieved in a wide variety of cells and organisms using different genetic approaches, but the fastest and most precise method currently available **is** CRISPR-Cas9 genome editing, which allows targeted disruption of specific genes with high efficiency.

PCR Primers

PCR primers are the key ingredient that makes one PCR different from another. PCR primers are short, single-stranded DNA molecules of about 18 to 25 nucleotides that are designed to be complementary to each of the ends of the target sequence that will be amplified. A standard PCR uses two primers, often called the "forward" and "reverse" primers. They are manufactured commercially and can be ordered to match any DNA sequence. DNA polymerase requires primers to initiate replication. In order to amplify a specific DNA sequence, the sequence of the primer annealing sites on the target DNA must be known. The forward and reverse primers are oriented on opposite strands of the DNA. After hybridization to a complementary region of the target DNA, the primers provide the 3′hydroxyl ends by which DNA polymerase mediated DNA synthesis proceeds. The DNA fragments to be amplified should not be greater than about 3 kb in length and ideally will be less than 1 kb. The length of the primer influences the rate at which it hybridizes to the template DNA, longer primers hybridizing at a slower rate.

Taq DNA Polymerase

For PCR, the enzyme of choice is *Taq* DNA polymerase extracted from the hyperthermophilic bacteria *Thermus aquaticus*, isolated first at a hot spring in Yellowstone National Park in 1976. It has been found even in water heaters. This enzyme is heat-tolerant (thermostable). *Taq* polymerase has its optimum enzymatic activity at 75°C to 80°C, commonly a temperature of 72°C is used with this enzyme, and substantially reduced activities at lower temperatures. The enzyme is not destroyed at the elevated temperatures required to unwind a DNA double helix. Most DNA polymerases are denatured and permanently lose their activity at such temperatures. This enzyme lacks the 3′ → 5′ exonuclease activity (proofreading ability) resulting in relatively low replication fidelity.

Polymerase Chain Reaction: the *in vitro* Version of DNA Replication

Polymerase chain reaction (PCR) was discovered by Dr. Kary B. Mullis in 1983. PCR has revolutionized the field of molecular biology to such an extent that its creator, Kary B. Mullis, was awarded the Nobel Prize in Chemistry in 1993. PCR is an indispensable tool for *in vitro* amplification (and thus isolation) of a defined DNA molecule with a specific DNA sequence. With this technique a target sequence of DNA can be amplified a billion-fold in several hours. The only requirement is that the sequence of nucleotides on either side of the sequence of interest be known. That information is needed to construct primers on either side of the sequence of interest. Once that is done, the sequence between the primers can be amplified. PCR has become a very powerful tool in molecular biology. PCR provides a sensitive, reliable, efficient, selective, convenient, and extremely rapid means of amplifying any desired sequence of DNA. PCR copies a DNA molecule without restriction enzymes, vectors, or host cells. It is faster and easier than conventional cloning. DNA to be amplified by PCR does not have to be purified. It can be used to amplify whatever DNA is present, however small in quantity or poor in quality. DNA sequence as short as 50-100 bp and as long as 10 kb can be amplified. One can start with a single cell, hair follicle, or spermatozoon and amplify the DNA sufficiently to allow for DNA analysis and a distinctive band on an agarose gel.

The following components are needed to perform PCR in the laboratory:

1. Target DNA: Contains the sequence to be amplified.
2. Two primers: Complementary to the ends of amplified region, forward and reverse.
3. A thermostable DNA polymerase (like *Taq* DNA polymerase or Vent polymerase): Enzyme that catalyses the reaction.
4. Water: The medium for all other components.
5. Buffer solution (Basic components: 20 mM Tris-HCl p^H 8.4, 50 mM KCl, 1.5 mM MgCl$_2$): Maintains p^H and ionic strength of the

reaction solution and provides a suitable chemical environment for the DNA polymerase.

6. All four dNTPs (deoxyribonucleoside triphosphates): DNA building blocks – dATP, dGTP, dCTP, dTTP, from which the DNA polymerase builds the new DNA. All four must be mixed at equal ratios. Always use balanced solutions of all four dNTPs to minimize the error rate. The final concentration of each dNTP should be between 50 and 500 μM, the most commonly used concentration is 200 μM.

7. Thermocycler or PCR machine: The device has a thermal block with holes where tubes with the PCR reaction mixtures can be inserted. This instrument has the capability of rapidly switching between the different temperatures in discrete, pre-programmed steps. Thus, the reactions can be set up, placed in the thermal cycler and the technician can return several hours later and obtain the product and proceed from that point.

8. Mg^{2+} ions: Mg^{2+} is required as an essential co-factor for thermo-stable DNA polymerase, *e.g.*, *Taq* polymerase. Its concentration can have significant effects on the outcome of the PCR. Since Mg^{2+} ions form complexes with dNTPs, primers and DNA templates, the optimal concentration of $MgCl_2$ has to be selected for each experiment. The concentration of free Mg^{2+} ions depends on the concentrations of compounds like dNTPs, free pyrophosphates (PPi) and EDTA (*e.g.*, from TE buffer). These compounds bind to the ions via their negative charges. Therefore, the concentration of Mg^{2+} should always be higher than the concentration of these compounds. The most optimal concentration should be determined empirically and may vary from 1 mM to 5 mM. The most commonly used $MgCl_2$ concentration is 1.5 mM, with a dNTP concentration of 200 μM each. Increase concentration of Mg^{2+} when increasing the concentration of dNTPs. Excess Mg^{2+} in the reaction can increase non-specific primer binding and increase the yield of non-specific products and promote misincorporation. Too little Mg^{2+} in the reaction can result in a lower yield of the desired product. Increases in dNTP

concentrations reduce the concentration of free Mg^{2+} thereby interfering with the activity of the polymerase enzyme.

The DNA, DNA polymerase, buffer, dNTPs, and primers in vast excess are placed in a thin-walled tube and then these tubes are placed in the PCR thermal cycler.

Cycling Profile for standard PCR

The PCR process consists of a series of twenty to thirty-five cycles. Each consists of three steps. The three main steps of PCR are:

Step 1: Denaturation

Initial denaturation

It is very important to denature the template DNA completely by initial heating of the PCR mixture. Normally, heating for 2 minutes at 94°-95°C is enough to denature complex genomic DNA. *Taq*-polymerase is also activated by this step.

Denaturation during cycling

Denaturation at 94°-95°C for 20 to 30 seconds is usually sufficient but must be adapted for the thermal cycler and tubes being used.

If the denaturation temperature is too low, the incompletely melted DNA "snaps back", preventing efficient primer annealing and extension or leading to "self-priming" which can lead to false positive results.

Use a longer denaturation time or higher denaturation temperature for GC rich template DNA.

Never use a longer denaturation time than absolutely required, unnecessary long denaturation times decreases the activity of the polymerase.

Step 2: Primer annealing

The temperature is then lowered to 40°C-65°C for twenty seconds. The lower temperature promotes base-pairing between primers and the ends of the template strands. The two primers bind to their complementary sequences on single-stranded DNA. This step is called annealing. The temperature for this step varies depending on the size of the primer. Primers are chosen such that one is complementary to the one strand at one end of the target sequence and that the other is complementary to the other strand at the other end of the target sequence. The annealing temperature can affect the specificity of the reaction. DNA-DNA hybridization is temperature-dependent, if the temperature is too high, then no hybridization/annealing takes place, instead the primers and templates remain dissociated. If the temperature is too low, non-specific annealing will increase dramatically.

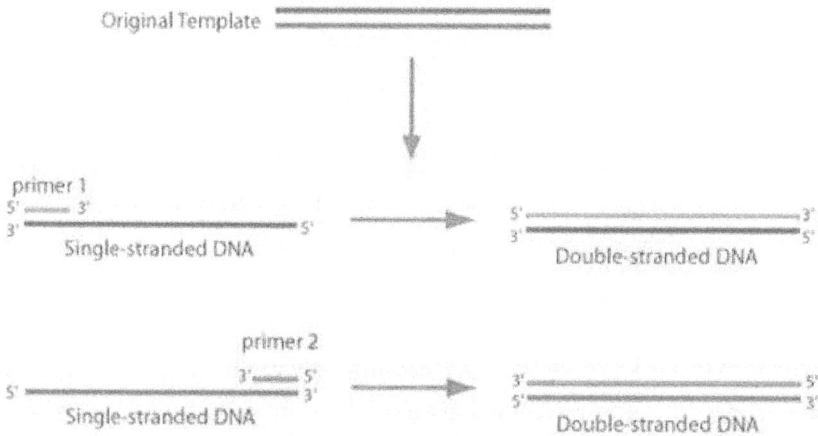

Step 3: DNA Synthesis

The temperature is then raised to 72°C for twenty seconds. *Taq* DNA polymerase extends the primer by its polymerase activity in the 5' to 3' direction, attaching the appropriate dNTPs that are complementary to the template. This results in the synthesis of new DNA strands complementary to the parent template strands. The extension time depends both on the DNA polymerase used and on the length of the DNA fragment to be amplified. As a rule-of-thumb, at its optimum temperature, the DNA polymerase will polymerize a thousand bases per minute.

These three steps make up one PCR cycle. The next cycle will begin by denaturing the new DNA strands formed in the previous cycle. Under optimum conditions, *i.e.*, if there are no limitations due to limiting substrates or reagents, at each extension step, the amount of target DNA is doubled, leading to exponential (geometric) amplification of the specific DNA fragment. As long as the primers are in excess, they will bind preferentially to the opposite DNA template strand. Then DNA polymerase can extend the primers and synthesize new DNA. Twenty cycles provide an amplification of 10^6 (*i.e.*, 2^{20}) and 30 cycles, 10^9 (2^{30}). Each cycle takes 5-10 minutes. In an optimal reaction less than 10 template molecules can be amplified in less than

40 cycles to a product detectable by gel electrophoresis. Most PCRs should only include 25-35 cycles. As cycle number increases, non-specific products can accumulate.

The basis of PCR is temperature changes and the effect that these temperature changes have on the DNA. The various stages in the cycle are controlled by changes in temperature since the temperatures for denaturation, primer annealing, and DNA replication are different. The entire cycling process of PCR is automated and can be completed in just a few hours. It is directed by a machine called a thermocycler, which is programmed to alter the temperature of the reaction every few minutes to allow DNA denaturing and synthesis.

Final extension

Usually, after the last cycle, the reaction tubes are held at 72°C for 5-15 minutes to promote completion of partial extension products and annealing of single-stranded complementary products. After the final extension, immediately put the samples on ice or store at 12°C.

At the end of a PCR experiment, a sample of the reaction mixture is analysed by agarose gel electrophoresis.

Types of PCR

- ❖ Real-time PCR
- ❖ Nested PCR
- ❖ Inverse PCR
- ❖ Reverse transcription PCR
- ❖ Asymmetric PCR
- ❖ Multiplex PCR

Reverse Transcription PCR (RT-PCR)

Polymerase chain reaction (PCR) amplification of complementary DNA (cDNA) can also be used to detect specific transcripts in RNA sample. In RT-PCR, reverse transcriptase is used to reverse transcribes all of the mRNAs in

RNA sample into cDNA. This single stranded cDNA can then be amplified by PCR using primers that anneal to a specific sequence. The amplified DNA fragments that are produced can be analyzed by agarose gel electrophoresis. The amount of amplified fragment produced is proportional to the amount of target mRNA in the original RNA sample. RT-PCR is extremely sensitive and is widely used in expression profiling, to determine the expression of a gene or to identify the sequence of an RNA transcript, including transcription start and termination sites. If the genomic DNA sequence of a gene is known, RT-PCR can be used to map the location of exons and introns in the gene.

Applications of PCR

Polymerase chain reaction (PCR) is a powerful technique that amplifies specific regions of DNA, allowing analysis even from extremely small samples. This makes PCR invaluable in **forensic science**, where only trace amounts of DNA may be available, as well as in the study of **ancient DNA** from specimens tens of thousands of years old. PCR is also widely used in **quantitative analyses**, such as estimating the amount of a specific DNA sequence in a sample. Techniques like **real-time PCR** measure DNA accumulation after each cycle, enabling precise quantification of gene expression levels.

PCR can also amplify **RNA molecules** once they are converted into complementary DNA (cDNA) using **reverse transcriptase**. This **RT-PCR** approach is particularly useful for measuring mRNA levels in different tissues or at different times, reflecting the activity of the corresponding genes.

In medicine, PCR enables the **early diagnosis of malignancies**, such as leukemia and lymphomas, by detecting translocation-specific malignant cells with sensitivity up to **10,000-fold higher** than conventional methods. PCR is also extensively used for **virus detection**, allowing identification soon after infection, often before symptoms appear. Quantitative PCR can measure **viral load** in patients, which is critical for monitoring disease progression and treatment.

PCR can compare **genomic differences** between organisms using techniques such as **random amplified polymorphic DNA (RAPD)**, where closely related organisms produce more similar banding patterns than distantly related ones.

Overall, PCR is indispensable in many laboratory applications, including:

- ❖ Mapping techniques used in the Human Genome Project.
- ❖ Genetic fingerprinting for forensic and paternity testing.
- ❖ Diagnosis of diseases and genetic disorders.
- ❖ Detection of bacteria and viruses in environmental samples.
- ❖ Analysis of microbial communities.
- ❖ DNA-based phylogeny and functional gene analysis.

DNA/Gene Library, Chromosome Library

DNA library or gene library is simply the collection of DNA fragments cloned into vectors and stored within host organisms so that researchers can identify and isolate the DNA fragments that interest them for further study. They contain either the entire genome of a particular organism or the genes that are expressed at a given time. There are basically two kinds of libraries: genomic DNA library and cDNA library.

Genomic DNA Library: A genomic library is the collection of entire genomic DNA fragments of an organism cloned in a suitable vector. A genomic library is prepared from the total DNA of a cell line or tissue. It contains all the genes and gene-related sequences and intergenic DNA sequences. Construction of a genomic DNA library starts with isolation and purification of genomic DNA. The genomic DNA is digested with a restriction enzyme resulting in DNA fragments of a specific size. The resulting DNA fragments are cloned into suitable vectors. These recombinant molecules are further transferred into the host cells to create a library. This library contains representative copies of all DNA fragments present within the genome. No library is perfect. However, a perfect genomic library would contain all of the DNA sequences in the entire genome.

Genomic libraries are suitable for a wide range of applications, including genome mapping and comparative genomics. It allows the study of regulatory elements and noncoding sequences that are important in gene expression and regulation. Genomic libraries also have certain disadvantages including the complexity and size of handling large DNA fragments and the resource-intensive process of creating and maintaining such libraries.

cDNA Library: A cDNA (complementary DNA) library is a collection of cDNA molecules derived from mRNA molecules in a tissue. Unlike genomic libraries, cDNA libraries represent only the coding region of expressed genes of an organism only, excluding non-expressed genomic regions such as introns and other noncoding sequences. cDNA libraries generally contain much smaller fragments than genomic DNA libraries, and are usually cloned into plasmid vectors. cDNA libraries are specifically created from eukaryotes to study the expressed genes. Prokaryotes do not contain introns. Therefore, creating cDNA libraries for prokaryotes is generally not necessary, as their genomic DNA directly corresponds to their mRNA.

cDNA libraries are useful for studying gene expression, protein functions, and producing recombinant proteins. Since cDNA libraries exclude noncoding regions, they provide a more focused view of the expressed genetic information.

The disadvantages of the cDNA library include the limitation of studying gene regulation due to the absence of regulatory sequences, limited gene diversity, and bias toward highly expressed genes.

Chromosome libraries: The collections of clones of fragments of individual chromosomes.

Analysis of Cloned DNA Sequences

After DNA has been cloned and amplified, it must be analyzed and characterized to confirm its identity and integrity. Two common methods for this purpose are **Southern blotting** and **DNA sequencing**, both of which rely on gel electrophoresis for initial separation of DNA fragments. Gel electrophoresis allows researchers to resolve DNA fragments by size, providing a foundation for further analysis, while Southern blotting and sequencing provide detailed information about the presence, structure, and exact nucleotide sequence of the cloned DNA.

Gel Electrophoresis

Gel electrophoresis is a widely used laboratory technique for separating charged biomolecules – such as DNA, RNA, and proteins – based on their size and electric charge. Molecules move through a gel matrix at speeds determined by their charge, size, shape, and the voltage applied. Agarose, a gelatinous substance derived from seaweed, is the most commonly used gel for DNA separation. Agarose gel electrophoresis effectively separates DNA fragments ranging from 100 base pairs (bp) to 25 kilobases (kb). Typically, gels range from 0.7% to 2% agarose: lower concentrations resolve larger DNA fragments, while higher concentrations resolve smaller fragments.

To perform agarose gel electrophoresis, an agarose-TAE solution is prepared and heated to dissolve the agarose. The solution is poured into a casting tray with wells, where it solidifies into a gel slab. The gel is then placed in an electrophoresis chamber filled with TAE buffer, with wells positioned near the negative electrode. DNA samples, often mixed with a tracking dye and a DNA ladder (a reference with known fragment sizes), are loaded into the wells. When an electric field (typically 50-150 V) is applied, negatively charged DNA migrates toward the positive electrode. Smaller DNA fragments move faster through the gel pores, while larger fragments migrate more slowly, resulting in size-based separation.

After electrophoresis, the gel is stained with a fluorescent dye, such as ethidium bromide or GelRed. Ethidium bromide intercalates between DNA bases and fluoresces under UV light, while GelRed, a safer alternative, also binds DNA and emits visible orange light under UV illumination. The DNA bands, visualized against the DNA ladder, indicate fragment sizes and allow for identification and analysis. Gel electrophoresis is commonly used to visualize DNA after PCR, enzymatic digestion, or other experimental treatments, providing a clear picture of molecular size and integrity.

Molecular/Genetic Marker

A **genetic marker** is a known DNA sequence within a genome that can be used to identify a specific location or locus. These markers represent positions in the DNA where variations occur, meaning multiple alternative sequences may exist at that site. Effective genetic markers are easily identifiable, linked to a specific locus, highly polymorphic, and inherited according to standard Mendelian principles. **Molecular markers** function as "tags" or "flags" on the DNA, enabling researchers to locate specific genes or regions of interest. They are valuable tools for examining genetic variation at the DNA level and have important applications in animal and plant breeding, as well as in mapping genes associated with desirable traits.

Advantage of Molecular Markers

Molecular markers provide a precise representation of an organism's genetic makeup at the DNA level. One major advantage is that they allow researchers to test for specific traits at very early stages, such as the embryonic stage in animals, eliminating the need to wait for the trait to become observable—a process that can sometimes take years. By using markers linked to genes of interest, breeding programs can shift from **phenotype-based** to **genotype-based** selection. This is particularly valuable for quantitative traits, where traditional selection relies solely on observable characteristics without knowing which genes are being targeted.

Molecular markers have a wide range of applications. They can be used to assess genetic diversity and relationships within and between populations, study biological processes such as mating systems, pollen movement, or seed dispersal in plants, and identify specific genotypes. Overall, they represent a powerful tool for improving breeding efficiency, genetic analysis, and understanding complex biological systems.

Gene *versus* Marker

Molecular markers should not be considered normal genes, as they usually do not have any direct biological effect. Instead, they serve as constant landmarks within the genome. In contrast, a gene of interest is directly responsible for producing specific proteins that determine particular phenotypes.

Markers, while genetically linked to genes, do not influence the trait itself. In many cases, genes responsible for a trait can be identified and directly assayed with a high degree of confidence. However, when a gene has not yet been isolated, molecular markers can be used to tag and track the gene of interest, facilitating studies of inheritance and selection.

Types of Molecular Markers

Different kinds of molecular markers exist. Molecular markers can be broadly classified into three categories, based on the chronological order of their development:

1. **First-generation markers** – These are based on hybridization techniques. A classic example is **Restriction Fragment Length Polymorphisms (RFLPs).**

2. **Second-generation markers** – These rely on **PCR-based methods** and include:
   - ❖ **Random Amplified Polymorphic DNA (RAPDs)**
   - ❖ **Amplified Fragment Length Polymorphisms (AFLPs)**
   - ❖ **Simple Sequence Repeats (SSRs or microsatellites)**
   - ❖ **Sequence Characterized Amplified Regions (SCARs)**

3. **Third-generation markers** – These are based on **DNA sequencing** and are the most recent development. The most common type is **Single Nucleotide Polymorphisms (SNPs).** Their detection requires sequence information. With advances in sequencing technology, SNPs have become increasingly popular in recent years.

The utility of molecular markers to breeders varies with the type of marker system used, as each has its own **advantages and disadvantages** depending on the context.

Restriction Fragment Length Polymorphisms (RFLPs)

Restriction Fragment Length Polymorphisms (RFLPs) were the first molecular markers used to assess genetic variability in organisms, though they are less commonly used today. RFLPs were invented in **1984 by the English scientist Alec Jeffreys** and are used to analyze unique patterns in DNA fragments generated by restriction enzymes, enabling genetic differentiation among individuals.

Principle: RFLPs detect DNA polymorphisms by exploiting differences in restriction sites – specific sequences in DNA recognized and cut by restriction enzymes. Variations in these sites between individuals result in DNA fragments of different lengths. Therefore, the distance between cleavage sites for a particular restriction enzyme differs among individuals or species, producing variable fragment lengths.

Procedure:

1. DNA is extracted from samples such as blood or saliva and purified.
2. The purified DNA is digested with specific restriction enzymes, generating thousands of DNA fragments of varying lengths.
3. These fragments are separated by **agarose gel electrophoresis** based on size.
4. The fragments are then detected using **Southern blotting** and visualized via **autoradiography** on X-ray film.

RFLPs were historically widely used in **Animal Genetic Resources (AnGR) assessments** and breeding program development because of their reliability and reproducibility, although newer marker technologies have largely replaced them due to efficiency and cost considerations.

Random Amplified Polymorphic DNA (RAPD)

Random Amplified Polymorphic DNA (RAPD) was first demonstrated by **Welsh and McClelland in 1990**. RAPDs are DNA fragments amplified by **PCR** using short, randomly generated synthetic primers, typically about **10 nucleotides long**.

Principle: Because the primers are short, PCR is performed at a **low annealing temperature (35-40°C)** to allow them to bind. However, this low temperature results in **non-specific binding**. These random oligonucleotides serve as both forward and reverse primers, amplifying multiple genomic fragments simultaneously. The amplified fragments, which are unique to each genotype, can be used to detect polymorphisms. Fragment sizes typically range from **500 bp to 5 kb**.

Procedure:

1. Genomic DNA is amplified by PCR using arbitrary 10 bp primers.

2. Because primers anneal at random sites, the number and size of PCR products vary among individuals, reflecting polymorphism.
3. The resulting **RAPD-PCR fingerprint patterns** are separated by **agarose gel electrophoresis** to visualize the banding profiles.

Applications: RAPD has been widely used to define genetic diversity among species. For example, RAPD fingerprints were used to generate unique genetic profiles for **ten animal species**, including wild boar, pig, horse, buffalo, cattle, dog, cat, rabbit, and kangaroo.

Advantages of RAPD:
1. No prior sequence information is needed to design primers.
2. Requires only a small amount of DNA, since PCR amplifies the template.
3. Simple, quick, and cost-effective compared to RFLP.

Disadvantages of RAPD:
1. **Poor reproducibility** – RAPD polymorphic profiles are not always consistent.
2. **Non-specific binding** of primers often leads to unreliable results.

Amplified Fragment Length Polymorphism (AFLP)

Amplified Fragment Length Polymorphism (AFLP) was first described by Vos *et al.* in 1995. It is considered a hybrid technique that combines features of both RFLP (digestion of DNA with restriction enzymes) and RAPD (amplification of DNA fragments by PCR).

Principle: AFLP detects genetic polymorphisms by generating DNA fragments through restriction digestion, attaching oligonucleotide adaptors to the fragment ends, and then selectively amplifying them using PCR primers complementary to the adaptors.

Procedure:
1. Digestion – Genomic DNA is digested with two different restriction enzymes.
2. Adaptor Ligation – Synthetic double-stranded adaptors (with known sequences) are ligated to the sticky ends of restriction fragments.
3. Selective Amplification – PCR is carried out using primers complementary to the adaptor sequences, often extended with selective nucleotides to reduce the number of amplified fragments.

4. Separation & Detection – Amplified fragments are separated (*e.g.*, by gel electrophoresis or capillary electrophoresis) and visualized, commonly through autoradiography or fluorescent labeling.

Advantages of AFLP:

- ❖ Not labor-intensive compared to RFLP.
- ❖ Faster and more efficient in generating large numbers of markers.
- ❖ Highly reproducible and reliable.
- ❖ Requires no prior sequence information.
- ❖ Ideal for studying population genetics, genome mapping, and genetic diversity.

Applications: AFLP has been widely used in:

- ❖ Detecting genetic polymorphisms.
- ❖ Evaluating and characterizing animal genetic resources.
- ❖ Genome typing and population structure analysis.

Compared to RFLP (slow, requires large DNA amounts) and RAPD (fast but less reproducible), AFLP is a balanced method: relatively fast, highly reproducible, and powerful for genome-wide studies.

Microsatellite Markers

Microsatellites, also known as **Simple Sequence Repeats (SSRs)**, are short tandemly repeated DNA sequences (1-6 base pairs in length) that are distributed throughout the genome. They have been widely used as molecular markers due to their **high level of polymorphism** and reproducibility.

Applications:

- ❖ Extensively used in **genetic diversity studies** and **population structure analysis**.
- ❖ Important in **mapping quantitative trait loci (QTLs)** in livestock.
- ❖ Applied in **marker-assisted selection (MAS)** programs.
- ❖ Widely utilized in **disease diagnosis** and **forensic investigations** due to their high discriminatory power.

Advantages of SSRs:

- ❖ **Highly polymorphic** compared to RAPD and AFLP, making them more informative.
- ❖ **Stable and reproducible**, with higher repeatability than RAPDs.
- ❖ Do not require the use of radioisotopes.

Disadvantages of SSRs:

- ❖ **Time-consuming** and **expensive** to develop.
- ❖ Requires prior sequence information for primer design.

Principle and Procedure:

1. Microsatellite loci are identified, often through sequence data.
2. **Primers** are designed to flank the repeat regions.
3. PCR is performed to amplify the microsatellite regions, producing alleles of different lengths depending on the number of repeats.
4. The amplified products are separated (usually by gel or capillary electrophoresis) to reveal **variation in repeat numbers**, which serves as the genetic marker.

Because of their **high variability and codominant inheritance**, microsatellites have been among the most widely used markers in the past decades for **population genetics, breeding programs, and QTL mapping** in both plants and animals.

Compared to **RAPD and AFLP**, microsatellites are **more polymorphic, more reliable, and more useful for long-term studies**, though they come at a higher cost and require more effort to develop.

Single Nucleotide Polymorphisms (SNPs)

A Single Nucleotide Polymorphism (SNP), pronounced *"snip"*, is the most common type of genetic variation. It occurs when a single nucleotide (A, T, C, or G) at a specific position in the genome differs among individuals in a population or between paired chromosomes in an individual. For a variation to be classified as an SNP, it must occur in more than 1% of the population.

Characteristics:

- ❖ SNPs may be located in coding regions, non-coding regions, or intergenic regions.
- ❖ If a SNP is within a gene, the gene may have multiple alleles.
- ❖ SNPs are the most abundant type of molecular marker in all organisms.
- ❖ In humans, 99.9% of DNA is identical, and most genetic differences are due to SNPs.

Biological and Practical Importance:

- ❖ SNPs explain much of the phenotypic variation between individuals.

- They are useful for parentage testing and diagnosis of genetic disorders.
- Most SNPs occur outside coding regions (since only 3-5% of the genome codes for proteins), but those within or near genes of interest are especially valuable, as they may be directly associated with specific traits.
- SNPs are highly informative for genetic diversity studies, selection, and disease association research.

Example: The cattle genome has approximately 3 billion nucleotides and scientists have identified around 40 million SNPs, many of which are associated with performance traits.

Advances in Technology: With improvements in whole-genome and gene sequencing, SNP detection has become faster and more accurate. Sequencing-based approaches allow researchers to capture a complete view of genetic variation, making SNPs one of the most powerful tools for characterizing genetic diversity and for use in marker-assisted selection.

Compared to older marker systems (RFLP, RAPD, AFLP, SSR), SNPs are far more abundant, stable, and directly linked to traits of interest, making them the preferred molecular markers in modern genomics and breeding programs.

Molecular Blotting Techniques

Identification of specific sequences of DNA, RNA, and proteins is important for various studies in molecular biology. Blotting is a molecular technique used to identify specific DNA, RNA, or protein sequences by transferring them from an electrophoresis gel onto a solid membrane, usually nitrocellulose or nylon, where they can be probed and visualized. The procedure involves three general steps. First, the mixture of molecules is separated by gel electrophoresis: DNA samples are run on agarose gels for Southern blotting, RNA on agarose gels for Northern blotting, and proteins on polyacrylamide gels for Western blotting. Second, the separated molecules are transferred from the gel onto a membrane that binds nucleic acids or proteins. The transfer can be achieved either by diffusion or by electrophoresis, and the membrane may be treated to covalently link the molecules to its surface.

Finally, the molecules of interest are detected by adding a probe or antibody that specifically binds to the target. For DNA or RNA, the probe is usually a complementary nucleic acid sequence labeled with a radioactive, fluorescent, or dye tag. For proteins, detection relies on antibodies, typically with a secondary antibody conjugated to an enzyme that produces a colorimetric or chemiluminescent signal in the presence of a suitable substrate.

Bands of DNA appear in an electrophoretic gel only if most DNA fragments are of the same size, such as after PCR amplification or plasmid digestion. However, digestion of chromosomal DNA produces a smear of fragments of many different sizes. In such cases, additional methods are needed to detect a specific DNA sequence within the smear, and this is accomplished using a Southern blot, in which a complementary labeled probe hybridizes to the sequence of interest. In summary, Southern blotting is used to detect DNA, Northern blotting to study RNA transcripts and gene expression, and Western blotting to analyze specific proteins.

Southern Blotting

Southern blotting is one of the central techniques in molecular biology and was the first to make a major impact on clinical molecular pathology. It was developed in 1975 by the British molecular biologist Professor Sir Edwin Mellor Southern at Edinburgh University, Scotland, and named after him. The method is used to determine the identity, size, and abundance of specific DNA sequences within a complex mixture of DNA. For example, it can be employed to locate a particular gene within an entire genome. In this process, genomic or other DNA samples are digested with appropriate restriction enzymes and separated by size through agarose gel electrophoresis. The resulting double-stranded DNA fragments are then denatured into single strands by treatment with a strong base such as NaOH. Following this, the DNA fragments are transferred from the gel onto a nitrocellulose or nylon membrane by capillary action, which immobilizes them while preserving their relative positions.

Once transferred, the single-stranded DNA fragments on the membrane are hybridized with a labeled single-stranded probe that is complementary to the sequence of interest. This probe may be radioactively, fluorescently, or chromogenically labeled, allowing it to bind specifically to its target DNA fragment and enable detection. After washing away unbound probes, the labeled hybridized DNA is visualized – by autoradiography using X-ray film in the case of radioactive or fluorescent probes, or by color development on the membrane in the case of chromogenic probes – producing a characteristic banding pattern, or DNA fingerprint. By running different DNA samples side by side on the same gel, the resulting fingerprints can be compared to identify shared or unique sequences.

Southern blotting has been widely applied in DNA fingerprinting, criminal investigations, victim identification, and paternity testing. It is also used to detect the presence of a DNA sequence within a mixture, to determine the size of restriction fragments, and to analyze fragments larger than those typically amplified by PCR. Although the invention of PCR has largely replaced Southern blotting in many applications due to its simplicity and speed, the Southern blot remains an important foundational technique. It also paved the way for the development of related blotting methods such as **Northern blotting** for RNA and **Western blotting** for proteins, both named in reference to Southern's original method.

Northern Blotting

Northern blotting is a laboratory method used to detect specific RNA molecules within a mixture of RNAs. The technique was developed by James Alwine and his colleagues in 1979 and is widely applied in the study of gene expression. Unlike DNA, RNA is already single-stranded and does not require restriction enzymes for fragmentation. Instead, the RNA is first denatured to remove any secondary structures and ensure that the strands remain unfolded. The denatured RNA molecules are then separated by size using agarose gel electrophoresis, where they often appear as streaks rather than distinct bands because of the presence of many small RNA fragments.

Following electrophoresis, the separated RNA fragments are transferred by capillary action from the agarose gel onto a nylon membrane or diazobenzyl oxy methyl (DBM) paper, which immobilizes the RNA. The membrane is then hybridized with a labeled, single-stranded DNA probe that is complementary to the RNA sequence of interest. This probe, which is tagged with either a radioactive isotope or a fluorescent dye, binds specifically to the target RNA, forming a DNA-RNA hybrid. After washing away unbound probes, the labeled hybrid allows the RNA molecule of interest to be detected among the many RNA species on the membrane.

Northern blotting is particularly useful for analyzing RNA from specific tissues or cell types in order to measure gene expression levels. It can also be employed in the diagnosis of diseases by detecting abnormal RNA transcripts or changes in gene expression patterns.

Western Blotting (Protein Immunoblotting)

Western blotting, also known as protein immunoblotting, is a widely used analytical technique for detecting a specific protein from among a complex mixture of proteins in blood, tissue, or cell extracts. It can also be used to evaluate the size of a protein of interest and measure its level of expression. The method was developed by George Stark's group at Stanford University in 1979 and has since become an essential tool in molecular biology, diagnostics, and medical research.

The process begins with preparation of the protein sample, which is treated with sodium dodecyl sulfate (SDS), a detergent that denatures proteins by unfolding them into linear chains and imparting a uniform negative charge. The proteins are then separated according to size using SDS-PAGE (sodium dodecyl sulfate-polyacrylamide gel electrophoresis) or, in some cases, native PAGE. Following separation, the proteins are transferred from the gel onto a blotting membrane, usually made of nitrocellulose (NC) or polyvinylidene difluoride (PVDF), where they become immobilized. To prevent nonspecific binding, the membrane is subjected to a blocking step before being incubated with a primary antibody that specifically recognizes the target protein, forming a protein-antibody complex. After unbound primary antibodies are washed away, the membrane is incubated with a secondary antibody, which binds specifically to the primary antibody. This secondary antibody is conjugated to a reporter enzyme that produces a detectable signal, either through color formation or light emission, thereby revealing the position of the target protein.

Detection can be performed in several ways. If the antibody is radioactively labeled, the protein band can be visualized by autoradiography on X-ray film. More commonly, enzyme-linked secondary antibodies are used in conjunction with chromogenic substrates, which generate visible colored bands at the antigen site. Alternatively, chemiluminescent substrates may be employed, producing light for highly sensitive detection.

Western blotting is a versatile method with numerous applications, including quantifying protein expression, detecting defective proteins, identifying bacterial or viral proteins in serum, and confirming the presence of antibodies to pathogens such as HIV. Because of its sensitivity and specificity, it remains a cornerstone technique in both research and clinical diagnostics.

DNA Sequencing

DNA sequencing is the process of determining the exact order of nucleotides in a DNA molecule. It reveals the precise arrangement of the four bases – adenine, guanine, cytosine, and thymine – within a strand of DNA. The nucleotide sequence is the most fundamental level of genetic information, serving as the blueprint that contains the instructions for building and maintaining an organism. Understanding genetic function, variation, and evolution would be incomplete without knowledge of DNA sequences. Sequencing an entire genome is a complex task that involves breaking the DNA into smaller fragments, sequencing those fragments, and then computationally assembling them into a long, continuous sequence called a consensus.

Over time, sequencing technologies have developed through at least three generations. The first-generation methods emerged in the 1970s, including the Maxam-Gilbert method, developed by Allan M. Maxam and Walter Gilbert (1976-1977), and the Sanger method (dideoxy chain termination), discovered by Frederick Sanger and his colleagues in 1977. These were the pioneering techniques that made DNA sequencing possible. The second generation introduced pyrosequencing, while the third generation brought high-throughput sequencing (HTS) or next-generation sequencing (NGS) techniques, which allow millions to billions of DNA molecules to be sequenced simultaneously with high speed and low cost. Illumina sequencing is the most widely used NGS technology, capable of running up to 500 million sequencing reactions on a single slide using fluorescently tagged nucleotides in a modified replication reaction.

One powerful approach, known as shotgun sequencing, is commonly used for sequencing entire genomes. This method involves randomly breaking DNA into fragments, sequencing them, and then assembling the overlapping sequences with the help of computational tools. Next-generation sequencing has dramatically accelerated genomic research, reducing costs while enabling large-scale studies of genetic diversity, disease, and evolutionary biology.

Sanger Sequencing: The Chain Termination Method

Sanger sequencing is based on the use of chain terminators, 2′, 3′-dideoxynucleotide triphosphates (ddNTPs) that are added to growing DNA strands and terminate synthesis at different points. Dideoxynucleotides (ddNTPs) were variants of deoxyribonucleotides (dNTPs).

Ingredients for Sanger sequencing

The classical chain-termination method (Sanger sequencing) involves making many copies of a target DNA region. Its ingredients include:

- ❖ A DNA polymerase enzyme.
- ❖ A DNA primer, which is a short piece of single-stranded DNA that binds to the template DNA and acts as a "starter" for the polymerase.
- ❖ The four DNA nucleotides (dATP, dTTP, dCTP, dGTP).
- ❖ The template DNA to be sequenced.
- ❖ Dideoxy, or **chain-terminating**, versions of all four nucleotides (ddATP, ddTTP, ddCTP, ddGTP). These ddNTPs will also be radioactively or fluorescently labeled for detection in automated sequencing machines.

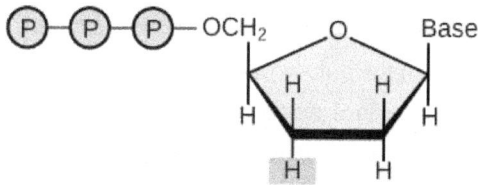

Dideoxynucleotide (ddNTP)

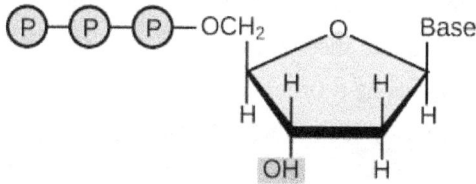

Deoxynucleotide (dNTP)

Dideoxy nucleotides are similar to regular, or deoxy, nucleotides, but with one key difference: they lack a hydroxyl group on the 3' carbon of the sugar ring. In a regular nucleotide, the 3' hydroxyl group acts as a "hook," allowing a new nucleotide to be added to an existing chain. Once a dideoxy nucleotide has been added to the chain, there is no hydroxyl available and no further nucleotides can be added. The chain ends with the dideoxy nucleotide, which is marked with a particular color of dye depending on the base (A, T, C or G) that it carries.

Method of Sanger Sequencing

The DNA sample to be sequenced is divided into four separate sequencing reactions, containing all four dNTPs at high concentration (dATP, dGTP, dCTP and dTTP), primer, and the DNA polymerase. To each reaction only one of the four dye-labeled, chain-terminating dideoxynucleoside triphosphates (ddNTPs; ddATP, ddGTP, ddCTP, or ddTTP) are added at low concentration than the ordinary nucleotides.

The mixture is first heated to denature the template DNA, and then cooled so that the primer can bind to the single-stranded template. Once the primer has bound, the temperature is raised again, allowing DNA polymerase to synthesize new DNA starting from the primer. DNA polymerase will continue adding nucleotides to the chain until it happens to add a dideoxy nucleotide instead of a normal one. At that point, no further nucleotides can be added, so the strand will end with the dideoxy nucleotide.

This process is repeated in a number of cycles. By the time the cycling is complete, it's virtually guaranteed that a dideoxy nucleotide will have been incorporated at every single position of the target DNA in at least one reaction. That is, the tube will contain fragments of different lengths, ending at each of the nucleotide positions in the original DNA. The ends of the fragments will be labeled with dyes that indicate their final nucleotide.

After the reaction is done, the fragments are run through a long, thin tube containing a gel matrix in a process called **capillary gel electrophoresis**. Short fragments move quickly through the pores of the gel, while long fragments move more slowly. As each fragment crosses the "finish line" at the end of the tube, it's illuminated by a laser, allowing the attached dye to be detected.

The smallest fragment (ending just one nucleotide after the primer) crosses the finish line first, followed by the next-smallest fragment (ending two nucleotides after the primer), and so forth. Thus, from the colors of dyes registered one after another on the detector, the sequence of the original piece of DNA can be built up one nucleotide at a time. The data recorded by the detector consist of a series of peaks in fluorescence intensity. The DNA sequence is read from the peaks in the chromatogram.

Uses and limitations

Sanger sequencing gives high-quality sequence for relatively long stretches of DNA (up to about 900900900 base pairs). It's typically used to sequence individual pieces of DNA, such as bacterial plasmids or DNA copied in PCR. However, Sanger sequencing is expensive and inefficient for larger-scale projects, such as the sequencing of an entire genome or metagenome (the "collective genome" of a microbial community). Although genomes are now typically sequenced using other methods that are faster and less expensive, Sanger sequencing is still in widely used for the sequencing of individual pieces of DNA, such as fragments used in DNA cloning or generated through polymerase chain reaction (PCR).

The Maxam-Gilbert Method

Maxam-Gilbert sequencing rapidly became more popular, since purified DNA could be directly used. The basis of Maxam-Gilbert technique is cleavage of the existing DNA molecule using chemical reagents that act specifically at a particular nucleotide. So here a primer is also not needed.

The double-stranded DNA fragment to be sequenced is first labeled by attaching a radioactive phosphorus group to the 5' end of each strand. Dimethyl sulphoxide is then added and the labeled DNA sample is heated to 90°C. This results in breakdown of the base-pairing and dissociation of the DNA molecule into its two component strands. The two strands are separated from one another by gel electrophoresis, which works on the basis that one of the strands probably contains more purine nucleotides than the other and will therefore be slightly heavier. One strand is purified from the gel and divided into four samples, each of which is treated with one of the cleavage reagents. The modification and cleavage reactions are carried out under conditions that result in only one breakage per strand. Some of the cleaved fragments retain the ^{32}P label at their 5' ends. After electrophoresis, using the same special conditions as for chain termination sequencing, the bands visualized by autoradiography will represent these labeled fragments. The nucleotide sequence can now be read from the autoradiograph exactly as for a chain termination experiment.

Pyrosequencing Method

Pyrosequencing is a method of DNA sequencing based on the "sequencing by synthesis" principle, in which the sequencing is performed by detecting the nucleotide incorporated by a DNA polymerase. Pyrosequencing relies on light detection based on a chain reaction when pyrophosphate is released. It is based on detecting the activity of DNA polymerase with another chemo luminescent enzyme. Essentially, the method allows sequencing a single strand of DNA by synthesizing its complementary strand enzymatically, one base pair at a time, and detecting which base was actually added at each step. The template DNA is immobile, and solutions of A, C, G, and T nucleotides are sequentially added and removed from the reaction. Light is produced only when the nucleotide solution complements the first unpaired base of the template. The sequence of solutions which produce chemiluminescent signals allows the determination of the sequence of the template.

Automated DNA sequencing

Automated sequencing has been developed so that more DNA can be sequenced in a shorter period of time. In 1986, Leroy Hood and Lloyd Smith automated Sanger's method. In this new sequencing technology, radioactive markers are replaced with fluorescent ones. Each ddNTP terminator is tagged with a different color of fluorophore: red, green, blue, or yellow. Thus, instead of having to run four separate sequencing reactions, the reactions can be performed in a single tube containing all four ddNTP's, each labeled with different color dyes, one for each ddNTP. These dyes fluoresce at different wavelengths, which are read via a machine.

Applications/Uses of DNA Sequencing Technologies

The determination of a DNA sequence has numerous important applications. First, sequencing allows researchers to **identify and locate genes** within a DNA segment by screening for characteristic gene features. Second, DNA sequences from different organisms can be **compared to assess homology**, enabling the study of **evolutionary relationships** both within and between species. Third, sequencing provides the ability to **analyze genes for functional regions**, such as regulatory elements, coding regions, or motifs, which can inform studies of gene function, expression, and regulation.

RNA Sequencing

RNA sequencing refers to techniques used to determine the sequences of RNA molecules. It typically involves converting RNA into complementary DNA (cDNA) through reverse transcription, followed by high-throughput or shotgun sequencing, often using next-generation sequencing (NGS) technologies. RNA sequencing allows researchers to determine both the primary sequence and the relative abundance of each RNA molecule in a biological sample. The complete set of RNA molecules in an organism is called the **transcriptome**. Just as genome sequencing has been critical for understanding gene structure and function, sequencing the transcriptome provides insights into the diverse functions, regulation, and interactions of

RNA molecules. Compared to DNA sequencing, RNA sequencing is more challenging due to the relative instability of RNA molecules.

DNA Fingerprinting

For sexually reproducing species, no two individuals share exactly the same DNA sequence, except identical twins. In humans, about 99.9% of the DNA sequence is identical across individuals, and the remaining 0.1% contains variations unique to each person. These differences form the basis of **DNA fingerprinting**, also called DNA typing, DNA profiling, genetic fingerprinting, genotyping, or identity testing. It is a laboratory technique used to identify individuals based on their unique DNA patterns and is widely applied in forensics, paternity testing, and evolutionary studies.

The method was invented in 1984 by **Professor Sir Alec Jeffreys** at the University of Leicester, UK. It is called a "fingerprint" because, like physical fingerprints, it is extremely unlikely for two unrelated individuals to have identical DNA patterns.

Sources of DNA

Samples for DNA fingerprinting can be obtained from blood, semen, saliva, hair follicles, teeth, bones, tissues, buccal smears, amniotic fluid, chorionic villi, or skeletal remains. DNA is extracted from these cells and purified for analysis.

Original Method (RFLP)

Jeffreys' original technique used **restriction fragment length polymorphism (RFLP)**. In this method, DNA is:

1. Cut with restriction enzymes.
2. Separated by gel electrophoresis.
3. Transferred to a membrane.
4. Hybridized with labeled probes targeting polymorphic regions.

The resulting band pattern, visualized via autoradiography, produced a unique "fingerprint" for each individual (except identical twins). Although highly accurate, RFLP required large amounts of high-quality DNA and several days to complete, making it impractical for many forensic applications.

Modern Methods (PCR-based)

Today, DNA fingerprinting primarily uses **PCR-based techniques,** which are faster, more sensitive, and require much smaller DNA samples. Instead of restriction enzymes, these methods target **short tandem repeats (STRs) or variable number tandem repeats (VNTRs)** – highly polymorphic regions cataloged in genomic databases such as NCBI.

Using PCR, specific loci are amplified with primers, producing many copies of STR or VNTR fragments. The fragments are then separated by gel electrophoresis or capillary electrophoresis. STRs, being shorter, may require high-resolution methods, while VNTRs produce more distinct banding patterns.

By analyzing multiple loci, a **unique DNA profile** is generated for each person. In forensic and paternity testing, these profiles are compared to establish matches or exclusions, and statistical analysis is applied to calculate the probability of random matches among unrelated individuals.

Applications of DNA Fingerprinting

DNA fingerprinting has revolutionized forensics, genetics, and biology by providing an accurate and reliable method for **individual identification** and the study of **genetic relationships.** Its applications span multiple fields and continue to expand with technological advances, benefiting both scientific research and societal needs. In **forensic science,** DNA fingerprinting is used to identify criminals, victims, and resolve legal **paternity or maternity disputes.** It also helps in identifying **unidentified bodies** after natural disasters, accidents, or conflicts. In **agriculture and animal husbandry,** DNA fingerprinting is employed to identify and protect commercial varieties of crops and livestock. Additionally, it is a valuable tool in **population genetics,** enabling the study of genetic diversity, racial origins, historical migration patterns, and invasions. Beyond these, DNA fingerprinting is used in **medical genetics** to locate and study genes associated with hereditary diseases. Overall, the technique remains a cornerstone of modern genetics, forensic investigations, and conservation biology.

Genome Editing and CRISPR-Cas9

Genome editing (or gene editing) refers to a set of technologies that enable scientists to make precise changes to an organism's DNA. Using these tools, genetic material can be added, removed, or altered at specific locations in the genome. Genome editing has broad applications: it is used in agriculture for modifying crop plants and livestock, in laboratory organisms (such as mice) for research, and is being explored in medicine for the treatment and prevention of diseases. Current research focuses on both single-gene disorders (*e.g.*, cystic fibrosis, hemophilia, sickle cell disease) and complex conditions such as cancer, heart disease, HIV infection, and neurological or psychiatric disorders.

Several genome-editing methods exist, but the most widely used and revolutionary is **CRISPR-Cas9** (Clustered Regularly Interspaced Short Palindromic Repeats and CRISPR-associated protein 9). This method is faster, cheaper, more accurate, and more efficient than earlier techniques, which is why it has transformed biological research.

Natural Origin of CRISPR-Cas9

CRISPR-Cas9 was adapted from a naturally occurring defense mechanism in bacteria. When infected by viruses, bacteria incorporate fragments of viral DNA into special regions of their genome known as **CRISPR arrays**, which act as a genetic memory of past infections. If the virus attacks again, the bacteria transcribe these arrays into RNA molecules that recognize the viral DNA. The RNA guides a nuclease enzyme, **Cas9**, to cut the viral DNA, thereby neutralizing the infection.

How CRISPR-Cas9 Works in Gene Editing

Scientists harness this system for targeted genome editing:

1. A **guide RNA (gRNA)** is designed with a short (~20-base) sequence complementary to the target DNA.
2. This gRNA binds to the **Cas9 enzyme**, forming a complex.
3. When introduced into a cell, the gRNA directs Cas9 to the specific DNA site.

4. Cas9 acts as a pair of **molecular scissors**, cutting across both DNA strands at the target site.

5. The cell's own **DNA repair machinery** is then triggered. Researchers exploit this process to:

1. **Disable a gene** → by introducing small insertions or deletions that disrupt its function.

2. **Insert new DNA** → by supplying a repair template carrying the desired sequence.

3. **Correct mutations** → by replacing faulty DNA segments with the normal sequence.

Key Components

Cas9 enzyme: Cuts the DNA at the targeted location.

Guide RNA (gRNA): Directs Cas9 to the specific DNA sequence. It has two parts:
1. A scaffold region that binds Cas9.
2. A customizable sequence (~20 bases) that matches the target DNA.

Significance

CRISPR-Cas9 has become the most versatile and precise genetic manipulation tool available. Its simplicity and power have revolutionized genetics, biotechnology, agriculture, and medicine. While research is ongoing to ensure safety and minimize unintended effects, its potential for treating human diseases and advancing science is extraordinary.

Microarray

The word **"array"** means an orderly arrangement. A **DNA microarray** (also called DNA array, DNA microchip, DNA chip, or biochip) is a **high-throughput, hybridization-based technology** that can detect the expression of thousands of genes simultaneously. It allows researchers to study which genes are active (expressed) and in what quantities, by measuring the mRNAs transcribed in a sample. Because of this, it is also known as **expression profiling**.

A microarray consists of a small plastic, glass, or silicon chip (sometimes a nylon membrane) containing thousands of microscopic spots. Each spot holds a short, single-stranded DNA fragment (probe) corresponding to a known gene. When target DNA or RNA from a sample is applied to the chip, only complementary sequences hybridize with the probes. Hybridization is detected using fluorescent labels or other detection methods. Unbound fragments are washed off, leaving behind only the paired ones.

Applications

- **Gene expression analysis** → measures the transcriptome (all expressed genes) in a sample.
- **Genotyping** → detects sequence variations, such as SNPs (single nucleotide polymorphisms).
- **Proteomics (antibody arrays)** → uses antibodies on the chip to study proteins.

How a Microarray is Made and Used

1. **Preparation of the chip**: Manufacturers attach DNA probes (fragments of known genes) to glass slides. Depending on the experiment, these may represent the entire genome or a selected subset of genes.
2. **Conversion of mRNA to cDNA**: Messenger RNA (mRNA) from the sample is reverse-transcribed into complementary DNA (cDNA).
3. **Labeling**: The cDNA is tagged with fluorescent dyes. For example:
   - cDNA from **tumor cells** → red dye.
   - cDNA from **healthy cells** → green dye.
4. **Hybridization**: Both samples are mixed and applied to the microarray. cDNA fragments compete to bind to complementary DNA probes on the chip. Non-complementary sequences are washed off.
5. **Detection**: The color of each spot indicates relative gene expression:
   - **Red** → higher expression in tumor cells.
   - **Green** → higher expression in healthy cells.
   - **Yellow** → equal expression in both samples (red + green overlap).

Outcome

The final result is a glass slide covered with thousands of colored spots. Computer analysis of the fluorescence patterns provides a detailed map of which genes are expressed, and to what extent, in different conditions (*e.g.*, diseased vs. healthy cells).

Applications of DNA Microarray Technology

Microarray technology has a wide range of applications and holds great promise for both research and clinical use. Two of the most important and widely used applications are:

Gene Expression Profiling

1. Used to measure and compare the expression levels of thousands of genes across different cell populations.
2. Helps answer key biological questions such as *when, where, and to what extent* specific genes are expressed.
3. Useful in identifying faulty or abnormally expressed genes, which can serve as potential **targets for disease diagnosis and therapy** (*e.g.*, cancer research).

Comparative Genomics

1. Applied to study genomic alterations, such as **sequence variations** and **single nucleotide polymorphisms (SNPs)**.
2. Helps in understanding genetic diversity, evolutionary relationships, and identifying genetic risk factors for diseases.

DNA Topology

DNA topology refers to the three-dimensional structural properties and spatial arrangement of the DNA double helix, particularly the **degree of supercoiling, twisting, and knotting** of DNA strands. These topological states are regulated by a group of enzymes known as **DNA topoisomerases**.

Topoisomerases maintain DNA topology by **temporarily breaking and rejoining DNA strands**, thereby relieving or introducing supercoils. This prevents the accumulation of torsional stress that arises during processes such as DNA unwinding.

Proper control of DNA topology is crucial for:

- **DNA replication** – allowing smooth unwinding of the helix.
- **Transcription** – enabling RNA polymerase movement along DNA.
- **Recombination and repair** – facilitating strand exchange and correction.
- **DNA packaging** – compacting DNA into chromosomes without tangling.

Thus, DNA topology plays a central role in **genome stability** and the accurate functioning of all cellular processes involving DNA.

READINGS AND REFERENCES

Abramsky, L., & Chapple, J. (1997). 47, XXY (Klinefelter syndrome) and 47, XYY: estimated rates of indication for postnatal diagnosis with implications for prenatal counselling. Prenatal Diagnosis: Published in affiliation with the International Society for Prenatal Diagnosis, 17(4), 363-368.

Agarwal, S., Cogburn, L., & Burnside, J. (1994). Dysfunctional growth hormone receptor in a strain of sex-linked dwarf chicken: evidence for a mutation in the intracellular domain. Journal of Endocrinology, 142(3), 427-434.

Allan, J., Cowling, G. J., Harborne, N., Cattini, P., Craigie, R., & Gould, H. (1981). Regulation of the higher-order structure of chromatin by histones H1 and H5. The Journal of cell biology, 90(2), 279-288.

Amir, R. E., Van den Veyver, I. B., Wan, M., Tran, C. Q., Francke, U., & Zoghbi, H. Y. (1999). Rett syndrome is caused by mutations in X-linked MECP2, encoding methyl-CpG-binding protein 2. Nature genetics, 23(2), 185-188.

Anso, I., Naegeli, A., Cifuente, J. O., Orrantia, A., Andersson, E., Zenarruzabeitia, O., Moraleda-Montoya, A., García-Alija, M., Corzana, F., & Del Orbe, R. A. (2023). Turning universal O into rare Bombay type blood. Nature Communications, 14(1), 1765.

Anthony, T. F., Jeffry, H.M., David, T. S., Richard, C. L. & Williams, M. G. (2000). An Introduction to Genetic Analysis. 7th edn. W.H. Freeman and Company, New York, pp860.

Bachtrog, D., Mank, J. E., Peichel, C. L., Kirkpatrick, M., Otto, S. P., Ashman, T. L., Hahn, M. W., Kitano, J., Mayrose, I., Ming, R., Perrin, N., Ross, L., Valenzuela, N., & Vamosi, J. C. (2014). Sex determination: why so many ways of doing it? PLoS Biol, 12(7), e1001899. https://doi.org/10.1371/journal.pbio.1001899

Bademkiran, S., Icen, H., & Kurt, D. (2009). Congenital recto vaginal fistula with atresia ani in a heifer: a case report. YYU Veteriner Fakultesi Dergisi, 20(1), 61-64.

Bateson, W., & Punnett, R. C. (1905-1908). *Experimental studies in the physiology of heredity*. Reports to the Evolution Committee of the Royal Society (Reports II-IV). In J. A. Peters (Ed.), *Classic papers in genetics* (pp. 42-60). Englewood Cliffs, NJ: Prentice-Hall, 1959.

Baur, E. (1907). Untersuchungen über die Erblichkeitsverhältnisse einer nur in Bastardform lebensfähigen Sippe von Antirrhinum majus. Berichte der Deutschen Botanischen Gesellschaft 25, 442-454.

BBC, 2024. World record broken for living thing with most DNA 31 May 2024. By Helen Briggs, @hbriggs, Environment Correspondent, BBC News.

Beukeboom, L. W., & Perrin, N. (2014). The evolution of sex determination. Oxford University Press & British Academy.

Brook, J. D., McCurrach, M. E., Harley, H. G., Buckler, A. J., Church, D., Aburatani, H., Hunter, K., Stanton, V. P., Thirion, J.-P., & Hudson, T. (1992). Molecular basis of myotonic dystrophy: expansion of a trinucleotide (CTG) repeat at the 3′ end of a transcript encoding a protein kinase family member. Cell, 68(4), 799-808.

Burnside, J., Liou, S. S., Zhong, C., & Cogburn, L. A. (1992). Abnormal growth hormone receptor gene expression in the sex-linked dwarf chicken. General and comparative endocrinology, 88(1), 20-28.

Caspersson, T., Farber, S., Foley, G. E., Kudynowski, J., Modest, E. J., Simonsson, E., Wagh, U., & Zech, L. (1968). Chemical differentiation along metaphase chromosomes. Experimental cell research, 49(1), 219-222. https://doi.org/10.1016/0014-4827(68)90538-7

Chase, C. D., & Gabay-Laughnan, S. (2004). Cytoplasmic male sterility and fertility restoration by nuclear genes. In Molecular biology and biotechnology of plant organelles: chloroplasts and mitochondria (pp. 593-621). Springer.

Consortium, I. H. G. S. (2004). Finishing the euchromatic sequence of the human genome. Nature, 431(7011), 931-945.

Daniels, G. (2013). Human Blood Groups, 3rd ed. Oxford. UK: Blackwell

Publishing.

Dean, L., & Dean, L. (2005). Blood groups and red cell antigens (Vol. 2). NCBI Bethesda.

Decuypere, E., Huybrechts, L., Kühn, E., Tixier-Boichard, M., & Merat, P. (1991). Physiological alterations associated with the chicken sex-linked dwarfing gene. *Critical Reviews in Poultry Biology, 3*(3), 191-221. https://doi.org/10.1080/001664809290190U

Dinh, C. T., Nisenbaum, E., Chyou, D., Misztal, C., Yan, D., Mittal, R., Young, J., Tekin, M., Telischi, F., & Fernandez-Valle, C. (2020). Genomics, epigenetics, and hearing loss in neurofibromatosis type 2. Otology & neurotology, 41(5), e529-e537.

Diribarne, M., Deretz-Picoulet, S., Allain, D., & Thepot, D. (2011). Inheritance of hair length and body size in Rex rabbits. World Rabbit Science, 19(4), 203-210.

Dodgson, J. B., & Romanov, M. N. (2004). Use of chicken models for the analysis of human disease. Current Protocols in Human Genetics, 40(1), 15.15. 11-15.15. 12.

Duan, D., Goemans, N., Takeda, S. I., Mercuri, E., & Aartsma-Rus, A. (2021). Duchenne muscular dystrophy. Nature Reviews Disease Primers, 7(1), 13. https://doi.org/10.1038/s41572-021-00248-3

Ekanem, E., Poozhikalayil, S., & Sinha, A. (2020). The Bombay blood group: how rare is it? A case report and a review of the literature. J Adv Med Res, 32, 24-29.

Ellinghaus, D., Degenhardt, F., Bujanda, L., Buti, M., Albillos, A., Invernizzi, P., Fernández, J., Prati, D., Baselli, G., Asselta, R., Grimsrud, M. M., Milani, C., Aziz, F., Kässens, J., May, S., Wendorff, M., Wienbrandt, L., Uellendahl-Werth, F., Zheng, T., . . . Karlsen, T. H. (2020). Genomewide Association Study of Severe Covid-19 with Respiratory Failure. N Engl J Med, 383(16), 1522-1534. https://doi.org/10.1056/NEJMoa2020283

Falconer, D. S. & Mackay T. F. C (1996). Introduction to quantitative genetics. 4th Edition. Longman Group Ltd., England.

Fricke, H., & Fricke, S. (1977). Monogamy and sex change by aggressive dominance in coral reef fish. Nature, 266(5605), 830-832.

Gardner, E., Simmons, M., & Snustad, D. (1991). Principles of genetics. John

Wiley and Sons, New York.

Garrido-Ramos, M. A. (2015). Satellite DNA in plants: more than just rubbish. Cytogenetic and Genome Research, 146(2), 153-170.

Gorgoni, B., & Gray, N. K. (2004). The roles of cytoplasmic poly (A)-binding proteins in regulating gene expression: a developmental perspective. Briefings in Functional Genomics, 3(2), 125-141.

Griffiths, A. J. F., Miller, J. H., Suzuki, D. T., Lewontin, R. C., & Gelbart, W. M. (1996). *An introduction to genetic analysis* (6th ed.). New York: W. H. Freeman and Company.

Guillaume, J. (1976). The dwarfing gene dw: its effects on anatomy, physiology, nutrition, management. Its application in poultry industry. World's Poultry Science Journal, 32(4), 285-305.

Haddad-Mashadrizeh, A., Mirahmadi, M., Yazdi, M. E. T., Gholampour-Faroji, N., Bahrami, A., Zomorodipour, A., Matin, M. M., Qayoomian, M., & Saebnia, N. (2023). Introns and their therapeutic applications in biomedical researches. Iranian Journal of Biotechnology, 21(4), e3316.

Hahn, C., & Salajegheh, M. K. (2016). Myotonic disorders: A review article. Iranian journal of neurology, 15(1), 46.

Hartl, D., & Clark, A. (1989). Principles of population genetics. Sinauer Associates, Inc., Publishers, Sunderland, Massachusetts, U.S.A.

Herrick, J. B. (1910). Peculiar elongated and sickle-shaped red blood corpuscles in a case of severe anemia. Archives of internal medicine, 6(5), 517-521.

Horst, P. (1988). Using the major gene for feather restriction. Poultry Misset 4: 8-9.
https://doi.org/10.1038/266830a0
https://doi.org/10.1056/NEJMra1802338
https://doi.org/10.3389/fmicb.2015.00545

Ihara, T., Akashi, H., & Bishop, D. H. (1984). Novel coding strategy (ambisense genomic RNA) revealed by sequence analyses of Punta Toro phlebovirus S RNA. Virology, 136(2), 293-306.

Jost, A. (1953). Problems of fetal endocrinology: The gonadal and hypophyseal hormones. Recent Progress in Hormone Research, 8, 379-418.

Jukes, T. H., & Osawa, S. (1990). The genetic code in mitochondria and chloroplasts. Experientia, 46, 1117-1126.

Klug, W. S., Cummings, M. R., Spencer, C. A., & Ward, S. M. (2005). Essentials of genetics. Pearson Education.

Leonard, A. C., & Grimwade, J. E. (2015). The orisome: structure and function. Frontiers in Microbiology, 6, 545.

Li, X. (2011). Sex chromosomes and sex chromosome abnormalities. Clinics in laboratory medicine, 31(4), 463-479.

López-Flores, I., & Garrido-Ramos, M. (2012). The repetitive DNA content of eukaryotic genomes. Genome dynamics, 7, 1-28.

Luger, K., Mäder, A. W., Richmond, R. K., Sargent, D. F., & Richmond, T. J. (1997). Crystal structure of the nucleosome core particle at 2.8 Å resolution. Nature, 389(6648), 251-260.

Mangus, D. A., Evans, M. C., & Jacobson, A. (2003). Poly (A)-binding proteins: multifunctional scaffolds for the post-transcriptional control of gene expression. Genome biology, 4, 1-14.

Mann, J., Cahan, A., Gelb, A., Fisher, N., Hamper, J., Sanger, R., Tippett, P., & Race, R. (1962). A sex-linked blood group. The Lancet, 279(7219), 8-10.

Marigo, V., Roberts, D. J., Lee, S. M., Tsukurov, O., Levi, T., Gastier, J. M., Epstein, D. J., Gilbert, D. J., Copeland, N. G., & Seidman, C. E. (1995). Cloning, expression, and chromosomal location of SHH and IHH: two human homologues of the Drosophila segment polarity gene hedgehog. Genomics, 28(1), 44-51.

Martin, R. D. (1994). The specialist chick sexer. Bernal Publishing, Box Hill, UK.

Moore, J. K., & Haber, J. E. (1996). Cell cycle and genetic requirements of two pathways of nonhomologous end-joining repair of double-strand breaks in Saccharomyces cerevisiae. Molecular and cellular biology, 16(5), 2164-2173.

Najari, B. B. & Alukal, J. P. (2018). Reproductive Medicine: Reciprocal translocations. Skinner, MK (Editor). Encyclopedia of Reproduction. Academic Press.

Newton, C. R., & Graham, A. (1997). *PCR*. Oxford: BIOS Scientific

Publishers.

Nicholas, F. (1996). Introduction to veterinary genetics. Clarendon Press, Oxford.

Nicholl, D. (2008). An introduction to genetic engineering. Third Edition. Cambridge University Press.

Nielsen, J., & Wohlert, M. (1991). Chromosome abnormalities found among 34910 newborn children: results from a 13-year incidence study in Århus, Denmark. Human genetics, 87, 81-83.

Nogler, G. A. (2006). The lesser-known Mendel: his experiments on Hieracium. Genetics, 172(1), 1-6.

Ouyang, J., Xie, L., Nie, Q., Zeng, H., Peng, Z., Zhang, D., & Zhang, X. (2012). The effects of different sex-linked dwarf variations on Chinese native chickens. *Journal of Applied Genetics, 53*(2), 173-178. https://doi.org/10.1007/s13353-011-0079-0

Panch, S. R., Montemayor-Garcia, C., & Klein, H. G. (2019). Hemolytic Transfusion Reactions. N Engl J Med, 381(2), 150-162.

Patel, J. N., Donta, A. B., Patel, A. C., Pandya, A. N., & Kulkarni, S. S. (2018). Para-Bombay phenotype: A case report from a tertiary care hospital from South Gujarat. Asian Journal of Transfusion Science, 12(2), 180-182.

Pereira, E., Felipe, S., de Freitas, R., Araújo, V., Soares, P., Ribeiro, J., Dos Santos, L. H., Alves, J. O., Canabrava, N., & van Tilburg, M. (2022). ABO blood group and link to COVID-19: A comprehensive review of the reported associations and their possible underlying mechanisms. Microbial Pathogenesis, 169, 105658.

Pesole, G., Mignone, F., Gissi, C., Grillo, G., Licciulli, F., & Liuni, S. (2001). Structural and functional features of eukaryotic mRNA untranslated regions. Gene, 276(1-2), 73-81.

Phillips, J. E., & Corces, V. G. (2009). CTCF: master weaver of the genome. Cell, 137(7), 1194-1211.

Pierce, B. A. (2008). Genetics: A conceptual approach. 3rd Ed. W.H. Freeman and Company.

Plohl, M., Meštrović, N., & Mravinac, B. (2012). Satellite DNA evolution. Genome Dyn, 7(126), 10.1159.

Qadir, H., Larik, M. O., & Iftekhar, M. A. (2023). Bombay blood group phenotype misdiagnosed as O phenotype: a case report. Cureus, 15(9), e45555.

Radostits, O. M., Blood, D. C., & Gay, C. C. (1997). *Veterinary medicine: A textbook of the diseases of cattle, sheep, pigs, goats and horses* (9th ed.). London: Baillière Tindall.

Ranum, L. P., Rasmussen, P. F., Benzow, K. A., Koob, M. D., & Day, J. W. (1998). Genetic mapping of a second myotonic dystrophy locus. Nature genetics, 19(2), 196-198.

Richmond, T. J., Finch, J. T., Rushton, B., Rhodes, D., & Klug, A. (1984). Structure of the nucleosome core particle at 7 Å resolution. Nature, 311(5986), 532-537.

Rosenfeld, R. G., Rosenbloom, A. L., & Guevara-Aguirre, J. (1994). Growth hormone (GH) insensitivity due to primary GH receptor deficiency. Endocrine reviews, 15(3), 369-390.

Ruiz-Ruano, F. J., López-León, M. D., Cabrero, J., & Camacho, J. P. M. (2016). High-throughput analysis of the satellitome illuminates satellite DNA evolution. Scientific Reports, 6(1), 28333.

Silva, D. M. d. A., Utsunomia, R., Ruiz-Ruano, F. J., Daniel, S. N., Porto-Foresti, F., Hashimoto, D. T., Oliveira, C., Camacho, J. P. M., & Foresti, F. (2017). High-throughput analysis unveils a highly shared satellite DNA library among three species of fish genus Astyanax. Scientific Reports, 7(1), 12726.

Stewart, M. D., Merino Vega, D., Arend, R. C., Baden, J. F., Barbash, O., Beaubier, N., Collins, G., French, T., Ghahramani, N., & Hinson, P. (2022). Homologous recombination deficiency: concepts, definitions, and assays. The Oncologist, 27(3), 167-174.

Storry, J. R., Clausen, F. B., Castilho, L., Chen, Q., Daniels, G., Denomme, G., Flegel, W. A., Gassner, C., de Haas, M., & Hyland, C. (2019). International society of blood transfusion working party on red cell immunogenetics and blood group terminology: report of the Dubai, Copenhagen and Toronto meetings. Vox sanguinis, 114(1), 95-102.

Sturtevant, A. H. (2001). *A history of genetics* (reprint ed.). Cold Spring Harbor, NY: Cold Spring Harbor Laboratory Press.

Tamarin, R. (2022). Principles of genetics. McGraw-Hill Book Co., Boston,

USA.

Tarpy, D., Nielsen, R., & Nielsen, D. (2004). A scientific note on the revised estimates of effective paternity frequency in Apis. Insectes Sociaux, 51(2), 203-204. https://doi.org/10.1007/s00040-004-0734-0

Van Vleck, L. D., Pollak, E. J., & Oltenacu, E. A. (1987). Genetics for the animal sciences. W. H. Freeman and Co. New York, USA.

Vanier, M. T., & Millat, G. (2003). Niemann-Pick disease type C. Clin Genet, 64(4), 269-281. https://doi.org/10.1034/j.1399-0004.2003.00147.x

Vos, P., Hogers, R., Bleeker, M., Reijans, M., Lee, T. v. d., Hornes, M., Friters, A., Pot, J., Paleman, J., & Kuiper, M. (1995). AFLP: a new technique for DNA fingerprinting. Nucleic acids research, 23(21), 4407-4414.

Wallace, B. (1981). Basic population genetics. Columbia University Press, New York, U.S.A.

Wallace, J. A., & Felsenfeld, G. (2007). We gather together: insulators and genome organization. Current opinion in genetics & development, 17(5), 400-407.

Weaver, R., & Hedrick, P. (1997). Genetics, 3rdedn. Wm. C. Brown Publishers, England, 45, 297.

Weiler, K. S., & Wakimoto, B. T. (1995). Heterochromatin and gene expression in Drosophila. Annual review of genetics, 29(1), 577-605.

White, C. L., Suto, R. K., & Luger, K. (2001). Structure of the yeast nucleosome core particle reveals fundamental changes in internucleosome interactions. The EMBO Journal, 20(18), 5207-5218. https://doi.org/10.1093/emboj/20.18.5207

Zanotto, P. d., Gibbs, M. J., Gould, E. A., & Holmes, E. C. (1996). A reevaluation of the higher taxonomy of viruses based on RNA polymerases. Journal of virology, 70(9), 6083-6096.

Zheng, J., Liu, Z., & Yang, N. (2007). Deficiency of growth hormone receptor does not affect male reproduction in dwarf chickens. Poultry Science, 86(1), 112-117.

Zhou, B.-R., Jiang, J., Feng, H., Ghirlando, R., Xiao, T. S., & Bai, Y. (2015). Structural mechanisms of nucleosome recognition by linker histones. Molecular cell, 59(4), 628-638.